COMPOSTING OF AGRICULTURAL AND OTHER WASTES

Proceedings of a seminar organised by the Commission of the European Communities, Directorate-General Science, Research and Development, Environment Research Programme, held at Brasenose College, Oxford, UK, 19–20 March 1984

COMPOSTING OF AGRICULTURAL AND OTHER WASTES

Edited by

J. K. R. GASSER
Agricultural and Food Research Council, London, UK

ELSEVIER APPLIED SCIENCE PUBLISHERS
LONDON and NEW YORK

ELSEVIER APPLIED SCIENCE PUBLISHERS LTD
Crown House, Linton Road, Barking, Essex IG11 8JU, England

Sole Distributor in the USA and Canada
ELSEVIER SCIENCE PUBLISHING CO., INC.
52 Vanderbilt Avenue, New York, NY 10017, USA

British Library Cataloguing in Publication Data

Composting of agricultural and other wastes:
proceedings of a seminar organised by the
Commission of the European Communities,
Directorate-General Science, Research and
Development, Environment Research Programme:
held at Brasenose College, Oxford, 19–20 March 1984.—
(EUR; 9419)
1. Compost. 2. Agricultural wastes.
I. Gasser, J. K. R. II. Environment Research Programme.
631.8′75 S661

ISBN 0-85334-357-8

WITH 56 TABLES AND 92 ILLUSTRATIONS

Publication arrangements by Commission of the European Communities, Directorate-General Information Market and Innovation, Luxembourg

EUR 9419

LEGAL NOTICE

Neither the Commission of the European Communities nor any person acting on behalf of the Commission is responsible for the use which might be made of the following information.

Printed in Great Britain by Galliard (Printers) Ltd, Great Yarmouth

PREFACE

The Contact Co-ordination Group on Composting was established by the Commission of the European Communities to co-ordinate work on composting in the European Communities relevant to the Secondary Raw Materials Programme. At early meetings of the Group, the two major areas for composting identified were urban waste including sewage sludge and agricultural wastes. The problems of sewage treatment and disposal have received attention over many years under the COST 68 programme. There is a large amount of technical information available, much of which has been discussed at Conferences. In contrast, the composting of agricultural, horticultural and forestry wastes appears to have received less attention and the information is more widely scattered in the scientific literature and part has either not been published or exists as reports.

The Group therefore discerned a need to collect together for discussion information on composting agricultural wastes and the relevant experience from composting urban wastes with the objectives of assessing the present position, identifying gaps in our knowledge and making recommendations for future action. These needs and objectives led to the organisation of this Seminar on "Composting Agricultural and Other Wastes", with participants mainly from countries of the Communities and some welcome guests from other places.

J.K.R. GASSER

CONTENTS

POSTERS

INTRODUCTION

Composting - Progress to date in the European Economic Community and prospects for the future

COMPOSTING - PROGRESS TO DATE IN THE EUROPEAN ECONOMIC COMMUNITY AND PROSPECTS FOR THE FUTURE

G.L. FERRERO, P. L'HERMITE
Commission of the European Communities
Directorate-General for Science, Research and Development
200, rue de la Loi
B - 1049 BRUSSELS
BELGIUM

SUMMARY

A general description is given of the R&D programme "Recycling of urban and industrial waste" with particular reference to activities to be coordinated in the field of composting. A number of composting research activities are currently in progress concerning inter alia : the specifications of the product, the definition of parameters and analytical measurements applicable to the composting process, the recovery of energy.

The work of the seminar and the conclusions it arrives at are to be used to ascertain interest in the research topics under the heading of activities to be coordinated and should provide the input required for continuing with this activity. They should also help to determine whether there are any other research topics where there is currently sufficient general interest for action at Community level.

Composting comes within the framework of research and development activities carried out by the Commission of the European Communities into areas of priority concern like energy, raw materials, the environment, etc. and is one of the activities to be coordinated in the programme on the recycling of urban and industrial waste.

The programme was originally adopted by the Council of Ministers on 12 November 1979. It was allocated 9 million ECU and was to run for four years, thus finishing on 30 October 1983. It has been extended for another two years by the Council Decision of 12/13 December 1983 and has been allocated another 2 million ECU until 31 December 1985.

On 25 May 1982 a non-Community country - Sweden - signed a cooperation agreement with the EEC relating to this programme and is thus participating on an equal footing with the other Member States in research and coordination work. This cooperation agreement has been extended for another two years.

The programme includes action of two kinds :

(i) indirect actions, i.e. shared-cost contracts, the Community contributing a maximum of 50 % of the total cost of research; contracts may be signed with companies, university laboratories, regional, provincial or municipal authorities, national research centres, etc.;

ii) activities to be coordinated - composting comes under this heading, as we have already mentioned; the aim is to coordinate at Community level between national research projects being carried out in Member States and subsidized by public funds and research being carried out under contract and financed by EEC funds; these coordinated activities also include the organization of meetings between experts appointed by the Member States on specific and topical subjects of common interest and the dissemination of scientific information at seminars, colloquia, symposia, etc. With all these coordinated activities, the Community's sole contribution is to organize the meetings.

The programme was based on the findings of a number of specific studies carried out by national experts which identified those areas of research where a Community-level research scheme would be desirable and worthwhile.

This takes account of the pressing problems posed by the huge quantities of waste of all types arising in the Member States of the Community. A breakdown of quantities is given in Table 1.

The programme also takes account of the need to :

(i) save raw materials and energy in order to increase our level of self-sufficiency;

(ii) reduce the quantities of waste to be disposed of, thus effectively helping to protect the environment.

The R&D programme on the recycling of urban and industrial waste is subdivided into four main research areas :

I - Sorting of household waste

II - Thermal treatment of waste

III - Fermentation and hydrolysis

IV - Recovery of rubber waste

Each of these research areas is divided in turn into research topics.

Table 2 shows the whole programme, distinguishing between indirect actions and activities to be coordinated.

As can be seen from Table 2, composting is included in research area III and is a coordinated activity only, in other words there is no provision for contracts with a financial contribution from the Commission, only for the coordination of existing projects through the organisation of meetings of experts and the like.

In fact, the potential for composting waste is enormous if one considers the quantities of agricultural waste which arise in the Member States of the Communities.

It is obviously in the interest of Society as a whole therefore to encourage the composting of agricultural waste so as to rationalize our use of resources and to improve protection of the environment.

In order to decide on those projects which, at Community level, could most effectively help to bring about the wider use of composting methods, it was decided to set up a Committee of experts to coordinate composting research who would be chosen by the Member States of the EEC and appointed by the members of the Advisory Committee on the Management of the programme on the recycling of urban and industrial waste. The Committee of experts (the Contact Coordination Group on Composting) was given the task of identifying priority research topics in respect of which specific projects of proven common interest could, in a relatively short time with regard to the length of the programme, lead to positive results.

Some of these specific projects are already under way and concern :

(i) drawing up a list of composting research projects being carried out in Member States - the list has been put on computer and for the last few months has been available to anyone who wishes to obtain the information;

(ii) recovery of energy from composting - work is being done at the moment on examining the stage research has reached and the results so far obtained;

(iii) comparing aerobic and anaerobic processes - comparing technical and economic feasibility to help choose the optium process;

(iv) the metal content of compost from urban waste - this project is being carried out in parallel with a similar project specifically on sewage sludge which is part of the COST 68 programme and is intended to result in a proposal for a Community directive;

(v) determining criteria for the production of compost - specifications of the product;

(vi) estimating the total quantity in each country of organic waste which is not currently used to produce compost but which could be so used in the future;

(vii) study being carried out into the use in agriculture of residues from anaerobic digesters;

(viii) definition of parameters and analytical measurements applicable to composting processes - this project is the extension to composting of the work already carried out on the anaerobic digestion processes which appeared in a recent publication;

(ix) study of the engineering of composting processes and preparing models of the process.

In addition composting of sewage sludges in view of their agricultural use is part of the activities of the Concerted Action COST 68 ter "Treatment and Use of Sewage Sludge" implemented within the framework of the Third Environmental Protection Research Programme, which will be extended from this year to "Liquid agricultural wastes". Composting is a long establisched method of stabilizing wastes which when applied to sewage sludges lead to a loss of solid matter due to gas release, a decrease in the moisture due to the increase fo temperature with a subsequent loss or elimination of pathogens and a stabilisation of the final product. The compost can be spread on land with the usual farming devices for manure and its organic value which contributes to the long-term maintenance of soil nitrogen constitutes its main interest.

The work done to date has prompted us to formulate a number of working hypotheses regarding what still remains to be done at Community level. One of the difficulties recognized by all concerned is the current lack of specifications for different types of compost. These would guarantee users a specific product quality and properties which remain standard over time and which would be tailored to the different ways in which compost is used. Because there are no standard specifications, it is difficult to market compost and there is also a serious effect on the product image and on market development potential. This applies both to compost from urban waste and to compost from agricultural waste, because, even though there are perhaps fewer problems with agricultural waste, there are still the same obstacles to marketing it more widely.

One of the priority tasks at Community level is therefore felt to be any action which would encourage the wider use of compost. Alongside this action, there is a need to provide information for potential users in order to help them decide on the product most suited to their particular needs (cultivation of : vegetables, flowers, fruit trees, timber crops, etc.).

It is also important to ensure that products are absolutely harmless (i.e. check their phytotoxicity levels) and to provide instructions for use and details on how to obtain the best results.

The need for action along these lines at Community level was stressed repeatedly at the symposium on "Biological reclamation and land utilization of urban wastes" held in Naples from 11 to 14 October 1983.

At the end of the symposium, a number of recommendations were put forward.

1. To encourage projects which can make a worthwhile contribution to international cooperation on the categorization of products, and selecting those parameters and analytical measurements which guarantee the best quality control for compost.

2. To encourage pilot research projects on the combined treatment and recycling of urban waste, where the particular method of composting determines the recovery of the organic fraction.

3. To encourage those research and coordination projects of common interest which set out to improve the quality of the product and ensure that it is used.

4. To encourage those research projects which can make a worthwhile contribution to a basic knowledge of those bio-oxidative stabilization processes typical of composting.

5. To encourage schemes to publish information on possible uses for compost, according to the different degrees of stabilization and the functions they may be required to perform in agriculture, particularly on dry southern terraces where organic substances can do much to prevent deterioration of the soil.

6. To encourage a detailed study to help determine the advantages, in energy terms, of organic manuring as a means of improving chemical and water balances, reducing disease and nutritional problems in plants, reducing the use of pesticides and chemical soil-correctives, reducing the need for mechanical working of soil and increasing the ground volume which can be penetrated by roots as a way of increasing crop intensity.

7. To encourage awareness of and the preparation of regulations on pollution by heavy metals. Encouraging a rapid turnover of organic substances in agriculture should not have heavy metal pollution as a side-effect and it should be a requirement that these processes do not bring about situations which in the future may be difficult to remedy. The introduction of heavy metals into our food chain could have disastrous consequences which at the moment it is totally impossible to forecast. It is essential therefore to distinguish between waste and toxic or dangerous residues which may prevent recovery of this waste. To this end, it is important to analyse the effect of accumulated quantities of heavy metal in the soil, in particular by carrying out studies on the effects which changes in the humus content of soil (as the result of irregular manuring) may have cyclically on the balance of heavy metals in those parts of the soil to which the roots penetrate.

One of the purposes of this seminar is to determine these needs in aspect of compost from agricultural waste as well.

So what results can be expected from this seminar ?

First of all, it should confirm the degree of interest which the countries of the Community and those outside the Community have in the composting of agricultural waste. We also expect to be able to ascertain the level of interest in the research topics which are currently being coordinated, and whether they are still topical, using the contributions of those present at the seminar to add to the information we already have so as to help us take as broad a view as possible which is an essential feature of our current activities.

Lastly, the gaps which still exist in this field show that that there is a need for more research projects and coordination with the emphasis particularly on those gaps which are still evident in primarily Community projects.

As we have mentioned, the current programme has been extended until 31 December 1985 although a new programme which is to be called "The treatment, recycling and reutilization of waste" is in preparation.

Composting is one of the research topics proposed for that programme. In fact, on the basis of the results so far achieved it will be possible to assess the value of what has already been done and the need for any further action for the new programme. It is hoped that this seminar will provide the input which is essential if our Community action is to continue.

TABLE 1

Quantities of waste arising in the Member States of the European Economic Community in millions of tonnes per year

Domestic waste	150
Agricultural waste	950
Industrial waste	160
Sewage sludge	300
Waste from extractive industries	250
Demolition waste and debris	170
Consumer waste (discarded cars, used tyres, etc.)	120
Other waste (street litter, dead leaves, etc.)	200

TABLE 2

RESEARCH AND DEVELOPMENT PROGRAMME

R E S E A R C H T O P I C S	Activities to be coordinated	Indirect Actions
RESEARCH AREA I		
Sorting of household waste		
1. Assessment of waste sorting projects	x	-
2. Methods for sampling and analysis of household waste	x	-
3. Evaluation of health hazards	x	-
4. Technology for the sorting of bulk waste	x	x
5. Materials recovery		
5.1 Paper	x	x
5.2 Plastics	x	x
5.3 Non-ferrous metals	x	-
6. Energy recovery	x	x
7. New collection and transport systems	x	-
RESEARCH AREA II		
Thermal treatment of waste		
1. Firing of waste derived fuel	See I.6	See I.6
2. Pyrolysis and gasification	x	x
3. Recovery of metal and glass from residue	x	-
RESEARCH AREA III		
Fermentation and hydrolysis		
1. Anaerobic digestion	x	x
2. Carbohydrate hydrolysis	x	x
3. Composting	x	-
RESEARCH AREA IV		
Recovery of rubber waste		
1. Retreading	x	-
2. Size reduction	x	-
3. Reclaiming and recycling of rubber powder	x	-
4. Pyrolysis	x	-

REFERENCES

- Utilisation énergétique des déchets municipaux - Rapport de synthèse 1980-1981 (ENV/178/79-FR)

- Etude sur les aspects techniques, écologiques, économiques de la production, de la commercialisation et de l'utilisation du compost dans les différents Etats membres de la Communauté européenne - Université Catholique de Louvain- Unité de Génie Biologique

- Fermentation - Hydrolysis processes for the utilization of organic waste materials

  L.O. Hopkins,R.J. Griffith, D.M. Malone, T. McManus
  C.E.C. Studies on Secondary Raw Materials IV - I (1979)

- Die Struktur und die sozio-ökonomische Bedeutung der Abfallwirtschat und der Recycling-Industrien in der Europäischen Gemeinschaft Euroconsult - XI/493/83 (November 1982)

- Proposal for the Definition of Parameters and Analytical Measurements Applicable to the Anaerobic Digestion Processes
  Agricultural Wastes 7 (1983) 183-193

- Disinfections of Sewage Sludge : Technical, Economic and Microbiological Aspects : Proceedings of a CEC Workshop held in Zürich, May 11-13, 1982. EUR 8024

SESSION I

PRODUCTION OF COMPOST

Principles of composting leading to maximization of decomposition rate, odour control, and cost effectiveness

Technological aspects of composting including modelling and microbiology

Forced aeration co-composting of domestic refuse and sewage sludge in static piles

Methods for the evaluation of the maturity of municipal refuse compost

Phytotoxins during the stabilization of organic matter

PRINCIPLES OF COMPOSTING LEADING TO MAXIMIZATION OF DECOMPOSITION RATE, ODOR CONTROL, AND COST EFFECTIVENESS

M..S. Finstein and F.C. Miller
Department of Environmental Science
Cook College, Rutgers University
New Brunswick, New Jersey USA

Summary

In composting for waste management purposes, odour control and cost-effective design and operation are both served by maximizing the rate of decomposition. This is accomplished through deliberate heat removal by ventilation, with reference to a 60°C operational ceiling. Implementation is via a blower actuated by means of temperature feedback control. This strategy, compared to a conventional approach, yielded about four-fold more waste treatment in half the processing time.

1. STRATEGY FOR MAXIMIZING DECOMPOSITION RATE

In composting for waste treatment purposes, odour control and cost-effectiveness are linked objectives, in that both are served by maximizing the rate of organic matter decomposition. We reported elsewhere on two fundamentally different strategies for management of the composting microbial ecosystem: one designed to maximize the rate of microbial decomposition; the other, owing to its design, results in a minimal rate (1). Reprints of this paper were sent in advance to the attendees of the conference. This permits us today to simply list the major conclusions from this paper, so that at the conference we can focus on the practical implications of these and related findings (2).

1.1 Major Scientific and Technical Conclusions Derived from Reference 1

i. The composting microbial ecosystem tends strongly to self-limit via excessive accumulation of metabolically generated heat, leading to inhibitively high temperature. The threshold to significant inhibition is approximately 60°C, and inhibition increases sharply at higher temperatures. Unless controlled through deliberate heat removal, composting masses typically peak at 80°C, at which point the rate of decomposition is extremely low.

ii. This self-limiting tendency must be countered if decomposition is to be fostered. Consequently, the central problem in the design and control of composting facilities is heat removal with reference to a 60°C operational ceiling.

iii. A practical means of removing heat from the composting mass is through ventilation. The main ventilation-associated mechanism of heat removal (ca.90%) is the vaporization of water (3). Ventilation also supplies O_2 for aerobic decomposition (main source of heat).

iv. The forced pressure-mode of ventilation removes heat more efficently than the vacuum-induced mode (4).

v. During composting the rate of heat generation varies with time.

Hence to maintain a given temperature this must be matched by correspondingly time-variable heat removal. Implementation is through temperature feedback actuation of a blower, using standard control equipment. Composting mass and blower thereby interact, to seek an assigned set-point temperature.

vi. To achieve the desired operational ceiling of 60°C, it might be necessary to assign a lower set-point (e.g. 45°C).

vii. The blower capacity (head and volume) must suffice to meet peak demand for ventilation. A fresh waste (e.g. raw sewage sludge) demands more ventilation than a partly treated one (e.g. digested sludge).

viii. A temperature gradient is established along the axis of airflow; drying is relatively uniform along this axis.

ix. Managed thus, decomposition and drying are related because the following chain of relations is established: decomposition generates heat; the heat vaporizes water; the vaporization causes drying. Production of water replenishes only ca. 10% of the water removed.

x. A consequence of this control strategy is that the composting mass is well-oxygenated (typical O_2 level, 17% v/v). This is because approximately 9x more air is needed to remove heat than to supply O_2 for aerobic respiration.

1.2 General Relevance of the Conclusions Derived from Reference 1

These conclusions are based on a field study in which composting ecosystem management strategy was isolated as the experimental variable. The experimental design was that two 36 tonne piles, consisting of a mixture of primary (raw) sewage sludge and woodchips, were constructed side-by-side. One of the piles (9A) was controlled according to the principles summarized above, to maximize decomposition rate through temperature control. The other pile (9B) was controlled according to the Beltsville Static Pile Process (5, 6) which, owing to its basic design, seeks inhibitively high temperatures (typical peak, 80°C).

Convenience required that the principles embodied in pile 9A, and others similary controlled, be given a name, and we selected that of "Rutgers Static Pile Process". At one level, therefore, the study can be viewed as a comparison between two (non-proprietary) processes. More importantly, it is a scientific comparative analysis between fundamentally different strategies for managing the composting microbial ecosystem. As such, the findings are generally relevant to composting.

2. ODOUR CONTROL

Composting facilities operated for waste managment purposes receive large amounts of putrescible material, constituting a potential for nuisance. Such material is readily decomposed microbially, by definition. Hence, accelerating decomposition at the outset of processing represents treatment of the problem at its source. However, treatment of the symptom is the more common response.

2.1 Treating the Symptom in Isolation from the Cause

The Beltsville Process is a well-developed case in point, serving as such to illustrate a widespread problem. In this spirit we review the aspects of this process intended to control odour, in the context of its basic approach to composting ecosystem management.

The stated operational goal is to maintain 5% to 15% O_2 inside the pile (5, 6), apparently based on the opinion that the presence of O_2 is a sufficient condition to ensure effective composting. In fact it is a necessary, but not a sufficient, condition.

The ventilation regimen prescribed to maintain an oxygenated condition is as follows. A 0.25 KW blower serves a 45 tonne pile; the blower is operated in the vacuum-induced mode; operation is on a fixed schedule, by timer; the usual schedule is for blower operation 20% of the time. This regimen is applied regardless of the strength of the waste, and succeeds in keeping the pile oxygenated.

One might well wonder how so little ventilation, rigidly scheduled, keeps such a large mass of sewage sludge oxygenated! It does so by fostering the composting ecosystem's self-limiting tendency (1.1, item i). The ventilation provided suffices to avert anaerobic conditions during the temperature come-up stage, but does not remove enough heat to prevent a temperature ascent to biologically inhibitive levels (1.1, item x). Consequently, at peak temperatures (typically 80°C) the need for O_2 is slight and is provided by the slight ventilation, thereby maintaining an oxygenated condition. The hidden price, however, is that the basic waste treatment objective of rapid putrescible (odour-causing) matter decomposition is sacrificed.

The vacuum-induced option was adopted as an odour-control measure, so that exhaust gas could be vented through a scrubber pile (5, 6). This necessitates placement of a condenser-water trap between the composting mass and the scrubber pile. Despite this measure, condensasation occurs in the scrubber pile, making it soggy and potentially malodorous. Moreover, the scrubber pile further decreases blower efficiency, and is itself a space-consuming management problem. In all, these measures represent a mistaken treatment of the symptom in isolation from the underlying cause.

2.2 Addressing the Problem's Cause

The Rutgers process control strategy addresses the odour-problem at its source, in that it is designed to maximize decomposition rate. This was demonstrated in the composting of primary (raw) sewage sludge from a mixed residential-industrial (mainly food processing) area. The sludge was received as a belt-filter press cake (moisture content, 76%; volatile solids content, 74%). Samples removed from the piles were evaluated by odor test panels on a scale of -5 to +5 (-5 = most unpleasant; 0 = neutral; +5 = most pleasant), as reported in detail (2, 7).

The first evaluation was a part of a field trial (7) in which set-point temperature was the experimental variable (Table I). A sample was removed from each pile on day seven, and fours hours later presented to a panel. After overnight storage in sealed jars the samples were re-evaluated. The material representing the highest composting temperature (set-point, 65°C; median pile temperature, 67°C) was judged most unpleasant, and that representing the lowest temperature (set-point, 45°C; median pile temperature, 48°C) least unpleasant. The temperatures characteristic of the Beltsville Process are even more suppressive of decomposition than those resulting from the 65°C set-point.

The second odor evaluation (2) was part of an extension of the comparative study of the two fundamentally different control strategies (1). Subsequent to the period of temperature feedback control (Rutgers), or the standard 21 day period of fixed-schedule ventilation (Beltsville), material was screened and formed into derivative pile for curing (Table II). During this phase of the study the Rutgers material suffered a prolonged period of dryness (moisture content, approximately 25%), presumably causing a hiatus in curing (1.1, item ix). Water was eventually added to this pile. The test panel results are reported chronologically, and also in adjusted

Table I. Effect of set-point temperature on odour of day 7 samples.*

	Set-point temperature (°C)	Period of feedback control (days)	Temperature of pile during period of feedback control (°C)				Odour evaluation (test panel mean score)	
			Median	Range	Observations No.	Observations % >60°C	Day of sampling†	After over-night storage‡
Pile 4A	45	0.5 to 14.7	48	25-63	1020	1.8	-1.35	-2.33
Pile 4B	55	1.1 to 8.9	62	18-78	641	58	-2.00	-3.30
Pile 4C	65	1.2 to 3.5	67	64-74	182	100	-3.00	-3.85

*Each pile consisted of 6 tonnes of a mixture of primary (raw) sewage sludge cake and woodchips. Judging from CO_2 levels (typically 1% to 4%), O_2 was at a high level in all of the piles.

†Seventeen member odour test panel.

‡Twenty-nine member odour test panel.

Table II. Effect of Process control strategy on odour of curing samples.

		Parent piles				Derivative curing pile samples (test panel mean score)	
		Pile temperature (°C)[†]				Total composting time[‡]	
				Observation			
	Strategy	Median	Range	No.	% >60°C	85 days[§]	90 days[$]
Pile 9A	Rutgers[*]	53	24-68	1755	13	-1.30	+0.66
Pile 9B	Beltsville	70	45-82	1519	91	-3.15	-2.29

[*]The set-point was 45°C for the first 44 hours of operation, and 48°C thereafter. (Set-point not applicable to Beltsville process.)

[†]The summary periods were: Rutgers, hrs. 10 to 380 (period of temperature feedback control); Beltsville, hrs. 100 to 500 (excludes temperature rise.

[‡]Inclusive of parent and curing stages. Adjusting for the period of dryness (see text), the -1.30 mean score relates to day 25, and +0.66 to day 30.

[§]Twenty member odour test panel.

[$]Thirty-eight member odour test panel.

form to take into account the presumed curing hiatus. Regardless, the control strategy designed to maximize decomposition rate yielded the more acceptable samples.

It is not suggested that rate maximization assures freedom from odour nuisance. Given the nature of the material received for treatment, the possibility of equipment or processing failure (8), and operational factors outside the boundaries of composting per se, no such assurance is possible. Nonetheless, maximizing the rate of decomposition is a basic step toward the goal of odour control.

3. COST EFFECTIVE PROCESSING

The premise of this section is that capital and operating costs at composting facilities are strongly influenced by decomposition rate ($\equiv$ reaction rate). In part this is because the need for facility space (or volume) is inversely related to rate, as is the weight and volume of process residue that must be handled. Moreover, for any given processing duration, the higher the rate the more stable and easily handled the residue. This facilitates storage, transport, and final disposition, preferably through some avenue of resource-recovery.

Data relating cost and decomposition rate are not available. Nonetheless, certain aspects of these and related matters can be usefully discussed based on fundamental principles and some known and estimable costs.

3.1 Matching Heat Removal to its Generation

The required matching of heat removal to heat generation, in reference to temperature, is achievable via feedback control of a blower (1.1, item v). Fixed-schedule blower operation plays a brief role, preventing anaerobic conditions prior to the onset of blower demand (via temperature feedback). The equipment for the system consists of a thermistor and temperature-controller (feedback), and a timer (schedule). Inclusion of an event-recorder is desireable to log the number and duration of "demand" events. A complete description of this system is available (9).

Each batch of composting material, regardless of size, requires one control system. The equipment (non-expendible) cost is approximately $450 (includes one channel of a multi-channel recorder).

3.2 Heat Removal Mechanisms

We now reexamine the conclusion (1.1, item iii) that forcing air through the mass is the best way to remove heat.

3.2.1 Different approaches to ventilation

Two basic approaches may be taken to ventilate a composting mass: the material can be forced to pass through air, or air can be forced to pass through the matrix. The first corresponds to agitation of the mass (e.g. windrowing) and the second to use of a blower. With respect to the two options for blower operation, the forced-pressure mode is preferable for its greater heat removal efficiency (4).

In view of the large amount of mechanical energy involved, it would undoubtly be costly to exercise effective control over temperature by means of agitation. Agitation also serves to mix and abrade the material, but these functions are outside the scope of the present discussion. The relevant point is that agitation does not lend itself to temperature control for rate maximization. This contrasts to the proven effectiveness of forced-pressure ventilation in conjunction with temperature feedback control.

3.2.1.1 Air requirement

Assuming feedback control, the required installed blower capacity is defined by the peak demand for ventilation exerted by the particular waste. Our experience is with one sludge (2.2), mixed with either woodchips, or with screened (woodchip-free) recycled compost (10). Peak demand for air ranged from 57 to 134 m^3/wet tonne-hour.

Total air usage is that provided on a fixed schedule (small amount prior to onset of demand) plus that demanded via feedback. Total usage ranged from 3,370 to 13,140 m^3/wet tonne.

Demand for ventilation ceased because of dryness (e.g. Reference 1, Fig. 5). In all of our experience, cessation occurrred after 4.3 to 15.8 days of composting.

Total air usage for pile 9B (Beltsville Process) was 2326 m^3/wet tonne. Thus, the Rutgers strategy has the larger ventilation requirement. The supplying of air by blower is only a minor part of overall costs (11).

3.2.1.2 Uniform air distribution: an unresolved problem

If the composting mass is in the form of a lengthy pile, and if the placement of air delivery holes is uniform along the length of the ductwork, and the holes are of equal size, the distribution of air is non-uniform in the longitudinal dimension. This is because more air exits from the part of the ductwork close to the blower than from the distant part.

We presented a solution to this problem (12) which, after publication, was discovered to be flawed. Although a correction is being prepared, this is not available at the time of writing.

3.2.2 Conduction

Conductive heat removal inevitably occurs at the interface of the composting-mass and its surroundings (ground, ambient air, vessel walls — if relevant). However, conduction does not lend itself to controlled removal. Basically, this is because the rate of heat generation ($\equiv$ decomposition) is potentially high relative to the thermal conductivity of the composting material. Hence, heat cannot be removed fast enough to arrest the temperature ascent short of inhibitively high levels. Consider the following example.

A mixture of raw sewage sludge and woodchips has a bulk density of 500 kg/m^3 and a volatile solids content of 85%. At the peak level of activity heat output amounts to 0.0138 watts/g volatile solids (13). This translates to 5865 watts/m^3. Further, assume that the mixture has a thermal conductivity intermediate between that of wood (0.17 watts/m·C°) and that of water (0.556 watts/m·C°) (14). The analysis need not be continued to conclude that slow heat removal relative to its generation would result in inhibitively high temperatures in the bulk of the composting material.

Enhancing conduction through a heat exchanger system does not seem feasible, for at least three reasons. First, an extensive system of tubes, for example, throughout the composting mass would be necessary to meet the peak demand for heat removal. Second, condensation and localized wetting would occur near the exchanger walls. Third, it would still be necessary to ventilate the mass to supply O_2, thereby competing with conduction for heat removal. Management of the two heat removal mechanisms in reference to a desired operating temperature would be difficult. In all, there are difficult obstacles to deliberate removal of heat through conduction.

3.2.3 Direct water cooling

One other possibility might be noted briefly—that of spraying water onto the top of the composting mass while drawing off excess water at the bottom through vacuum-induced ventilation (Dr. Georg Raicov, personal communication). The idea of "water cooling" is intriguing in that it might be possible to realize the desired control of temperature and, unlike "air cooling", the removal of heat is unaccompanied by drying. In the absence of documentation, however, we are unable to evaluate this possibility. Also, dryness is often a goal of the composting process (see 4.1).

3.3 Height Limitation

Increasing the height of the composting mass seems to offer a means of decreasing the processing area required, but this approach quickly encounters physical and biological constraints. An obvious constraint is that frictional resistance, together with compaction, increasingly restricts airflow. But this is put aside in the following theoretical exercise.

Suppose compostable matrix fills a silo constructed of perfectly insulating material (no conductive loss to the surroundings). The silo is of infinite height, and the matrix offers no resistence to airflow. Air is forced up through the matrix at finite velocity. The intention is to prevent inhibitively high temperatures throughout the column. This proves impossible because air and matrix temperature increases with height, as long as the matrix is biologically active (1.1, item viii). At some point the temperature reaches a peak value, which corresponds to total suppression of biological activity (no heat generation), and this temperature persists throughout the rest of the column's height. The intention is defeated by the heat generation-temperature interaction (7).

Returning from this hypothetical exercise to the real world, we define "critical height" as the point in the vertical dimension at which the 60°C operational ceiling becomes established. Critical height depends on numerous factors, such as porosity, compactability, the heat generation characteristics of the waste, the enthalpy of the inlet air, and airflow velocity.

A partially defined example is provided by pile 7 (8), pile 8 (15), and pile 9A (1). Mean pile height was 2.0 meters, and mean ambient temperature was 21°C. During quasi-steady state operation (period of temperature feedback control) the temperature gradient in the upward vertical direction was roughly 23°C/meter. Thus, the pile height was barely consistent with the plan to maintain temperature $\leq$ 60°C in the bulk of the material.

3.4 Configuration and Strategy: Useful Distinction

It is useful to make explicit the distinction between the physical and conceptual levels of composting process organization. By physical we mean factors such as mass geometry, ventilation equipment, turning machinery; by conceptual we refer to composting ecosystem dynamics. Collectively, the physical attributes make up "process configuration"; a coherent concept for managing the composting ecosystem may be referred to as a "process control strategy". This distinction was basic to our field study (1), in which control strategy (Rutgers vs. Beltsville) was isolated as the experimental variable by holding configuration (static pile) constant (1.1).

3.5 Configurations: Unenclosed and "Enclosed"

During composting per se the useful physical manipulations are ventilation and agitation. One or both manipulations can be performed in various permutations, which comprise the configuration level of organization. Some examples follow. Example #1: the mass is in the form of a free-standing elongated pile, is not disturbed mechanically through turning or other agitation, and is ventilated by blower. This describes the static pile configuration (control strategy not specified). Static pile composting is usually unenclosed, but is sometimes roofed-over or enclosed in a shed. Example #2: the mass is in the form of an elongated free-standing pile, and is periodically turned. This describes the "windrow configuration", which also may be unenclosed, under a roof, or in a shed. Example #3: a hybird of # 1 and # 2 (ventilated by blower and turned). Example #4: the mass is in a silo-like vessel (single stage), the silo is loaded at the top and unloaded (via gravity) at the bottom, air is forced through the mass. Example #5: the mass is in a trench-like vessel, the material is agitated and moved via a digger or paddle arrangement, and is ventilated by blower. Example #6: the mass is in a rotating drum, which agitates and moves it from a point of entry to a discharge port. When a structure is involved, whether roof, shed, or vessel, this usually represents only a first composting stage. Typically, a subsequent curing stage is performed in an unenclosed area.

These examples demonstrate two points. First, at the level of configuration composting processes are highly individualistic. Commonly, a process is patented on the basis of some set of mechanical or structural characteristics. Second, regardless of configuration the same ecosystem dynamics are at play. Consequently, the control strategy designed to maximize decomposition rate can be implemented through many (though not all) configurations. Insofar as concerns structure, the least costly configuration is static-pile.

Nonetheless, enclosed, and especially invessel, composting is often credited with certain advantages relative to unenclosed composting, despite the usual absence of any coherent process control strategy. One purported benefit is protection against the elements. This is probably based on the notions that rainfall is harmful to the process and that conservation of heat is both enhanced and desirable. Neither idea withstands scrutiny. Moreover, vessel walls interfere somewhat with controlled heat removal through ventilation, by perturbing airflow patterns (16). Another disadvantage of enclosure is that the saturated exhaust air becomes a formal operational problem.

A common assumption is that enclosure improves odour control, but as was already developed, this objective is logically pursued through intervention at the level of ecosystem dynamics.

These structures are costly, whether consisting of a roof, a shed, or some type of vessel. The additional capital expenditure might be justified only if superior performance is demonstrable, based on objective criteria, relative to rationally controlled unenclosed composting.

3.6 Mass Balance

The data reported in Reference 1, and related data (2) were used to calculate mass balances (manuscript in preparation). This further extends the comparison between the two fundamentally different process control strategies, both being implemented in unenclosed static pile configuration. With respect to the mixture of sewage sludge and woodchips, the outcome is expressed in terms of Rutgers Process (designed to maximize decomposition rate) vs. Beltsville Process (minimal rate): composting time, 11.4 days vs. 21 days; total dry sludge solids decomposed, 38.1% vs. 10.4%; dry sludge volatile solids decomposed, 51.8% vs. 14.2%; water removed, 78.2% vs. 19.4%. Thus, approximately four-fold more waste treatment was accomplished in half the time through use of feedback control in reference to temperature.

Comparable data are available with respect to the use of recycled compost (woodchip-free) as the bulking agent, in static pile configuration using the Rutgers strategy (10). The following results represent a dry weight-recycle ratio of 1 part compost to 3 parts sludge: composting time, 9.5 days; total dry solids decomposed, 24%; dry volatile solids decomposed, 35%; water removed, 86%. The term solids refers to the sludge-compost mixture.

4. PRODUCT DEVELOPMENT

The composting field is traditionally product-oriented in that the dominant economic motive for operating facilities, at least until recently, was compost production. This orientation finds its fullest expression in the mushroom industry, which naturally seeks high quality feedstock for conversion to productive mushroom compost. The favoured feedstocks are usually agricultural wastes, which otherwise can be spread on soil without need for composting. Hence waste management is only peripherally involved. Production of compost for use as an organic soil amendment can be more closely linked to the needs of waste management, yet a product-orientation persists.

Waste management per se is a service-oriented enterprise dedicated to environmental quality. Many wastes or fractions thereof can be composted, and doing so may advance management goals even in the absence of productive use of the process residue (see 3.,4.1). It is desireable, of course, to find uses for this residue, as a negative factor (disposal) is thereby replaced with a positive one (resource-recovery). This might offset a minor part of the cost of the service. Regardless, management of the waste is the overriding motive, and as such takes precedence over any particular use of the process residue.

This perspective makes clear the distinction between process and product (17), and brings into focus the need of effectively integrating the two. The relationship between decomposition and drying intrinsic to the process frequently plays a decisive role, as developed below.

4.1 Drying per se as a Goal.

Removal of water from sewage sludge is often a major operational goal, as this improves ease of handling, storing, transporting and disposal. Composting, properly controlled, is an excellent means to this end, as the vaporization which causes the drying is driven by heat generated at the expense of putrescible (odour-causing) material. Two important goals are thus pursued in tandem. Moreover, since heat generation and vaporization are physically intimate, heat exchangers with their inevitable inefficiencies are not involved. Energy input in the form of ventilation is therefore moderate.

In our experience with raw sewage sludge the mean starting moisture content of the mixture of sludge and woodchips, or sludge and recycled compost, was 61%, and this decreased to 28% during 10.8 days of composting. (The amount of water removed was noted previously in section 3.6.) The composting was unenclosed, and as much as 13 cm of rain fell during the processing period, resulting in only superficial wetting. The cost of electricity to provide ventilation, assuming \$0.06/kw-hr, was \$0.38/tonne water removed. This performance may be compared to that of air drying (no biological heat generation), which was: 65% to 38% in 18.8 days, at a cost for ventilation of \$6.98/tonne water removed.

The much greater water removal through composting, per unit air, results from an increase in the air's enthalpy (heat content) with passage through the biologically active matrix. Biological heat generation accounts for approximatley 95% of the vaporization, with the remainder caused by unsaturation of the inlet air. In air drying the only force driving vaporization is unsaturation, hence its high cost.

It should be emphasized that the composting process performance summarized above was obtained through application of the Rutgers strategy, in static pile configuration. In the Beltsville Static Pile Process little water is removed (65% starting moisture content decreased to 61% in 20.8 days-also see 3.6), hence the goal of drying is not usefully advanced. This is manifested operationally in that the moisture content is frequently too high for screening for woodchip removal, sometimes causing an overall processing bottleneck. The prescribed remedy (6) is to remove water prior to screening through spreading and air drying.

4.2 Production of Solid Fuel

Composting's capacity to dry material could be exploited to convert moist wastes to fuel. In the course of the "biological drying" some of the fuel value is consumed, but without this preparatory step a much greater energy loss is incurred in the furnace. Projected advantages of separating fuel production from combustion in this manner, compared to combustion of the undried waste, are as follows: the fuel is more easily stored, transported, and fed to the furnace; the fuel is more homogeneous; fuel production can be decentralized; having been spared much of the evaporative burden, the furnace can be smaller; better combustion is obtained; furnace down-time does not disrupt waste management.

4.2.1 Direct heat recovery: poor prospects

The more usual suggestion for energy recovery through composting is via direct extraction of heat. We doubt the practicality of this approach, in part because the relationship between decomposition and drying leads to serious difficulties with heat exchangers (3.2.2). More fundamentally, the temperature of the exchanger medium (liquid or gas) can be elevated, on a sustained basis, only to a theoretical maximum of about 60°C. Though the composting mass can self-heat biologically to 80°C, heat removal at this temperature cannot be sustained (1.1, items i and ii; 7).

4.3 Production of Compost

Traditionally, composting (the process) has been used mainly to produce compost (a product). The vexed questions of defining compost and judging its suitability for soil and crop are outside the scope of our emphasis. For the present purpose we make the seemingly reasonable assumption that rate maximization speeds production of the desired product, however it might be defined.

This raises the possibility of premature drying — meaning that dryness comes to retard microbial action before the desired level of stabilization is reached (in part this is illustrated in Reference 1, Figs. 5 and 6). It is not possible to generalize about whether premature drying will occur, as this depends on the working definition of compost, the nature of the waste (heat generation characteristics and initial water content), the relative contribution of conduction to heat removal, and other factors. The tactic of conserving water by suppressing heat generation through inhibitive temperatures (e.g. Beltsville Process) is not attractive. The disadvantages are that compost production is slow; moreover, the conservation of water is necessarily accompanied by the conservation of putrescible (odour-causing) material. It is preferable to design and control the system for rate maximization and, if necessary to sustain biological activity, add water partway through the processing.

For the present limited purpose we define compost in terms of nitrification, as follows: when NO_2^- and/or NO_3^- appears in the composting material it is considered ready to be used as a compost. As part of the curing study (2.2, Table II), the material was periodically tested qualitatively for NO_2^- and NO_3^- (Table III). Even though the material derived from the Rutgers pile suffered a prolonged period of dryness, it was first to manifest nitrification. Based on this study it appears that raw sewage sludge can be converted to a thoroughly nitrified process residue in less than one month, provided there is timely addition of water to prevent premature dryness.

5. ROLE OF GOVERNMENT

We borrowed the term "cost-effectiveness" from a recent U.S. Government bulletin on composting (18). One of its stated purposes is to provide guidance to the Federal Construction Grants
facility design and operation. Nowhere in the bulletin, however, is cost-effectiveness defined, nor is the problem of its definition addressed. It must be added that the problem is a difficult one.

We are impressed by how frequently cost-related factors are affected by decomposition rate. This points to a useful area of government involvement. First, to evaluate basic, unifying, strategies for maximizing decomposition rate. Second, to consider suggestions for means of measuring decomposition rate in field practice (15). We believe that presently available data could then support "technology-forcing" guidelines.

TABLE III. Effect of process control strategy applied to parent pile on nitrification in derivative curing pile.

	Derivative curing pile (first appearance of oxidized N)			
	Total composting time (days)†		Total adjusted composting time (days)‡	
Parent pile (control strategy)*	NO_2^-	NO_3^-	NO_2^-	NO_3^-
Rutgers (9A)	86	99	25	38
Beltsville (9B)	113	123	NA	NA

*See Table II.

†Inclusive of parent and curing stages.

‡Adjusted = period of extreme dryness (ca. 25% moisture content) subtracted from total.

NA - not applicable.

ACKNOWLEDGEMENTS

This is the New Jersey Agricultural Experiment Station Publication No. 17513-1-84 and was supported by state funds, and by U.S. EPA (MERL/ORD Cincinnati) Project No. R806829010.

We thank Dr. P.F. Strom for many discussions on these matters.

REFERENCES

1. FINSTEIN, M.S., MILLER, F.C., STROM, P.F., MACGREGOR, S.T., PSARIANOS, K.M. (1983). Composting Ecosystem Management for Waste Treatment. Bio/Technology 1:347-353.
2. FINSTEIN, M.S., MILLER, F.C., MACGREGOR, S.T., PSARIANOS, K.M. Draft-final report to USEPA (Merl/ORD Cincinnati), submitted February, 1982.
3. FINSTEIN, M.S., CIRELLO, J., MACGREGOR, S.T., MILLER, F.C., PSARIANOS, K.M. (1980). Sludge Composting and Utilization: rational approach to process control. Final report to EPA, NJDEP, CCMUA, pp. 211. Rutgers, the State University of New Jersey. Available from U.S. Dept. Commerce, National Technical Information Service, Springfield, VA 22161, Accession No. PB82 136243.
4. MILLER, F.C., MACGREGOR, S.T., PSARIANOS, K.M., CIRELLO, J., FINSTEIN, M.S. (1982). Direction of ventilation in composting wastewater sludge. J. Water Pollut. Control Fed. 54:111-113.
5. EPSTEIN, E., WILLSON, G.B., BURGE, W.D., MULLER, D.C., ENKIRI, N.K. (1976). A forced aeration system for composting wastewater sludge. J. Water Pollut. Control Fed. 48:688-694.
6. WILLSON, G.B., PARR, J.F., EPSTEIN, E., MARSH, P.B., CHANEY, R.C., COLACICCO, D., BURGE, W.D., SIKORA, L.S., TESTE, C.E., HORNICK, S. (1980). Manual for composting sewage sludge by the Beltsville aerated-pile method. USDA/EPA. Report, EPA-600/8-80-022, (MERL/ORD Cincinnati), pp.65.
7. MACGREGOR, S.T., MILLER, F.C., PSARIANOS, K.M., FINSTEIN, M.S. (1981). Composting process control based on interaction between microbial heat output and temperature. Appl. Environ. Microbiol. 41:1321-1330.
8. MILLER, F.C., MACGREGOR, S.T., PSARIANOS, K.M., FINSTEIN, M.S. (1983). A composting processing failure: diagnosis and remedy. Indust. Waste: Proc. 15th Mid-Atlantic Conf., p. 463-471. Ann Arbor Science Book, Ann Arbor, MI.
9. MILLER, F.C., FINSTEIN, M.S. (1983). Equipment for control and monitoring of high rate composting. Proc. Intern. Symp. Biol. Reclamation Land Util. Urban Wastes, p. 551-560. (Eds., F. Zucconi, M. DeBertoldi, E.S. Coppola) Naples, 11-14 October 1983.
10. MILLER, F.C., MACGREGOR, S.T., PSARIANOS, K.M., FINSTEIN, M.S. (1982). Static-pile sludge composting with recycled compost as the bulking agent. Indust. Waste: Proc. 14th Mid-Atlantic Conf. p. 35-44. Ann Arbor Science Book, Ann Arbor, MI.
11. KASPER, V. JR., DERR, D.A. (1981). Sludge composting and utilization: an economic analysis of the Camden sludge composting facility. Final report to EPA, NJDEP, CCMUA pp. 342. Rutgers, the State University of New Jersey.
12. PSARIANOS, K.M., MACGREGOR, S.T., MILLER, F.C., FINSTEIN, M.S. (1983). Design of composting ventilation system for uniform air distribution. BioCycle 24(2): 27-31.
13. MILLER, F.C. (In preparation). Ph.D. Thesis, Rutgers University, New Brunswick, New Jersey.
14. HOLMAN, J.P. (1981). Heat Transfer, McGraw-Hill, Inc., New York, N.Y.
15. FINSTEIN, M.S., MILLER, F.C., STROM, P.F. (In press). Evaluation of composting process performance. Proc. Conf. Composting Solid Wastes and Slurries. The University of Leeds, 28-30 September 1983.
16. HOITINK, H.A.J., KUTER, G.A. (1983). Factors affecting composting of municipal sludge in a bioreactor. Draft-final report to EPA (CR-807791-01-0). Ohio Agric. Res. Devel. Center, the Ohio State University, Wooster.

17. FINSTEIN, M.S., MILLER, F.C. (1982). Distinction between composting (the process) and compost (the product). Letter to BioCycle 23(6):56.
18. Anonymous (1981). Composting processes to stabilize and disinfect municipal sewage sludge. pp. 43. EPA, Water Programs Operations (WH-547) 430/9-81-011, MCD-79. Washington, D.C.

DISCUSSION

BIDDLESTONE: Within the approach of 'waste management' how do you relate your strategy to the question of pathogen destruction? Is there an over emphasis on this aspect?

FINSTEIN: The need for pathogen control depends on the end use of the compost. Killing of pathogens results from a combination of time and temperature - a quick kill is not necessary. Pathogen control results in unwanted water removal when unnecessary attempts are made to achieve high temperatures, which are best at the end of the process, when required.

TECHNOLOGICAL ASPECTS OF COMPOSTING INCLUDING MODELLING AND MICROBIOLOGY

M. de BERTOLDI, G. VALLINI, A. PERA

Istituto di Microbiologia Agraria, Università degli Studi di Pisa, Italy
Centro di Studio per la Microbiologia del Suolo, C.N.R., Pisa

F. ZUCCONI

Istituto di Coltivazioni Arboree, Università degli Studi di Napoli, Italy

Summary

Composting is a bioxidative process leading to a higly stabilized organic product, which may then contribute directly to soil conditioning and fertility. Compared to other processes (fermentation or anaerobiosis), composting allows a fast, simple and safe approach to bulk treatment of organic wastes. It also produces less odour, and develops a thermophilic stage which decreases the concentration of animal and plant pathogens in the mass. On the other hand, fermentation and anaerobiosis lead to products which will undergo further stabilization (bioxidation) when exposed to air. For industrial composting of organic residues, maximum process efficiency is required, which in turn is a reflection of our capability to manage the microbial population growth and activity. Main factors which need to be optimized are: oxygen supply to the mass, moisture and temperature control, C/N ratio, balance of nutrients and pH. Also, fast diffusion and a proper succession of microbial populations, as separately required in the different stages, are most essential. Generally, substratum composition (carbon, nutrients, pH, moisture) and therefore microbial metabolism tend to become limiting factors when processing urban, agricultural and industrial waste. Process programming and plant design must then take into account the specific limitations in dealing with different organic materials. Management of the process of composting must take into account the end-product value and compatibility with plant cultivation. Production of phytotoxic metabolites characterizes the intermediate phases of organic matter degradation and may be used as a parameter to assess the degree of stabilization of the process. Use of organic soil conditioners in agriculture, may then involve different products in relation to the degree of stabilization, according to specific needs as well as considerations on costs and energy conservation.

1. INTRODUCTION

Composting as a means of waste disposal is increasingly used the world over. The concept of organic matter recovery for use in agriculture is becoming more popular than landfill and incineration. This is because composting not only permits energy recovery from waste but also guarantees disposal of the most highly polluting fraction: biodegradable organic matter.

More than one hundred plants are already in operation in the United States of America and many more have been constructed or have been planned (35).

As well as being an ecologically valid system, composting is not to be underrated agronomically. Indeed, composting provides large amounts of organic matter for agriculture, which can be used to replace that lost every year through normal agricultural practice. In Europe, where some farmland has been depleted for more than two thousand years, the problem of restoration of organic matter is becoming increasingly urgent.

Organic matter in soil is one of the fundamental necessities for biological fertility and when its content falls below certain levels there is no turning back in the process of what is known as "desertification".

It follows that composting is a solution to both problems of waste disposal and organic matter availability in agriculture.

2. WHY COMPOSTING?

Several different kinds of waste contain large amounts of organic matter which can be used in agriculture: solid urban waste, sewage sludge, agricultural waste, food-factory waste and some sorts of industrial waste. Some of these wastes require careful separation of organic matter from inert material (glass, plastic, metals, etc.). Moreover, all require size reduction and chemical-biological conditioning before using in agriculture. The organic fraction of wastes is heterogeneous (Table 1) and if introduced fresh in the soil, would undergo a rapid process of transformation by the soil microflora. Composting radically transforms the various organic substances; it mineralizes the simpler and easily assimilable and humifies more complex compounds.

Table 1. Organic fraction composition of urban solid waste.

ITEM	Percentage
Volatile matter	70-90
Protein	2-8
Lipids	5-10
Total sugar	5-7
Cellulose	30-60
Starch	2-8
Lignin	3-8
Crude fibre	35-40

Mature compost is made up of fairly homogeneous organic matter with high molecular weight and so more stable (1,2,36).

Addition of fresh organic matter to the soil is to be avoided because it results in a change in the ecosystem in which the crop is developing (28). Once the organic matter is placed in the soil, if it is not partially humified, it will be degraded by the microflora resulting in a production of intermediate metabolites which are not compatible with normal plant growth (27,37,38). Other disadvantages are competition for nitrogen between microorganisms and roots, a high carbon/nitrogen (C/N), and the production of ammonia in the soil (10,17).

Composting is therefore a way of obtaining a stable product from biological oxidative transformation, similar to that which naturally occurs in the soil.

3. PRACTICAL SYSTEMS OF COMPOSTING

3.1. Principles

There is still much confusion today over the meaning of the word compost. Composting is a microbial reaction of mineralization and partial humification of organic substances which under optimum conditions take place within a month. It is very difficult to decrease this time and it is not possible for composting to take place in a few days as many assert . Composting time depends on the biological cycles of the microorganisms involved. Their replication time is conditioned by environmental factors and genetical constitution of the microorganisms. Although environmental factors may be improved, genetic limits remain .

In nature, organic matter spontaneously undergoes several processes of microbial transformation according to the composition of the substrate and the chemical and physical environment which microorganisms inhabit. The end-products are completely different depending on whether conditions are aerobic or anaerobic. Composting requires that the process be mostly aerobic so that organic matter is partially mineralized and humified.

Spontaneous composting occurs in nature, for instance the transformation of plant litter and manure; however these processes are slow, discontinuous and heterogeneous. To make composting suitable for the industry of waste disposal requires three fundamental points to be met: (1°), brevity of the process and low energy consumption; (2°), to guarantee a standard end-product not only safe for agriculture use but also of satisfactory fertilizing value; (3°), hygienic safety of plant and end-products.

For these three requirements to be satisfied simultaneously, composting cannot be spontaneous. It must be controlled in order to guarantee low costs and high quality end-product.

This means that several different kinds of practical composting systems have been devised: these are reported in Table 2.

The aim of all these systems is basically that of creating the best conditions for the process. These conditions directly influence the growth

Table 2. Summary of composting systems for waste.

OPEN SYSTEMS

- Turned pile
- Static pile
  - air suction
  - air blowing
  - alternating ventilation (blowing and suction)
  - air blowing in conjunction with temperature control

CLOSED SYSTEMS

- Vertical reactors
  - continuous
  - discontinuous
- Horizontal reactors
  - static
  - with movement of material

and metabolism of microorganisms which carry out the process. Therefore the factors which directly influence the process of composting are those which condition microbial metabolism (3,5,6,8,11,13,21,22,23).

In composting, the main factor that can be most influenced by technology, around which system designs are developed, is the availability of oxygen. With respect to design, the equipment for providing aeration ranges from the relatively simple to the very complex. This range leads to the generalized classification of compost technology described earlier as open (windrow, pile) and closed (mechanical, in a vessel or container). In practical terms the important point is how the two classes of systems compare with respect to effectiveness in accomplishing aeration, and even more important how they meet the objectives. If no significant improvement of the process and its product over that with a windrow system accompanies the use of a mechanized system, then the windrow system should be adopted, because the economics of the mechanized approach are less favourable (11).

The main features of open and closed systems are reported below.

3.2. Windrow

A windrowed mass of composting material can be aerated by either or both of two methods namely turning and forced aeration.

3.2.1. Turned pile

Pile turning, though widely used because it is simple, has its limits. In the first place the pile is oxygenated only periodically by turning. Since the optimal level of oxygen must be kept constant to enhance biological oxidations, interrupted aeration may be unsuitable (7) (Fig. 1).

Pile turning always requires more space, whether the piles in turning are moved laterally or when special machines are used which leave the pile in the same place; because, in the second method, the pile must not be too high for the machine to turn it. Periodic turning for disease control is not completely satisfactory (7,20). It must be added that, during the final stages of composting, when the material is nearly dry, dust containing a lot of Aspergillus fumigatus spores is released into the air (30,31).

3.2.2. Static pile

The idea of aerating a static composting pile by forced air is perhaps the best, because it allows exact amounts of oxygen to be given and enables the control of other important parameters such as moisture and temperature. It also has the advantage of low costs, smaller areas and there is no periodic turning of piles with all its disadvantages.

Ventilation of piles can be done in several ways (see Table 2).

The suction system was devised at Beltsville by the U.S.D.A. (12,34) and is widely adopted particularly in the U.S.A. With this system, an air flow of about 0.2 cubic meters/min/tonne at the blower inlet is sufficient to provide an oxygen concentration of 15% in a windrow of combined sludge cake and wood chips. To reduce the problem of odours, the air can be passed through a small filter pile of mature compost.

The process devised at Rutgers University in New Jersey is based on blowing and temperature control (14). A temperature controller with an adjustable temperature set point, continuously receives and interprets a signal from a thermistor in the pile. When the signal indicates a temperature more than the set point, the controller actuates the blower until ventilation decreases the pile temperature. This system has two main advantages in comparison to the preceding: blowing enhances evaporation giving a low moisture end-product and guaranteeing high stability. Automatic temperature control avoids long periods of high temperature (>60°). Since most microorganisms involved in composting do not survive at temperatures above 60°C, it is advisable that this temperature should not be reached in composting so that microbial activity is at maximum efficiency. This is particularly true for fungi which are strongly inhibited at high temperatures and which are useful in degrading cellulose and lignin.

High temperatures are highly selective towards the microflora; very few sporigenous bacteria survive above 70°C. This means composting is arrested until the temperature falls and the microorganisms can reinvade the mass.

Although high temperatures inhibit microbial population, they have, however, a positive effect in reducing pathogens. For this reason several processes include an initial phase of suction composting which permits the temperature to rise for a few days; the air stream is then inverted and blowing, in conjunction with temperature control, continues the process.

3.3. Closed systems

3.3.1. Vertical reactors

These reactors, generally over 4 m high, can be continuous - composting material in one large mass - or discontinuous - mass arranged on different floors. Discontinuous vertical reactors contain, on different levels, piles no higher than 2 or 3 m and therefore have no disadvantages apart from the high cost of the plant and maintenance.

In continuous vertical reactors, which may contain one single mass up to 9 m high, the process is extremely difficult to control, because uniform oxygen cannot be obtained even when blowing from beneath is used. Since volume of air per unit surface area must be proportional to the height of the pile, the lower part of the mass is over-ventilated with excessive cooling and drying and the upper layers are insufficiently aerated, because as air passes through the composting mass, it changes composition losing oxygen and acquiring carbon dioxide. For this reason, forced ventilation in reactors and in windrows is not suitable for heights over 3 m (Fig. 2).

Continuous vertical reactors may also be difficult to manage if the discharging mechanisms break down or need maintenance and repair. Manual discharge of 1000-2000 m^3 reactors is rather difficult.

3.3.2. Horizontal reactors

Many consider the Dano biostabilizer cylinder and other similar cylinders as horizontal reactors for composting. Although valid equipment, they are not to be considered as reactors for composting. Their main function is the differentiation of the various components of waste, through biological and physical-mechanical means. This simplifies separation of inert materials from biodegradable organic matter. Retention times of 24 to 36 hours are not enough for true composting, and only allow an initial process of microbial degradation of organic matter, which is most important for the following reasons. First, all organic matter is physically differentiated from inert materials, making separation easier and improving yield because of the smaller particle size of organic matter (9). Second, because organic matter obtained from Dano-type biological reactors is physically more homogeneous, it is partially biodegraded and uniformly invaded by microorganisms. The issuing material is therefore ideally prepared for the final process of composting whether in windrow or in reactor.

Let us now consider various horizontal reactors for composting. Generally preselected organic matter (separated from inert material) is used in these reactors and the process requires from 15 to 30 days. Material inside these reactors may be static or periodically turned and piles should not be higher than three meters. In these reactors air is nearly always blown from the bottom. Some include temperature control in addition to blowing.

If this equipment is used correctly, it combines the advantages of both open and closed systems, giving better control over the process and odours.

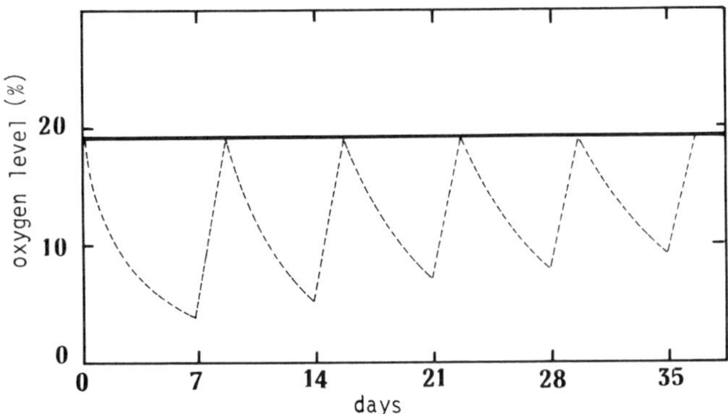

Fig. 1. The straight line represent the oxygen demand of a composting mass and the doted line the oxygen supplied with a periodic pile turning system.

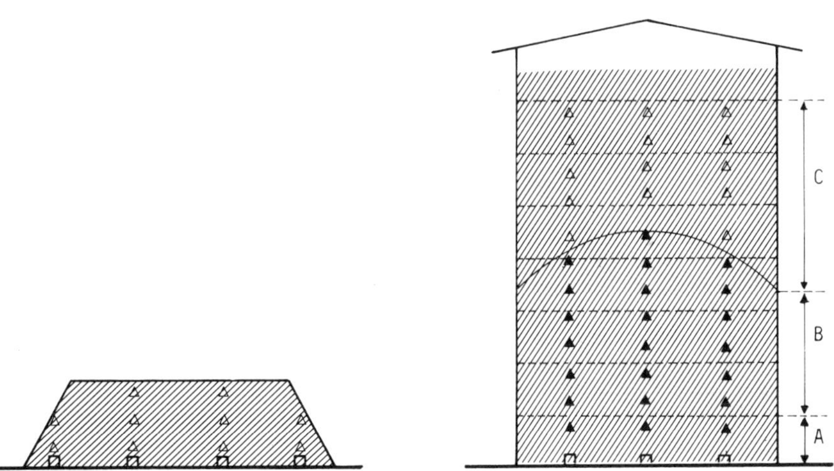

Fig. 2. On the left , a scheme for a static pile with forced aeration in which all the mass (not higher than 2.5 m) is correctly oxygenated. On the right, a scheme of a continuous vertical reactor with forced aeration from the bottom in which three different zones can be recognized: A, an overventilated and dried zone; B, zone with nearly correct oxygen supply; C, anaerobic zone.

The same principles as those for vertical reactors apply to this type: simple function and cheapness, otherwise they are not suitable.

4. THEORETICAL CONSIDERATION ON COMPOSTING

When discussing composting a careful distinction must be made between questions regarding the plant and those concerning the process itself. By plant it is meant the system of machinery, buildings and equipment for composting. By process we mean the correct application of the biological transformation with optimization of all the parameters and conditioning factors.

Today well made and sophisticated plants are available; much study has been dedicated to both experimental and theoretical aspects of the process, but the results have little practical industrial application. Civil and sanitary engineers who design and construct composting plants and microbiologists who analyse and devise methods for composting have been working independently and separately. Closer collaboration would certainly improve the practical results.

The theoretical principles underlying composting are quite simple; they must however be scrupulously respected during composting if the quality and the stability of the final product are to be guaranteed Unfortunately, some industrial plants have been constructed with no consideration for these principles, with the consequence that low quality phytotoxic and unstable composts have been put on the market. This can both be dangerous and make marketing of good quality compost difficult. The situation is further complicated by the absence of any legal and commercial definition of what is meant by the term compost. This means that series of products of different quality are available on the market; all are called compost, but have little in common.

4.1. Operational parameters

Many detailed analyses of the main parameters which govern composting have already been reported (3,5,7,8,11,14,15,16,29,32,33). Since composting is an exclusively biological process, all those factors which influence, whether directly or indirectly, microbial metabolism, affect the process.

From the point of view of the process itself, the optimum conditions mean producing a good quality compost in a short time. This result can only be obtained by working with all the parameters which control the process at an optimum level.

Optimization of these parameters depends on the system used. These parameters will vary according to whether the process is carried out in windrows or in a reactor and whether air flow is blown or sucked. For this reason it is not possible to define the ideal condition of raw materials without considering the composting system to be adopted.

The following is an analysis of the principal factors that control composting.

4.1.1. Composition of the mass

The preparation of organic matter for composting is at present carried out in two ways: (1°), mechanical; (2°), combined biological and mechanical. Mechanical processes involve size reduction by shredding, followed by separation of inert materials by screening, dynamic separation, air-flow separation, electromagnetic separation, etc. In combined biological-mechanical processes, waste is placed in biological reactors for a few (1-3) days, where the process of biological degradation begins together with mechanical size reduction. After this phase in the reactor, the biodegradable organic fraction has undergone drastic conditioning; it emerges macerated and broken-down so that separation from inert material is easily effected mechanically. Furthermore, the organic fraction has undergone microbial attack and so is better conditioned for the following process of composting; in addition to reduction to optimum size, it is more homogeneous and is uniformly invaded by microorganisms. This means that the physical and chemical characteristics of matter for composting can vary according to the preparation to which it has been submitted.

Organic matter for composting is composed of solids, water and gas with a constant interchange among the three fractions. Solid matter consists of ash, inert material and biodegradable organic matter containing water. The individual relationships among these components are extremely important for the evolution of the process and the quality of the end-product (25,26). These parameters can vary, as will be seen, within a limited range if the process is to proceed correctly. Microbial transformation of the organic fraction into compost is essentially an aerobic oxidative process. This means that the surface:volume ratio of the particles has a direct influence on the manner and speed of degradation; the air:water ratio in the particle interspace is equally important. Water and oxygen are indispensable for microbial activity and when the proportion is lower than the critical level, microbial metabolism and respiration slow down and stop; this means that the composting process slows or stops too.

4.1.2. Aeration of the mass

The air contained in the interspaces of the composting mass at the beginning of microbial oxidative activity varies in composition. The carbon dioxide content gradually increases and the oxygen level falls. The average $CO_2 + O_2$ content inside the mass is about 20%; oxygen concentration varies from 15 to 20% and carbon dioxide from 0.5 to 5% (29). When the oxygen content falls below this level, anaerobic microorganisms begin to exceed aerobic ones. Fermentation and anaerobic respiration processes take over. It is obvious therefore that the microorganisms must have a constant supply of fresh air to maintain their metabolic activities unaltered. In composting, oxygen is not only necessary for aerobic metabolism and respiration of microorganisms, but also for oxidizing the various organic molecules present in the mass.

One of the main functions of aeration is that of supplying oxygen so that a low level does not limit the process. Since the optimum level of oxygen must be constantly maintained to enhance bioxidative transformations, discontinuous aeration, such as periodic turning, is not suitable in this respect (7).

Oxygen consumption during composting is directly proportional to microbial activity. Therefore there is a direct relationship between oxygen consumption and temperature. Temperatures which enhance microbial activity are in the range 28-55°C, with a highest consumption of oxygen (15,24).

4.1.3. Temperature

Too often, high temperatures have been considered a necessary condition for good composting. High temperatures result from biological activity: heat liberated through respiration of microorganisms decomposing organic matter builds up within the pile since dispersion is low due to the natural insulation of solid urban waste (8,13). From the ecological point of view, Finstein calls this process "microbial suicide" (15). Excessively high temperatures inhibit growth in most of microorganisms present, thus slowing down decomposition of organic matter. Only a few species of thermophilic sporigenous bacteria show metabolic activity above 70°C: *Bacillus stearothermophilus*, *Bacillus subtilis*, *Clostridium* sp., and non-spore forming bacteria, gram-negative, aerobic: genus *Thermus* (8).

For rapid composting, high temperatures for long periods must be avoided. An initial thermophilic phase may be useful in controlling thermosensitive pathogens.

The problem of temperature control is solved by using forced ventilation throughout the process. Its advantage is that evaporative cooling is induced in the region of the pile that is most highly insulated and heat is carried to the outer layers. Another important feature in controlling temperature is the use of a temperature controlling unit responding to a temperature sensor placed in the pile. A ceiling temperature is established in which aeration time is automatically regulated by heat output, which is of course a function of biological activity (7,14,15,29).

4.1.4. Moisture

Moisture content and aeration are closely interrelated in terms of displacement of air in the interstices by water and promotion of aggregation and lowering of the structural strength of the material (18). Optimal moisture content in composting varies and essentially depends on the physical state and size of the particles and on the composting system used. Too little moisture means early dehydration of the pile which arrests the biological process giving a physically stable but biologically unstable compost. Too much interferes with the aeration by clogging the pores (5,7,19).

4.1.5. Carbon/nitrogen ratio

Low C/N ratios will slow decomposition and increase nitrogen loss. If the initial C/N ratio is greater than 35, the microorganisms must go through many life-cycles, oxidizing off the excess carbon until a more convenient C/N ratio for their metabolism is reached; low values would lead to a nitrogen loss through ammonia volatilization expecially at high pH and temperature values.

After extensive experiments on composting it was demonstrated that the general optimum C/N ratio was 25 in the starting material; higher values slowed the rate of decomposition, lower ones resulted in nitrogen loss (4,7).

4.1.6. pH level

Generally, organic matter with a high range of pH (from 3 to 11) can be composted. However, optimum values are between pH 5.5 and 8.0. Whereas bacteria prefer a nearly neutral pH, fungi develop better in a fairly acid environment.

In practice, the pH level in a pile cannot easily be changed. Generally, the pH begins to drop at the initiation of the composting process, as a consequence of the activity of acid-forming bacteria which break down complex carbonaceous material to organic acid intermediates.

High values of pH in the starting material in association with high temperatures can cause a loss of nitrogen through volatilization of ammonia.

5. CONCLUSIONS

The quantity of organic waste increases every year and disposal problems increase in proportion.

Many traditional disposal systems still in use are not sufficiently safe either ecologically or hygienically.

The present energy crisis makes it wasteful to throw away potentially recyclable refuse. Ecological factors and energy conservation oblige us to be increasingly careful about waste disposal.

Composting biodegradable organic waste is one of the best solutions to these problems because it guarantees complete disposal of waste and allows efficient recovery of energy in the form of organic fertilizer to be used in agriculture.

In order to avoid irreversible pollution of our agricultural land, every possible precaution must be taken from the start:

(1), avoid spreading on land non biodegradable toxic compounds, such as heavy metals, pesticides, plastics, etc.;

(2), good quality stabilized compost must be supplied to agriculture to prevent damage to growing crops;

(3), compost must be free from pathogens before use in agriculture;

(4), guidelines and legislation in this field must be introduced into Europe to define standards of quality for compost.

REFERENCES

1. CHANYASAK, V. and KUBOTA, H. (1981). Carbon organic ratio in water extract as measure of composting degradation. J. Ferm. Technol., 59, 215-219.
2. CHANYASAK, V. HIRAI, M., and KUBOTA, H. (1982). Changes of chemical components and nitrogen transformation in water extracts during composting of garbage. J. Ferm. Technol., 60, 439-436.
3. DE BERTOLDI, M., CITERNESI, U. and GRISELLI, M. (1979). Microbial studies on sludge processing. In "Treatment and use of sewage sludge". Proceedings of C.E.C. First European Symposium, Cadarache, 13-15 Febr. 1979 (D. Alexandre & H. Ott Eds.) p. 77, C.E.C., Bruxelles.
4. DE BERTOLDI, M. , CITERNESI, U. and GRISELLI, M. (1980). Bulking agents in sludge composting. Compost Science and Land Utilization, 21 (1), 32-35.
5. DE BERTOLDI, M. and ZUCCONI, F. (1980). Microbiologia della trasformazione dei rifiuti solidi urbani in compost e loro utilizzazione in agricoltura. Ingegneria Ambientale, 9, (3), 209-216.
6. DE BERTOLDI, M., CITERNESI, U. and GRISELLI, M. (1982). Microbial population in compost process. In "Composting" (The staff of Biocycle), p. 26. The JG Press, Emmaus, PA, USA.
7. DE BERTOLDI, M., VALLINI, G., PERA, A. and ZUCCONI, F. (1982). Comparison of three windrow compost systems. Biocycle, 23, (2), 45-50.
8. DE BERTOLDI, M., VALLINI, G. and PERA, A. (1983). The biology of composting: a review. Waste Management & Research, 1, 157-176.
9. DE BERTOLDI,M., PERA, A. and VALLINI, G. (1983). Principi del compostaggio. Proceedings of the international Symposium on "Biological reclamation and land utilization of urban waste". Napoli, 11-14 October 1983, Ed. by F. Zucconi, M. de Bertoldi and S. Coppola.
10. DE VLEESCHAUWER, T, VERDONCK, O. and ASSCHE, P., VAN (1982). Phytotoxicity of refuse compost. In "Composting" (The Staff of Biocycle) p. 54, The JG press, Emmaus, PA, USA.
11. DIAZ, L.F., SAVACE, G.M. and GOLUEKE, C.G. (1982). Resource recovery from municipal solid wastes, vol. II. C.R.C. Press, Boca Raton, Florida.
12. EPSTEIN, E., WILLSON, G.B., BURGE, W.D., MULLEN, D.C. and ENKIRI, N.K. (1976). A forced aeration system for composting waste water sludge. J. Water Poll. Contr. Fed., 48, 688-693.
13. FINSTEIN, M.S. and MORRIS, M.L. (1975). Microbiology of municipal solid waste composting. Advances in Applied Microbiology, 19, 113-151.
14. FINSTEIN, M.S., CIRELLO, J., MACGREGOR, S.T., MILLER, F.C. and PSARIANOS, K.M. (1980). Sludge composting and utilization; rational approach to process control. Final report to USEPA, NJDEP, CCMUA, Rutgers University,

New Brunswick, NJ, USA.

15. FINSTEIN, M.S., CIRELLO, J., MACGREGOR, S.T., MILLER, F.C., SULLER, D.J. and STROM, P.F. (1980). Discussion of Haug R.T. "Engineering principles of sludge composting" J. Water Poll. Contr. Fed., 52, 2037-2042.
16. FINSTEIN, M.S., MILLER, F.C., STROM, P.F., MACGREGOR, S.T. and PSARIANOS, K.M. (1983). Composting ecosystem management for waste treatment. Biotechnology, 1, 347-353.
17. GOLUEKE, C.G. (1977). Biological reclamation of solid wastes. Rodale press, Emmaus, PA, USA.
18. GOLUEKE, C.G. (1982). Selection and adaptation of a compost system. In "Composting" (The Staff of Biocycle) p. 85, The JG Press, Emmaus, USA.
19. GOLUEKE, C.G., LAFRENZ, D. CHASER, B. and DIAZ, F.L. (1980). Composting combined refuse and sewage sludge. Compost Science & Land Utilization, 21, (5), 42-48.
20. GOLUEKE, C.G. and MACGAUHEY, P.H. (1953) Reclamation of municipal refuse by composting. Tech. Bull. 9, Sanitary Engin. Res. Lab., Univ. Of California, Berkeley.
21. GRAY, K.R. and BIDDLESTONE, A.J. (1973). Composting process parameters. The Chem. Engineering, 2, 42-48.
22. GRAY, K.R. and BIDDLESTONE, A.J. (1974). Decomposition of urban waste. In "Biology of Plant Li tter Decomposition" (C.H. Dickinson and G.J.F. Pugh Eds.) p. 743, Academic Press, London, U.K.
23. GRAY, K.R., SHERMAN, K. and BIDDLESTONE, A.J. (1971). A review of composting, Part 1. Process Biochemestry, 6, (6), 32-36.
24. HAUG, R.T. (1979). Engineering principles of sludge composting. J. Water Poll. Cont. Fed., 51, 2189-2195.
25. HAUG, R.T. (1980). Compost Engineering: principle and practice. Ann Arbor Science, Ann Arbor, Michigan, USA.
26. HAUG, R.T. (1982). Modeling of composting process dynamics. In "Proceedings of the national conference on composting of municipal and industrial sludges". May 24-26 1982, Washington, D.C.
27 KATAYAMA, A., CHANYASAK, V., HIRAI, M.F., MORI, S., SHODA, M. and KUBOTA, H. (1983). Effect of compost maturity on komatsuma (*Brassica rapa* var. *previdis*) growth in Neubauer's pot. Proceedings of the International Symposium on "Biological Reclamation and Land Utilization of Urban Waste", Napoli 11-14 October 1983, Ed. by F. Zucconi, M. de Bertoldi and S. Coppola.
28. KONONOVA, M.M., NOWAKOWSKI, T.Z. and NEWMAN, A.C.D. (1966). Soil organic matter. Pergamon Press, Oxford, U.K.
29 MACGREGOR, S.T., MILLER, F.C., PSARIANOS, K.M. and FINSTEIN, M.S. (1981). Composting process control based on interaction between microbial heat output and temperature. Applied Environ. Microbiol., 41, 1321-1330.
30. MILLNER, P.D., D., BASSETT, D.A. and MARSH, P.B. (1980). Dispersal of *Aspergillus fumigatus* from sewage sludge compost piles subjected to mechanical agitation in open air. Applied Environ. Microbiol., 39, 1000-1009.

31. MILLNER, P.D., MARSH, P.B., SNOWDEN, R.B. and PARR, J.F. (1977). Occurrence of Aspergillus fumigatus during composting of sewage sludge. Applied Environm. Microbiol., 34, 765-772.
32. MILLER, F.C., MACGREGOR, S.T., PSARIANOS, K.M., CIRELLO, J. and FINSTEIN, M.S. (1982). Direction of ventilation in composting wastewater sludge. J. Water Pollution Control Federation, 54, 111-113.
33. SULER, D.J., and FINSTEIN, M.S. (1977). Effect of temperature, aeration and moisture on CO_2 formation in bench-scale, continuously thermophilic composting of solid waste. Applied Environm. Microbiol., 33, 345-350.
34. WILLSON, G.B., PARR, J.F., EPSTEIN, E., MARSH, P.B., CHANEY, R.L.,COLACICCO, D., BURGE, W.D., SIKORA, L.J., TESTER, C.F. and HORNICK, S. ((1980). Manual for composting sludge by the Beltsville aerated pile method. USDA, EPA 600/8-80 002, Cincinnati, Ohio, USA.
35. WILLSON, G.B., and DALMAT, D. (1983). Sewage sludge composting in the U.S.A. Biocycle, 24, (5), 20-23.
36. YOSHIDA, T. and KUBOTA, H. (1979). Gel chromatography of water extract from compost. J. Ferm. Techn., 57, 582-584.
37. ZUCCONI, F., FORTE, M., MONACO, A. and DE BERTOLDI, M. (1981). Biological evaluation of compost maturity. Biocycle, 22; (4), 27-29.
38. ZUCCONI, F., PERA, A., FORTE, M. and DE BERTOLDI, M. (1981). Evaluating toxicity of immature compost. Biocycle, 22 (2), 54-57.

This work was partially supported by Consiglio Nazionale delle Ricerche (C.N.R.), Progetto Finalizzato "Energetica 2".

DISCUSSION

LYNCH: I am concerned that you mention *Aspergillus fumigatus* spore clouds can be liberated from the composts you describe. As this is potentially a serious medical problem, would you please elaborate on it.

de BERTOLDI: *Aspergillus fumigatus* conidia are normally present during composting organic material; because of its thermo tolerance, it is not destroyed at temperatures of 55 - 60^{C} and therefore is still present in the end product. This does not constitute a risk because *A fumigatus* is normally present in all soils.

However, there may be some risks in composting plants for the worker, which are related to the composting system used. Enclosed systems are quite safe. Also static pipe with forced ventilation does not give too many problems in this context (personal experience and the work of Millner *et al*).

The problem may be serious with periodical turning especially when the material is nearly dry. The workers, if hypersensitive or having other pulmonary disease, may have a severe allergic response. In some people the inhaled conidia may germinate and the fungus may invade lung-parenchyma to produce typical aspergillosis.

BIDDLESTONE: If the air flow through a static pile is increased, is the problem of spore liberation exacerbated?

de BERTOLDI: I agree it may be but I am uncertain of the extent of the problem.

FLEGG: Experiments with mushroom composts show that oxygen levels in stacks drop initially, but then recover due to thermal effects drawing air into the stack through the sides. We find that stack turning is required more to mix the constituents of the stack than for oxygenating it.

de BERTOLDI: With a two metre high pile turning twice weekly is inadequate to provide a satisfactory oxygen level. Comparative experiments have demonstrated that the biological decomposition of organic matter during composting with the turning system is slower than with forced ventilation.

FORCED AERATION CO-COMPOSTING OF DOMESTIC REFUSE AND SEWAGE SLUDGE IN STATIC PILES

E.I. STENTIFORD, D.D. MARA AND P.L. TAYLOR
Department of Civil Engineering, The University of Leeds

Summary

Unsorted domestic refuse from the Doncaster area of South Yorkshire was shredded and mixed with sewage sludge (4-5% solids) to form the structural material for static pile composting. Positive and negative pressure aeration were used but the former was preferred from the point of view of temperature control. The microcomputer-based control programmes developed were used to maintain maximum temperatures in the range 50-60°C. Oxygen concentrations in the pile were measured, in addition to temperature, to improve control during the initial and end phase of composting.

In limited growth trials the Leeds compost was comparable with commercially available material. The addition of small amounts of compost to clay, pulverised fuel ash and shale produced a marked increase in grass yields. The production costs for compost do not make it economically viable where low cost landfill alternatives exist for refuse. It is economically attractive when compared with reactor-based systems.

1. INTRODUCTION

The composting of domestic refuse, with or without sewage sludge, has been practised for many years on various scales. In the last 50 years in particular a wide variety of schemes has been implemented (4) with varying degrees of success. These plants have either been capital intensive reactor-based systems or the low capital cost windrow type installations. A major disadvantage with a windrow system is the lack of any real control of the composting process unlike the more expensive reactor-based plants.

The composting of sewage sludge using forced aeration in static piles has been developed in the U.S.A. in the last ten years (1,2). The recent developments have included extensive improvements in process control (2). In the majority of full scale operations woodchips have been used as a bulking agent to give the required open structure to ensure adequate aeration. The woodchips are relatively inert and are not changed greatly during composting. In order to make these processes economically viable a high rate of woodchip recovery is required during the final compost screening.

The natural progression seemed to be a controlled forced aeration static pile system using domestic refuse in admixture with sewage sludge. This would not only have a level of control comparable with reactor-based systems but its economics would not be dependent on the recovery of the bulking agent. The object of the research work at the University of Leeds was to evaluate the possibilities of this composting process.

2. MATERIALS AND PROCESSING

The work reported in this paper relates to the use of domestic refuse and sewage sludge from the Doncaster area of South Yorkshire, England. Typical analyses for these materials are shown in Tables I and II. The sewage sludge showed only small variations around these reported values over the particular two year period concerned. However, the refuse not only varied markedly on an annual but also on a daily basis depending on the area of the town from which the collection was being made. The values presented in Table I should therefore only be used to give an approximate value for refuse composition. The heterogeneous nature of the material makes research work on a small scale a difficult undertaking if valid conclusions are to be drawn. It was for this reason that pile sizes between 20 and 50 tonnes were used.

The refuse as it arrived at the receiving station was not ideally suited to immediate composting due to its bulky nature. A simple refuse cutter/shredder was used (Mono Muncher type H27-60-2, Mono Pumps Ltd., Manchester) to break down this material into strips approximately 30 mm wide and up to 150 mm long. This gave a material which:

(i) had a relatively large surface area per unit volume, thus enhancing the breakdown rate of the refuse and making a greater area available for coating with sewage sludge;

(ii) had sufficiently large particles to give some degree of structural rigidity to the compost piles, thus ensuring reasonably good air transfer through the piles; and

(iii) had received some degree of mixing prior to adding sludge which at least meant that the material within each pile was relatively evenly distributed.

It was not within the scope of this project to evaluate the optimum size for the average particle of refuse used in a static pile composting system. Whilst the size used gave reasonable results further work is needed to assess whether changes in size would improve the overall process performance. However, the machine used did have the following characteristics which were thought to be desirable:

(i) it accepted all domestic refuse without presorting;

(ii) it was relatively inexpensive;

(iii) it was simple to install and operate; and

(iv) it produced very little dust.

3. PILE STRUCTURE

The initial work with refuse/sludge piles used a negative pressure (sucking) system mode of fan operation similar to the method developed at Beltsville in the United States (7). This was not particularly successful due to pile collapse restricting airflows in the lower sections of the pile. This collapse was most marked in the first week after pile construction (Figure 1) but continued progressively over the whole of the composting period. The initial collapse was due mainly to physical settling/consolidation of the pile with movements in the latter stages resulting from breakdown of the large organic fractions in the refuse. Similar changes have not been reported with sludge/woodchip piles due probably to the relatively inert nature of the bulking agent. Where shredded refuse is used it is not only the bulking agent but also a large proportion of the pile substrate.

Modifications to the pile structure, in particular using straw as the initial covering layer for the aeration pipes (Figure 2) produced better results with negative pressure operation. However, an important aspect of our work was to establish good control over the process, especially from a temperature point of view, and this was not found to be possible using this method.

Table I Typical Analysis for Doncaster Refuse

REFUSE FRACTIONS (AS RECEIVED)	% BY WEIGHT	% BY VOLUME
< 20 mm	15	5
Vegetables and Putrecibles	28	15
Papers	24	49
Metals	11	13
Textiles	4	5
Glass	11	2
Plastics	5	9
Unclassified	2	2
Total	100	100
Compostable	52 - 67	64 - 69
Moisture Content	35% (approx)	

Table II Typical Analysis for Sandall (Doncaster) Sewage Sludge

PARAMETER	AMOUNT
pH	5.7
Moisture Content %	96 *
Total Solids %	3.9*
Volatile Solids %	74.5
Ash %	25.5
Total Kjeldahl Nitrogen %	4.0
Total Organic Carbon %	41.4
Potassium (K) %	0.17
Total Phosphorus (P) %	1.71

*Based on the wet weight, all other values refer to dry solids.

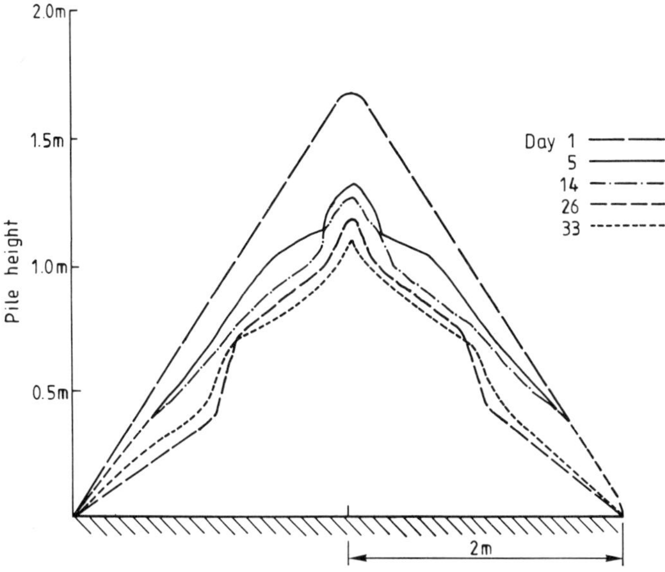

Figure 1 - Changes in pile cross-sectional dimensions as composting progresses (refuse/sludge mixture).

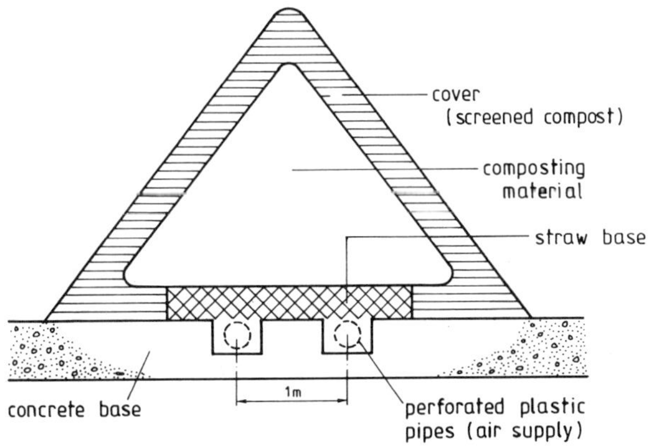

Figure 2 - Cross-section showing the most effective pile structure used.

Subsequent work during the process control and development phase utilised the positive pressure (blowing) mode of operation as developed at Rutgers University (2). It was possible using this method to hold temperatures down at the desired control values which were mostly set in the range 50–60°C. This control was possible due to the much greater heat loss associated with the relatively high rate of moisture evaporation.

4. PROCESS CONTROL

In controlling the composting process the major aim was to maximise the rate of biodegradation under aerobic conditions. The activity of the microorganisms, which are necessary for the early stages of composting, is inhibited to a marked extent when the temperature exceeds 60 °C (2). The optimum temperature for refuse/sludge mixtures appears to be in the range 45–60°C. A series of thermocouples placed throughout the pile were used both to monitor pile temperatures and to supply data for the control process. It is possible to achieve a form of control using one thermocouple well-positioned in the pile but more even conditions are achieved if a series of probes along the length of the pile core is used. In most cases the control systems used at Leeds were based on data from at least 5 thermocouples.

The use of temperature data only to control aeration rate, and hence temperature profiles, could cause problems in phases 1 and 3 of the typical process (Figure 3). In phase 1, before the desired control temperature is reached, excessive aeration could inhibit the required rapid temperature increase, whilst insufficient aeration could set-up anaerobic areas. In phase 3, excessive aeration with depleted substrate availability would bring about rapid cooling thereby reducing residual microorganism activity rates. Therefore in this phase sufficient air is required to maintain aerobic conditions whilst at the same time not producing a sharp temperature decrease. To reduce these problems associated with phases 1 and 3 a probe was used to monitor oxygen concentrations in the pile core.

Temperature and oxygen data were used by a microcomputer to control the rate of aeration applied to the pile. The system layout is shown schematically in Figure 4. The effect of using temperature and oxygen data to control aeration rates is shown in Figure 5. In this case 5 thermocouples were used for control together with one oxygen probe. The selected control temperature was 55 °C and this was reached between days 2 and 3. Prior to this the fan/airblower (used for aeration) "on-time" per 20 minute cycle was controlled by the oxygen concentration. In the period day 3 to day 20 the temperature was the controlling factor. In this particular trial the pile was removed from the composting pad at day 21 when oxygen concentration had once again become the controlling criterion for aeration.

Many different control programmes, based on temperature and oxygen data, were developed in this project some of which are discussed by Leton (5). On a commercial basis the use of an oxygen probe system might present operational difficulties so information from the research piles, controlled by temperature and oxygen feedback, has been used to produce a more sophisticated control programme using temperature input only. This not only gives good temperature control during phase 2 but also recognises phases 1 and 3 and supplies air accordingly.

In piles which were run on a fixed aeration cycle the typical temperature distributions achieved with positive and negative pressure aeration were as shown in Figure 6. In the case of negative pressure the highest temperatures were obtained in the core of the pile near the aeration pipes and with postive pressure the highest temperatures were near the apex.

In order to inactivate as many potential pathogens as possible the maximum amount of pile material should be subjected to the highest temper-

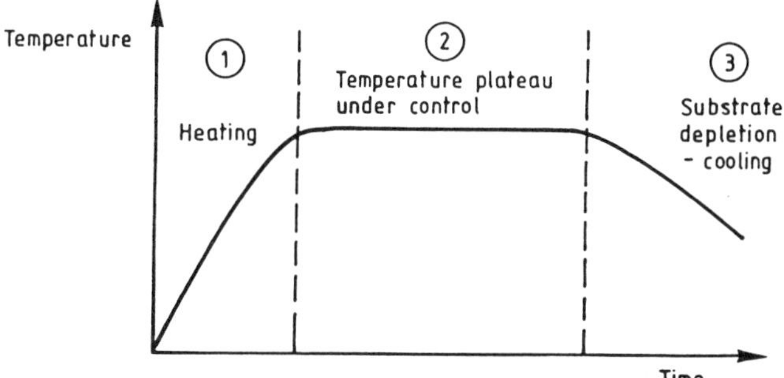

Figure 3 - Three phases of composting identified in a static pile system.

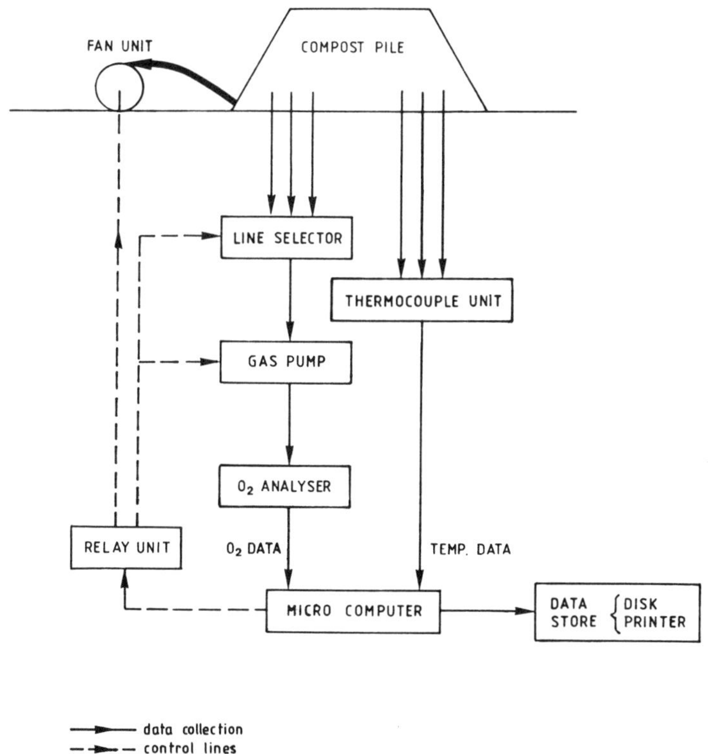

Figure 4 - Schematic showing system layout for the control of pile aeration.

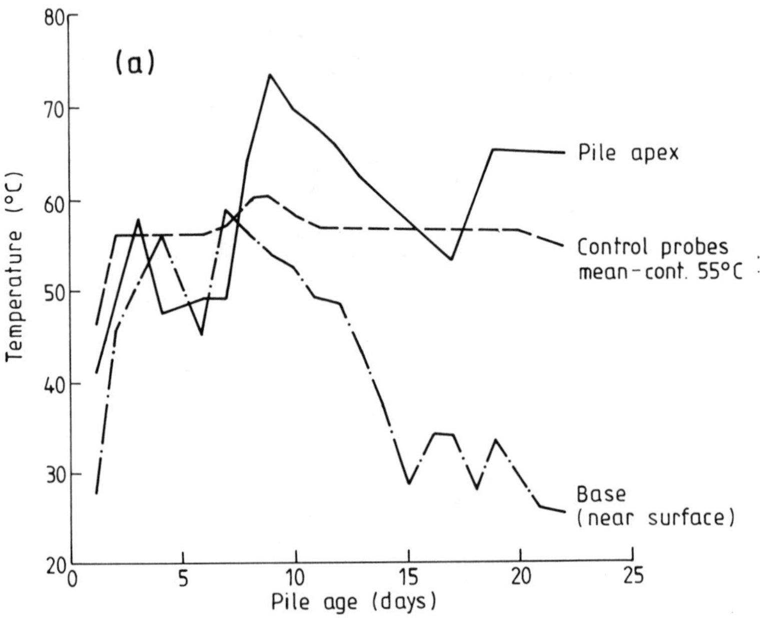

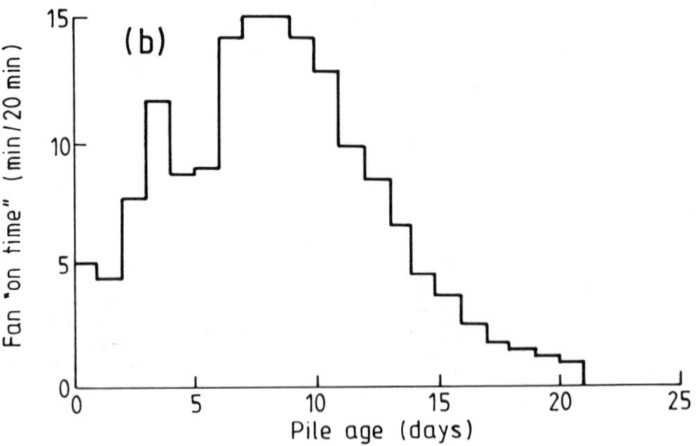

Figure 5 - Variation with time in a pile, controlled at 55°C, of
(a) temperature; and
(b) fan on-time

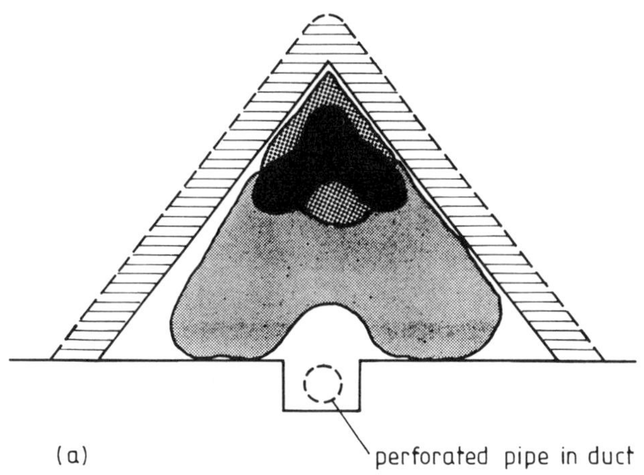

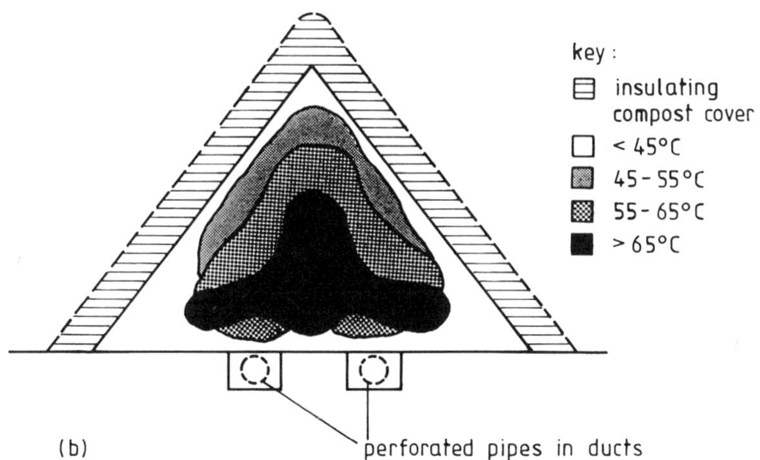

Figure 6 - Typical temperature cross-sectional distributions in a pile with (a) positive pressure (blowing) aeration; and (b) negative pressure (sucking) aeration.

ature for the longest period of time which is compatible with effective controlled composting. Although other factors, such as antibiotic production within the pile (6), are probably involved in this inactivation they are not readily controllable as is the temperature-time profile. It has been suggested that exceeding 55 °C for at least three days throughout the composting material should be sufficient to produce a pathogenically harmless material (3). This is not possible in all static pile systems since the outer layers have a temperature gradient from pile to ambient conditions. This ambient temperature is generally considerably less than 55 °C. The outer layers might consist of already composted material but they could still act as a microorganism reservoir from which the pile core could be reinvaded as it cools.

The negative pressure (sucking) system produces high core temperatures but relatively low peripheral temperatures whereas the reverse applies with the positive pressure (blowing) system. The best pathogen inactivation system might be a hybrid of the two with an initial negative pressure period producing a high core temperature. At the time when it was no longer possible to hold the desired control temperature (55-60 °C) the air flow would be reversed. This would enable control temperatures to be maintained whilst at the same time causing the high temperature area to move out from the core of the pile towards the surface layers. By suitable control it should be possible not only to produce rapid, controlled composting but also to submit a much greater amount of the compost material to temperatures around 55 °C than would be possible with either system on its own.

5. GROWTH TRIALS

The refuse was not presorted prior to composting so the inert material was unchanged by the process. In order to make a more visually acceptable product all the material used in the growth trials had passed through a 6 mm sieve. This compost was very similar in composition to others derived from similar raw materials. The limited growth trials which were carried out on the material were to confirm results found by others with this type of compost. The yields related to the growth of tomatoes produced in a greenhouse are shown in Table III. In this particular case there is no significant difference in yields between the two composts.

In the South Yorkshire area, mining and quarrying has necessitated much reclamation of spoiled tracts of land producing a demand for topsoil and compost. The effect on the yield of grass of adding Leeds compost to various waste heap materials was studied in a series of pot trials. Figure 7 shows the effect on monthly yields of various additions of compost to clay, pulverised fuel ash (PFA), and shale respectively. In all cases the addition of small amounts of compost produced an increase in yield which was particularly marked with PFA.

In all the growth experiments the compost used had been allowed to mature for at least 4 months after forced aeration had ceased. The use of seedling emergence tests, along the lines of those used by Zucconi (8), showed that a shorter maturing time than this inhibited seedling emergence. No such inhibition appeared to exist after 4 months.

6. PROCESS ECONOMICS

There are two levels of design which can be considered for a process of this type.

(i) A large scale processing operation run on factory-type lines as a direct competitor with more capital intensive fixed reactor vessel

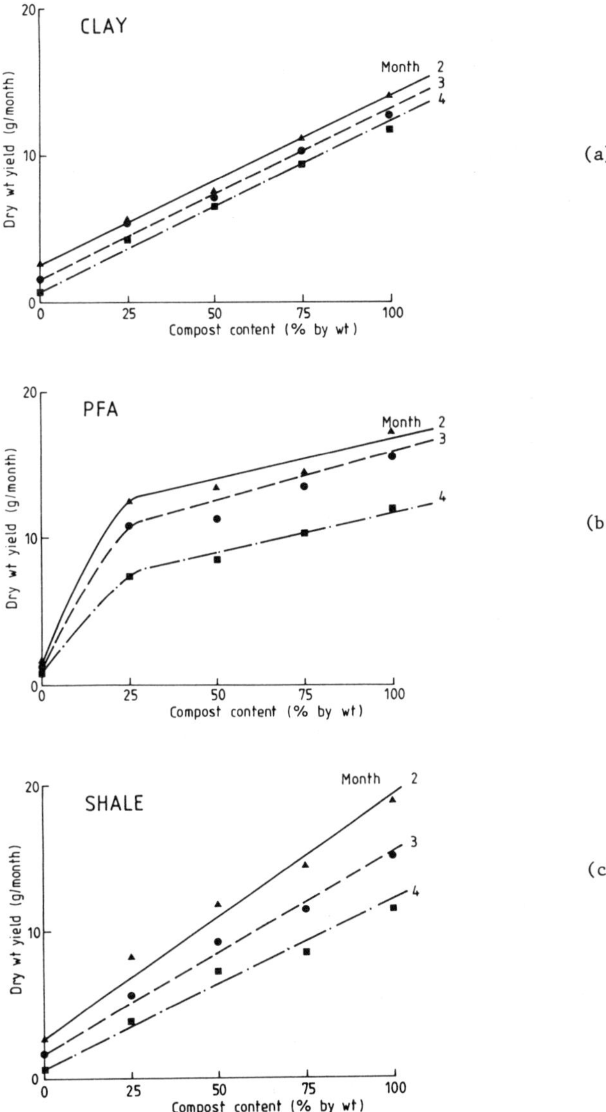

Figure 7 - Effect on grass yield of the addition of different amounts of Leeds compost to :
(a) clay;
(b) pulverized fuel ash; and
(c) shale

systems.

(ii) A small scale operation which runs on a batch basis with a minimum of mechanical plant.

In the first instance the type of costs involved are as indicated in Table IV.

The current bulk selling rate for this type of compost in the U.K. is £8 per tonne which does not make this factory-like system self sustaining from a cost point of view. The smaller plant throughput causes an increase in costs which becomes very marked as the rate falls below 100 tonnes/day. However, if we consider a reactor-based system, the production costs for example for a 150 tonne/day plant become £19/tonne refuse and £44/tonne compost.

In smaller systems which do not require the continuous building of compost piles but which might only require 1 or 2 per week the process could be run in parallel with other operations such as landfilling. This would substantially reduce the costs by making more effective use of operators and machinery.

It should be stressed that the majority of the costs used are estimates and should only be seen as approximations. However, their relative values should give a valid comparison between operation sizes and systems used.

In comparing high capital costs plants with less capitally intense system, such as static pile operations, a major argument in favour of the former has been the far more effective use of land space. However, this has to be considered alongside the requirement both systems have for allowing at least 4 months further maturing after initial composting. In the case of static pile systems this calls for an area 2-3 times the size of the composting pad. Thus the composting pad, or reactor plant, occupy a minor proportion of the total site area.

A major economic obstacle to the use of composting for refuse treatment in the U.K. is the very low cost of landfilling operations in many parts of the country. However, current problems associated with leachate and methane production from certain landfill sites could put more pressure on these disposal authorities to consider other methods of treatment and disposal. In some city areas in the U.K. refuse has to be transported a substantial distance to landfill sites and the cost of this makes composting an economic possibility.

7. CONCLUSIONS

This work concentrated for the most part on domestic refuse with sewage sludge, at 4-5% solids, being essentially only a nutrient-enhanced supply of moisture. To make a true comparison with sludge/woodchip systems the work needs to be extended to admixtures of refuse and sewage sludge cake (20-40% solids). The conclusions which can be made on the work carried out within the project to date are:

(i) the use of aerated static pile systems to produce compost from unsorted refuse and sewage sludge is viable practically on the scale examined;

(ii) close control of pile temperatures can be achieved using a positive pressure aeration system controlled by a microcomputer;

(iii) in growth trials the Leeds compost has compared favourably with commercially available material. In trials with waste materials small additions of compost substantially increased grass yields;

and (iv) on a current cost basis, with cheap landfill alternatives in the U.K., it is unlikely that refuse disposal authorities would install a composting system. However, should they consider this option then a forced aeration static pile system would appear to be more economic than reactor-based systems.

Table III Tomato yields for Leeds compost and a commercial product

	Leeds Compost	Commercial Compost
Number of fruit in sample	145	54
Average wt. of individual fruit (g)	59.2	55.7
Range of fruit wt. (g)	7.0 - 170.3	7.3 - 145.3

Table IV Cost (x£10^3) of construction and operation of a forced aeration static pile composting system (on annual cost basis)

Item	Average refuse throughput (tonnes/day)[(1)] 50	100	150	200
Civil Works[(2)]	61	104	150	196
Mechanical/Elect. equip[(3)]	75	80	112	130
Operation and Maintenance	140	165	230	282
Total annual costs	276	349	492	608
Savings[(4)]	28	55	82	110
Total net costs	248	294	410	498
Cost/tonne refuse (£)	19.8	11.8	10.9	10.0
Cost/tonne compost (£)[(5)]	45.6	27.0	25.1	22.9

Notes
(1) Total throughputs based on 50 weeks/annum and a 5 day working week.
(2) Based on 12% interest with 20 year payback.
(3) Based on 12% interest with 10 year payback.
(4) Includes income from the Water Authority for taking sewage sludge at £2/m^3 and the saving in landfill requirement at £1/m^3 of space.
(5) Assumes 0.45 tonnes sewage sludge added per tonne of refuse and a yield of 30% of the total as compost.

ACKNOWLEDGEMENTS

The author wishes to thank the Science and Engineering Research Council (SERC) who provided the funds for this project, the Yorkshire Water Authority (YWA) for their enthusiastic support and the South Yorkshire County Council (SYCC) for their cooperation.

REFERENCES

1. EPSTEIN, E., WILLSON, G. B., BURGE, W. D., MULLEN, D. C. and ENKIRI, N. K. (1976). A Forced Aeration System for Composting Wastewater Sludge. Jour. Wat. Pollut. Cont. Fed. Vol. 48, No. 4, p. 688-694, 1976.
2. FINSTEIN, M. S., CIRELLO, J., MACGREGOR, S. T., MILLER, F. C. and PSARIANOS, K. M. (1980). Sludge Composting and Utilization: Rational Approach to Process Control. USEPA Report No. C-340-678-01-1.
3. GOLUEKE, C. G. (1983). Epidemiological Aspects of Sludge Handling and Management. Biocycle, July-August, 24(4), p. 50-59, 1983.
4. GRAY, K. R., SHERMAN, K. and BIDDLESTONE, A. J. (1971). Review of Composting - Part 2: The Practical Process. Process Biochem. October, 1971.
5. LETON, T. G., TAYLOR, P. L., STENTIFORD, E.I. and MARA, D. D. (1983). Temperature and Oxygen Control of Refuse/Sludge Aerated Static Pile Systems. Paper presented at conference on "Composting of Solids Wastes and Slurries", University of Leeds, Sept. 1983.
6. WILEY, J. S. (1962). Pathogen Survival in Composting Municipal Waste. Jour. Wat. Pollut. Cont. Fed., Vol. 34, No. 1, p. 80-90, 1962.
7. WILLSON, G. B., PARR, J. F., EPSTEIN, E, MARSH, P. B., CHANEY, R. L., COLACICCO, D., BURGE, W. D., SIKORA, L. J., TESTER, C. F. and HORNICK, S. (1980). Manual for Composting Sewage Sludge by the Beltsville Aerated-Pile Method. USEPA Report No. EPA-600/880-022, May 1980.
8. ZUCCONI, F., FORTE, M., MONACO, A. and de BERTOLDI, M. (1982). Biological Evaluation of Compost Maturity. In "Composting - Theory and Practice for City, Industry and Farm". Publ. by J. G. Press, p. p. 34-40, 1982.

DISCUSSION

BRUCE: Have you experienced any problems associated with the presence of glass in the refuse component of the compost?

STENTIFORD: There have been no problems of damage to persons or plants due to the presence of glass. Damage can be caused to the tyres of front loaders used to handle the refuse. However, when marketing the product, the presence of glass can cause customer resistance.

VAN FAASEN: You used a minimum oxygen level of 5% for controlling aeration; others suggested that it should be at least 15%. The point where you measure the oxygen level in the pile will also be important for the range of O_2 concentrations throughout the pile. The same idea holds for temperature feed back control.

STENTIFORD: The minimum level of 5% oxygen was the value at which aeration commenced. If the pile was under oxygen control then aeration would be continued until the set upper limit was reached (generally 12 - 15%). The point of measurement is important and is the reason for a 5% lower limit because this should mean that, allowing for uneven distribution of O_2, all parts of the pile will contain sufficient O_2 to prevent anaerobic activity developing.

The positioning of temperature probes (thermocouples) is important and we think the results of our work (using up to 36 thermocouples per pile) has identified the best areas of the pile to use with perhaps only one thermocouple, to obtain the most representative reading to use in control.

LE ROUX: Sewage sludge and urban refuse contain heavy metals. Are they present in the compost and in what concentrations?

STENTIFORD: The metals are present in the compost, generally at concentrations which would be expected from those in the raw materials. One problem with metals is that as the material decomposes and mass is lost, the metal concentration increases when expressed in terms of a unit weight of compost. The range of values obtained for four piles started in May, July, September and October 1982 were:

Metal	concentration (mg/kg of dry solids)	
	Range	Mean
Cu	300 - 400	375
Cd	3.2 - 4.1	3.7
Ni	86 - 336	159
Pb	466 - 572	508
Zn	636 - 987	794

HOVSENIUS: We used the same methods 10 years ago but changed because of problems with plastics. These are now removed in a pretreatment which also removes paper. There can be problems with oxygen supply and temperature resulting from variation in the texture of waste. How would you scale up your model to serve say a population of 10 - 20,000 people?

STENTIFORD: The piles used contained 50-60 tonnes of material and the required number of heaps can be made. We take any type of refuse and have used both fine and coarse materials. The problems of plastics are reduced by mixing.

METHODS FOR THE EVALUATION OF THE MATURITY OF MUNICIPAL REFUSE COMPOST

J.L. MOREL*, F. COLIN**, J.C. GERMON***, P. GODIN****, C. JUSTE*****

* Ecole Nationale Superieure d'Agronomie et des Industries Alimentaires Phytotechnie (E.N.S.A.I.A.) Nancy.
** Institut de Recherches Hydrologiques (I.R.H.) Nancy.
*** Laboratoire de Microbiologie des Sols - I.N.R.A. Dijon.
**** Ministere de l'Environnement - Paris.
***** Station d'Agronomie - I.N.R.A. - Bordeaux.

Summary

Compost maturity is important in assessing the quality of the end-product and its possible uses. Some uses require well matured compost (horticulture, market gardening, plant nurseries,...), whereas other uses require fresh compost (hot beds in the greenhouse, mushroom production...).
Fresh compost allows heat recovery and matured compost is necessary to avoid adverse effects on plants due to the heat of decomposition high C/N ratio and phytotoxic compounds.
Several investigations have been made in order to determine compost maturity :
a - C/N ratio
b - polysaccharide analysis
c - Adenosine Tri Phosphate (A.T.P.) measurements
d - chromatographic test (EAWAG test, Switzerland)
e - colorimetric test, after extraction of humic components
f - respiratory activity test
g - phytotoxicity tests (germination and growth tests)
Some methods allow rapid determination of the maturity of intermediate and end-products of composting (d, e, f). The phytotoxicity tests (g) are easily interpreted by those who employ them but require several days. Chemical analysis (a) and more sophisticated methods (b, c) are more reliable for explaining the bio-oxidative decomposition processes and are necessary for ascertaining the value of rapid tests.
The principles, difficulties of achievement and benefits of these methods are compared in the text.

1. INTRODUCTION

Treatment of municipal refuse by composting gives an organic material which is valuable for plant production and may be used a) as an organic soil amendment for intensive cultivation or vineyards b) to form a growing medium or hot bed for horticulture and market gardening c) as a substrate for mushroom production.

However, following encouraging development, the commercialization of municipal waste compost has considerable problems with product-quality, mainly : a) physical aspects (sifting, and removal of items without agricultural value, many dangerous) and b) fertilizing value in which the maturity of the compost plays an important part, because its potential utility depends on this. Mushroom production and the preparation of hot beds in horticulture both require fresh compost. On the other hand, use of compost which is insufficiently matured can cause adverse effects on crops and significant loss of yields (phytotoxicity of compost, over-heating, depressive effects due to nitrogen immobilization).

The improvement of the commercial viability of compost demands that the stage of composting can be accurately determined, so that risks of adverse effects can be eliminated. It is therefore necessary to employ methods of testing the maturity of compost, not just in laboratories where the quality-control for products and experiments for new processes are carried out, but also on the composting plant.

Bearing this in mind the ministry responsible for the Environment has supported research in earlier years, aimed at trying to establish the characters evolved during composting of municipal refuse and to formulate methods to provide rapid information about compost maturity. This research was undertaken by different laboratories (I.N.R.A., I.R.H., E.N.S.A.I.A.) each working on the same range of compost which allowed better comparisons to be made between methods, results and coordination among the research teams.

As well as greenhouse experiments, which remain the best reference method but which take a long time to give positive information other methods have become available. Some are based on well-known parameters (e.g. C/N, respiration, state of organic matter). Others employ criteria which accurate analytical techniques have made possible (A.T.P., enzyme activity, polysaccharide measurements).

At present, as well as the traditional tools for investigating decomposition (C/N, temperature, humidity) other methods are available from the most sophisticated, which may be employed only in well equipped laboratories to the most basic which can be adapted to the immediate needs of compost treatment plants.

2. Methods based on the study of the evolution of biomass parameters

These methods are, in part, based on the initial hypothesis that the maturity of the compost may be assessed by the biological stability of the product. Estimating maturity therefore, amounts to measuring the activity of the microbial biomass or determining those constituents which are easily biodegradable and susceptible to degradation. Three kinds of methods may be distinguished : a) respirometry b) analysis of biodegradable constituents : total organic carbon, polysaccharides c) study of biochemical parameters (ATP, enzyme activity).

2.1. Respirometric methods

This type of method, described by SPOHN and KNEER (1968), CHROMETZKA (1968) was employed in studying the respiration of compost (absorption of oxygen or emission of carbon dioxide) either in the pure state (BENISTANT, 1978 ; NICOLARDOT, 1979) or in admixture with soil in proportions compatible with use in agriculture (MOREL *et al.*, 1979).

The results of these respiratory measurements over a large range of composts, with wide variations in age and origin, were compared with those from plant tests. Fig. I shows the rate of cumulative consumption of oxygen during a period of 28 days for composts in several stages of maturation, and indicates a marked drop in respiratory activity with ageing of the compost. Fig. II shows the production of ryegrass on a soil treated with composts related to the consumption of oxygen over 7 days by the same composts.

These respirometric tests together with tests on plants permit an empirical definition of the respiratory rate limits corresponding to the appearance of depressive effects in the production of dry-matter in rye grass, corresponding to the maturity limits.

Using these results, two principal respirometric techniques have been developed; both provide a means of determining maturity of the compost according to the degradability of the organic products it contains (NICOLARDOT, 1979 : MOREL et al., 1979). These techniques, however require specialized and well-equipped laboratories (use of respirometers) and can only be considered for occasional use (product control, development of new processes of compost-making, standardization of simple tests for the determination of maturity).

On the other hand, each of these techniques has been simplified and adapted for use for analysis in compost plants. For example, simplification has led to a maturity test which is practicable on the spot and involves measuring the consumption of oxygen by composts during three days, using material which is commonly available (GERMON et al., 1980). The oxygen consumption by compost is estimated from the measurement of the pressure drop caused by incubation of the sample in a leak-proof incubator, (carbon dioxide produced is absorbed by sodium hydroxide).

One kg of compost (with moisture content corresponding to 80% of waterholding capacity measured with the aid of a salad drainer) is placed in a 5 l graduated plastic vessel. Another 500 ml vessel contains 250 ml of 2N sodium hydroxide solution.

These two vessels are placed inside an air-tight incubator e.g. a 22 l pressure-cooker. The incubator is then attached to a FOURNIER-EUVRARD tensiograph by a vacuum pipe, previously regulated and standardized to measure and record a drop in pressure from 0 to 200 mm of mercury. The incubator and measuring device are then placed in a room where the temperature is constant at about 20°C.

Incubation lasts a maximum of 3 days, sometimes less if the product respires very actively. At the end of the incubation, the compost sample is taken out of the incubator and its volume is checked in the graduated vessel after carrying out a manual settling of the product in the vessel.

Fig. III gives an example of curves which may be obtained for a compost at several stages of maturity. After a lag phase, whose duration depends on the product itself, pressure decreases inside the incubator; the rate of decrease depends on how advanced degradation is and the biological activity of the compost.

Oxygen consumption is calculated using the following formula.

$$Q = v \times \frac{DPm}{76} \times \frac{32}{24.04} \times \frac{1000}{Pc} \times \frac{1000}{t}$$

Q = mg of oxygen/kg dry compost/hour
V = volume of incubator minus volume of compost and volume of NaOH
DPm = maximum drop attained (in cm of mercury)
Pc = dry weight of incubated compost in g
t = time taken for consumption of oxygen, that is the time from the end of the lag phase to the maximum drop attained (in hours)
24.04= volume (in litres) of a gramme molecule of oxygen at 20° and at 760 mm of mercury

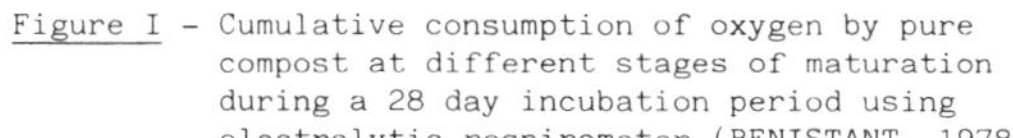

Figure I - Cumulative consumption of oxygen by pure compost at different stages of maturation during a 28 day incubation period using electrolytic respirometer (BENISTANT, 1978)

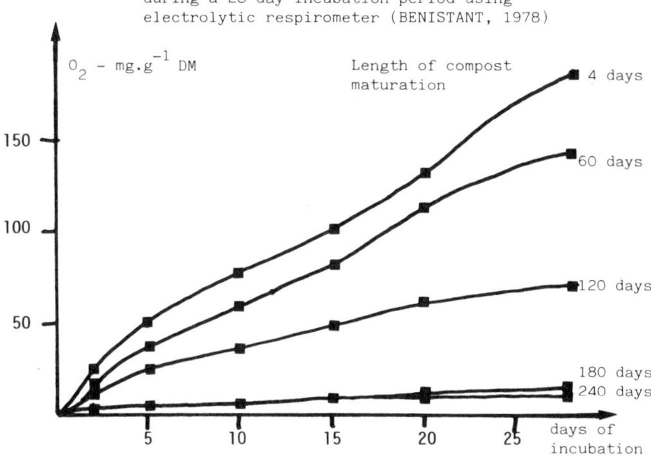

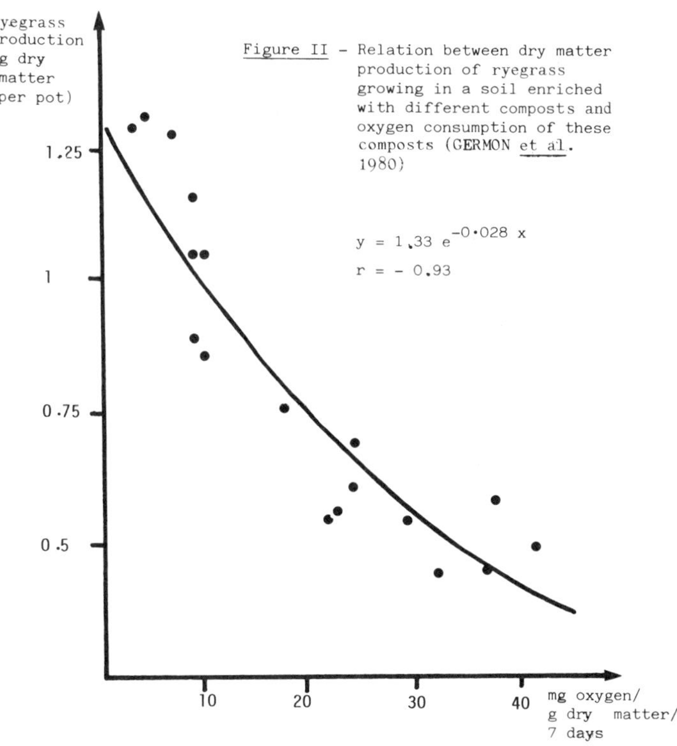

Figure II - Relation between dry matter production of ryegrass growing in a soil enriched with different composts and oxygen consumption of these composts (GERMON *et al.* 1980)

Figure III - Pressure drop in the incubator obtained for an urban refuse compost at different stages of maturity (GERMON et al., 1980)

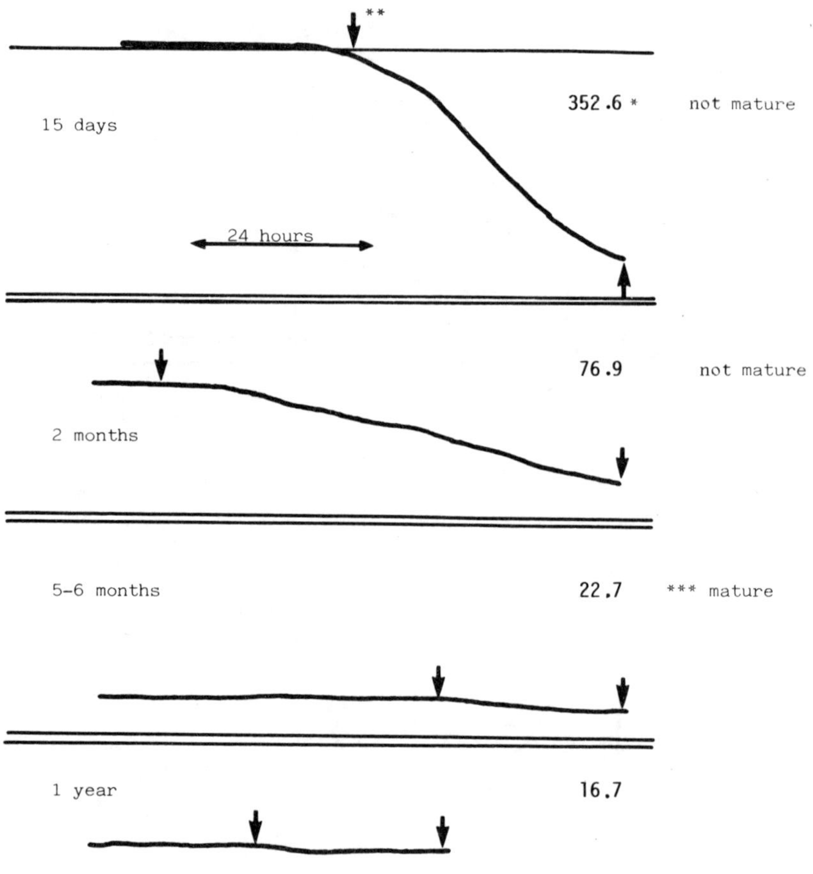

* oxygen consumption : expressed in mg/kg dry matter per hour

** arrows indicate the interval of time used for calculations

*** oxygen consumption is lower than 40 mg/kg dry matter/ hour thus the compost can be use as an organic amendment without any risk of depressive effet on plant production

With the help of laboratory electrolytic respirometers and the results of growth tests on ryegrass using eighteen kinds of compost from four different plants, it has been shown that a product may be considered mature when its level of oxygen consumption is less than 40 mg per kg of dry matter per hour. A high correlation between the oxygen consumption measurements was obtained by using both electrolytic respirometers and the test proposed here, which permits us to conclude that a consumption of 40 mg of oxygen measured in the simplified device constitutes a maturity limit.

This simple dependable test offers the possibility of quality control for urban refuse and the smooth running of the composting process. Its essential value is that the effects of the compost on plants may be predicted because the information it provides is directly related to the agricultural value of the compost. However, it does not allow for stunting effects due to other causes, such as phytotoxicity induced by pesticides in the products. The presence of pesticides can only be shown through direct growth tests on plants which will be dealt with in chapter 4.

The other simplification of respirometric techniques consists of looking for parameters which explain degradability in compost, these parameters are easily found through physico/chemical analysis. This technique will be dealt with in the next paragraph.

2.2. Assessing compost maturity through determination of degradable organic substances

Respiration of compost alone or mixed with soil is an efficient way of assessing the maturity (biological stability) of the product. Taking into account the length of time required for this kind of procedure, it was more realistic to envisage an assessment of compost maturity by indirect and more rapid means, which rely on a simple physiochemical characterization of the material (MOREL et al., 1979). In parallel with the respirometric study on compost samples, we also analysed the compost. Various parameters appeared to be useful due to the high correlation between them and the compost degradability. It was, therefore, possible through multiple regression analysis, by selecting variables which assured maximum precision, to anticipate compost behaviour in soil (Fig. IV). The relationship we found, shows very clearly how necessary it is to know how old the compost is, in order to estimate its maturity, but also, how inadequate this value itself is, as it must be accompanied by two variables which characterize clearly the degree of degradability of refuse i.e. the total organic carbon (representing all available organic substances) and how much water soluble sugars are contained (consisting of those most rapidly fermentable).

The correlation between the respiration of a soil + compost mixture and the dry matter production of ryegrass in a similar mixture, allows us to define two classes of maturity (as above). With a value of less than 2.4 which corresponds to a low rate of respiration of the soil + compost mixture, the compost may be considered mature. At the other end of the scale, above 2.7 the product is still highly degradable and will stunt vegetable growth, it is, therefore, not mature.

As in all predictive equations, there is a risk that the estimation may be erroneous. However, when employed in conjunction with the simplified respirometry method this technique enables us to overcome the problems inherent in the study of compost in which there are toxic substances. These have an artificially stabilizing effect on the product and may lead to the conclusion that the product is mature, even though there is much very degradable material and without allowing the presence of toxic substances to be detected.

2.3. Analysis of biochemical parameters of the biomass of compost (COLIN, 1978)

In parallel with these respirometric tests which provide overall estimation of compost degradability, it would appear desirable to look for more precise tests in order to measure evolution of the biomass in compost. Thus, we have shown that parameters of a biochemical nature, such as, the concentration of A.T.P., the hydrolytic enzyme activity (proteolytic, amylasic, cellobiasic activities) constitute a selected method for studying phenomena involved in the composting of waste matter, for both the accelerated composting stage and the maturation stage in stock piles.

The measurement of biochemical parameters was adapted and applied to follow the evolution of urban refuse during composting in a plant, consisting of an accelerated stage, carried out in a fermentation tower made up of six successive cells and maturation in a stock pile. The characteristics of the material during composting shows that each of the two operations corresponds to a distinct process:

- Accelerated composting in the tower decreases the different biochemical parameters characteristic of biological evolution (fig. V) either immediately or within a short space of time i.e. a day. This decrease occurs during the thermogenesis stage and is almost complete when the temperature reaches between 50° and 70°C. The duration of this stage is about 4 days, at the end of which, biological activity appears to be very low.
- Stock piling of material from the composting tower is accompanied by a second stage, completely distinct from the preceding stage (Fig. VI). This stage is accompanied, after a time by an increase in the different parameters characteristic of biological activity.

These parameters usually reach values greater than those measured in the composting tower. These high levels are reached after a period of 70 to 140 days in the stock-pile. Even after 190 days, development of the compost is still far from complete.

As well as application for the control of existing plants, the following method which has been developed, could be used for new composting techniques for ordinary urban refuse or for the application of classic techniques to new organic waste.

3. Chemical analysis methods for rapid assessment of maturity in urban refuse compost

During each of the composting stages, the organic constituents of urban refuse undergo transformations which lead to more biologically stable components. Apart from the loss of total organic carbon and easily degradable components an increase in highly polymerized organic products may also be found. The latter are less affected by immediate biodegradation and may be compared with humic compounds in the soil.

Techniques for assessing maturity in compost dealt with in the following section are based on modifications of the state of the organic matter which take place during composting. Total organic carbon, and C/N ratio the polysaccharide content and the state of humic substances can be

Figure IV - Estimation of degradability and maturity of urban compost by physico-chemical analysis (MOREL et al., 1979) AGE : days of maturation
COT : total organic carbon
PHs : extractable sugars by hot water

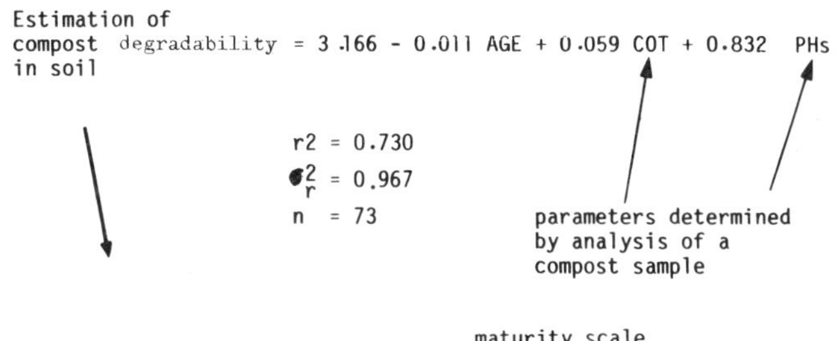

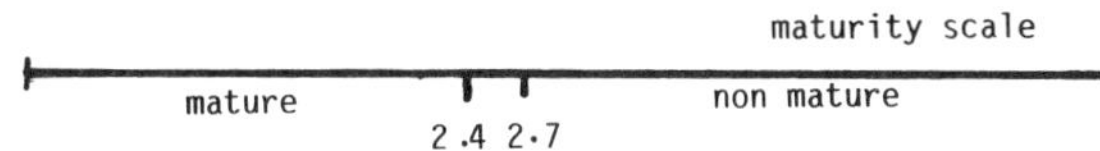

limits of maturity classes define as a result of growing test with ray grass

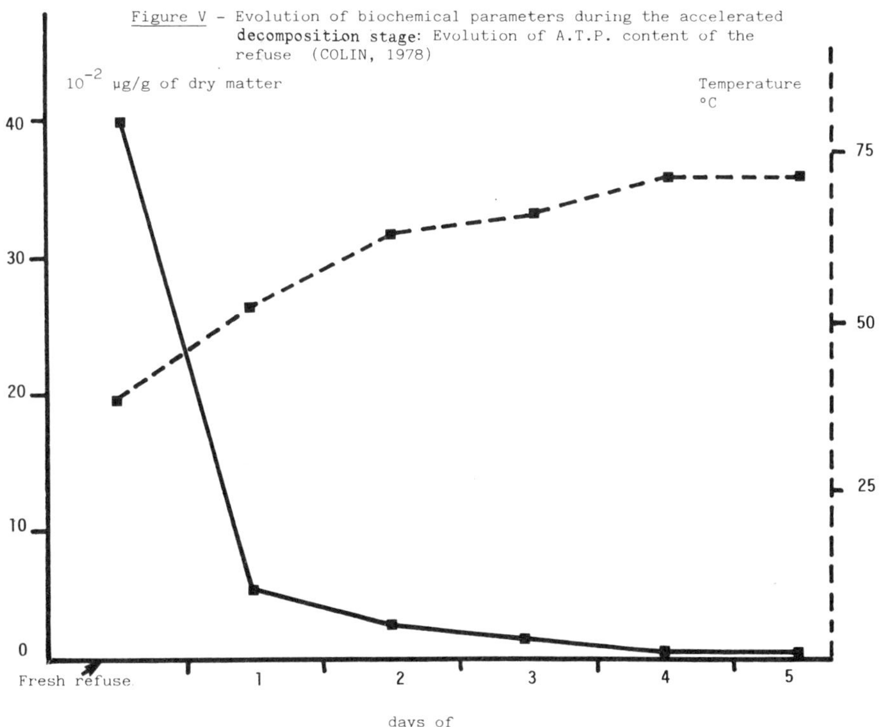

Figure V - Evolution of biochemical parameters during the accelerated decomposition stage: Evolution of A.T.P. content of the refuse (COLIN, 1978)

Figure VI - Evolution of enzymatic activities during stock piling of compost (COLIN, 1978)

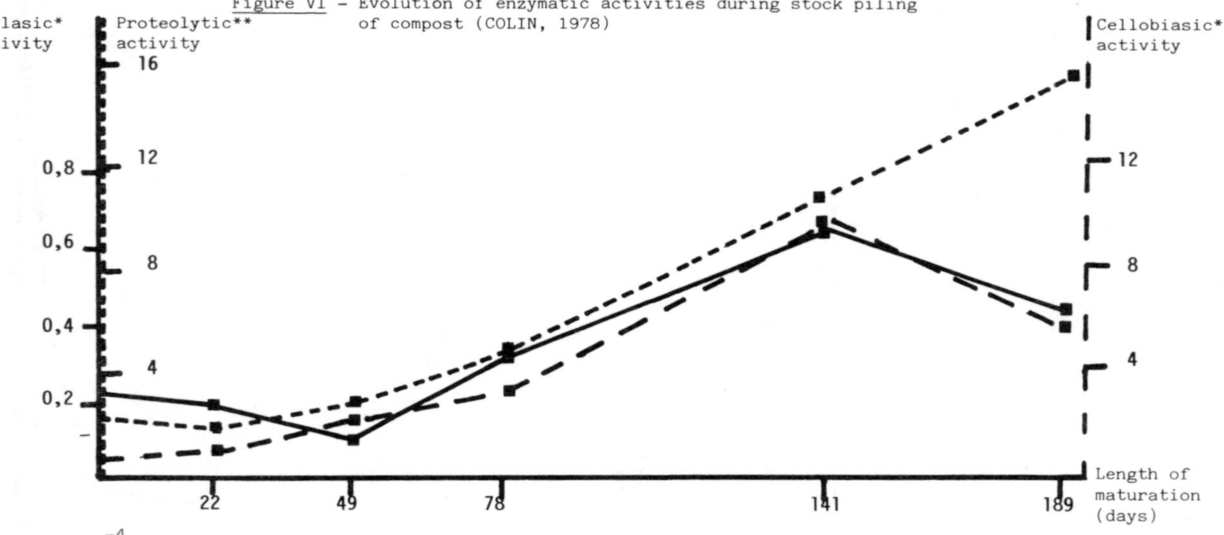

* 10^{-4} mole of reducing groups/g of dry matter/day
** mg of hydrolyzed caseine/g of dry matter/day
*** 10^{-7} mol of glucose formed/g of dry matter/day

considered to measure the development of compost. Note that the CEC, the water-holding capacity, the ash content may also be useful parameters for determining compost maturity.

3.1. Total organic carbon and C/N ratio

The C/N ratio is used mainly to define the quality of urban compost. As indicated in Fig. VII, the C/N changes with the ageing of the compost, reaching values, which are characteristics of a stable organic material. The normal french definition of maturity of compost (norme AFNOR U 44051) defines two classes of maturity according to the organic matter/nitrogen ratio :

Semi-mature urban compost	$OM/N \leq 60$
Mature urban compost	$OM/N < 50$

If the C/N ratio is a good indicator of the biological stability of compost, it would seem essential to interpret this parameter according to the initial characteristics of the product. Mixing urban refuse with waste such as sewage sludge, rich in nitrogen or slurry greatly influences the C/N ratio of the compost and can lower its values to those corresponding to a mature compost even though it has not yet degraded.

Therefore, as JUSTE and POMMEL (1977) proposed it is preferable to keep a constant check on the changes in the C/N ratio during fermentation, rather than measuring it occasionally. Thus, determination of the C/N when the compost is stockpiled proves very useful and the ratio (final C/N)/(initial C/N) will provide a better idea of the maturity of the material. Because of slight variations in the nitrogen level during composting (small increase) and the necessary delay for its determination, total N may be considered as a check only on the total organic carbon.

However, the maturity limits remain to be determined and in this respect because of the numerous analytical techniques available it is necessary to specify the technique for measuring the carbon content of compost.

3.2. Content of polysaccharides

Polysaccharide content, determined by the method of DUBOIS *et al.*, (1956) after extraction from compost using the GUCKERT (1973) technique for soil, enables the state of stability of compost to be assessed. This has already been considered during estimations of degradability in compost based on indirect measurements (paragraph 2.2.).

While the carbon level of compost decreases, all components of the organic matter are affected (cellulose, hemicellulose, lignin, soluble organic components, amino acids, simple sugars). Fig. VIII shows the variation in the polysaccharide content of different kinds of compost which may be extracted using hot sulphuric acid, which varies according to the maturation. In the beginning the simple polysaccharide content of urban refuse is high (20 % of the total organic material), but this total shrinks over the following period of time and represents only 4 % to 10 % after 240 days.

Water soluble polysaccharides undergo a similar evolution. Simple polysaccharide components therefore, are progressively decomposed by the microflora which develops during the course of different stages of thermogenesis and maturation. Water soluble, consisting mainly of

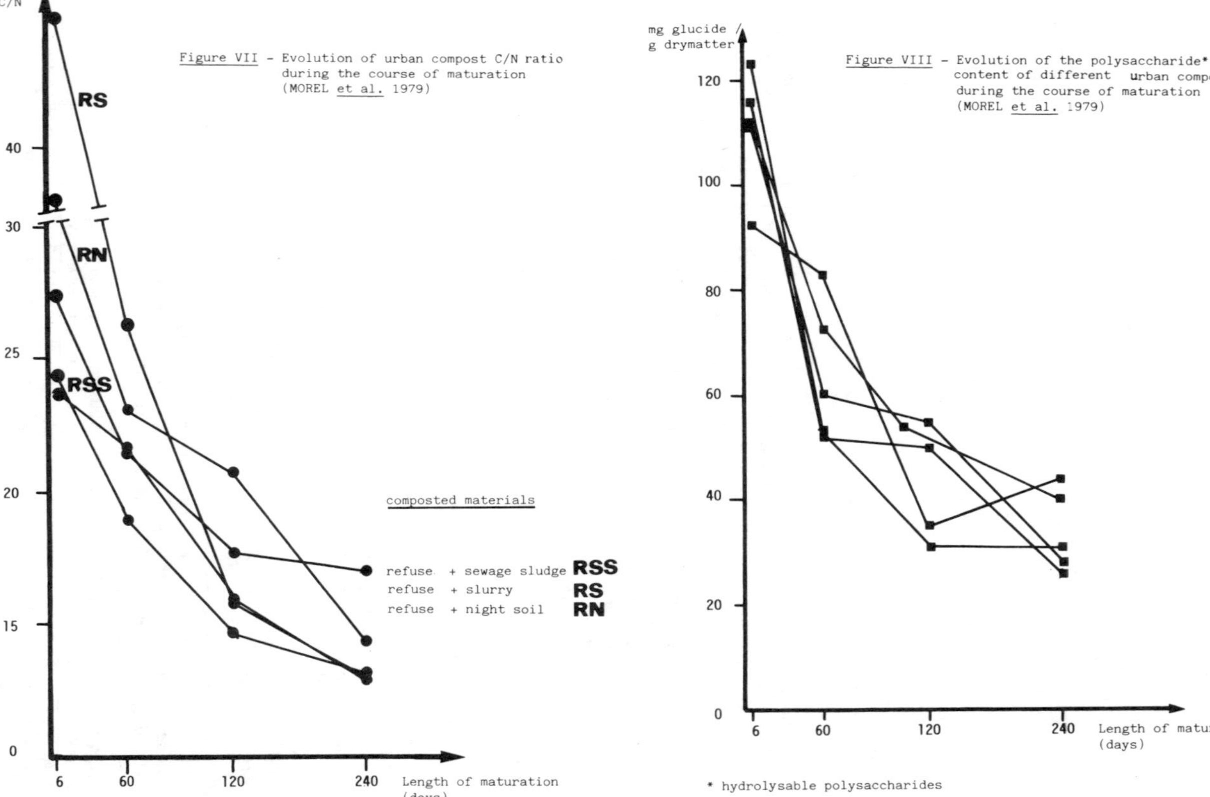

Figure VII - Evolution of urban compost C/N ratio during the course of maturation (MOREL et al. 1979)

Figure VIII - Evolution of the polysaccharide* content of different urban composts during the course of maturation (MOREL et al. 1979)

* hydrolysable polysaccharides

mono- or di-saccharides, sugars disappear much faster than hydrolysable sugars. On the other hand, genuine polysaccharides take far longer to decompose because of their structure (cellulose). At the end of composting, the quantity of polysaccharides extractable in acid is relatively high and seems to correspond with newly synthesised stabilized microbial components rather than those fractions which are not yet decomposed.

3.3. Overall analysis of the state of organic matter in compost

These methods are based on the modification of the state of organic matter contained in waste during composting. This change in state may be revealed by the alkaline extraction of the organic fraction (humic components) a procedure which is commonly used in the study of organic matter in soil.

3.3.1. Circular chromatography test (HERTELENDY, 1974)

This technique, previously described by PFEIFFER, was employed by the EAWAG research workers to determine the degree of maturity of compost. The humic substances extracted from the compost are isolated using a filter paper. The very mobile slightly polymerized components move towards the periphery of the filter paper, while the highly polymerized components stay in the centre. Therefore, simple observation of the chromatogram permits classification of the compost and an estimation, with reference to crop tests, of its probable effect on plant production.

The suggested technique is : 5 g of fresh compost is put into 50 ml of 1% NaOH. After 5 hours of agitation and centrifugation, the chromatography of the solution may be carried out on filter paper which has been pretreated with AgNO3 0.5%, by adding solution to the centre of the filter paper by a wick in the extraction solution. Chromatograms are read by the shape and colour of three separate zones, Schematically:

Chromatogram	Fresh compost	Mature compost
Centre	white - pink	red-violet
Transition zone	rings	irregular outline
Periphery	brown	clear with jagged edges

This technique is essentially qualitative and indicates the state of maturity of compost quickly. However, the interpretation of chromatograms may be different sometimes. In some cases it is quite simple to classify compost from the same plant but, in many situations a complete reversal of the chromatogram pattern of the type developed by HERLENDY (1974) may be found. This difficulty which we came up against, may be due to the fact that samples used at EAWAG were probably much more homogeneous (60 kg fresh compost put into composter) than those which we used (from compost plants where homogeneity is, for the most part, mediocre). However, this test could be very suitable for compost control when the initial product is known to be homogeneous (wood shavings, bark, straw,...). Composting treatment of these materials is used more and more.

3.3.2. Colorimetric method (MOREL, 1982)

Alkaline extraction of organic material from compost gives a brown hue which provides a range of shades, depending on the age of the compost. This hue is light at the beginning of composting and tends to darken as it matures. Such colour variations connected to the age of the compost, have provided a basis for the development of a test to determine maturity, which consists of periodic measurements of the optical density of alkaline extracts from the product. This test, while holding on to the notion of maturity - humification proposed by Hertelendy, allows problems connected to chromatogram observation to be avoided. Estimation of maturity is shown by a quantitative measurement, which is much more objective. This technique does not require sophisticated equipment and is, therefore readily available for composting plants.

For a given sample of compost, the procedure is as follows: put a quantity of compost, equivalent to 40 g of dry matter, in a container (capacity 1.5-2 l) which contains 1 l of sodium pyrophosphate 0.1 M. After having stood for 6 hours, during which it is agitated at regular intervals, the suspension is filtered twice so that all solid particles are removed. The optical density is read at 450 nm. In the absence of a colorimeter, a range of colours obtained from dilutions of an alkaline extract, may be used. Results obtained from both methods are comparable. Finding the stage of maturity requires constant control by colorimetric analysis of compost maturation. Although there is a highly significant correlation between the optical density of the alkaline extract and compost degradability in soil, one optical density limit which is common to all types of compost cannot be identified corresponding to the transition from non-maturity to maturity. Practically speaking, monthly samples should be taken in a composting plant, for colorimetric analysis.

The change in optical density of the alkaline extract with time reveals maturity as soon as a plateau occurs. We have shown that for a given plant, one optical density limit can be found corresponding to the start of maturity. This limit only applies locally, of course, but once it has been found, it simplifies the procedure greatly.

This test remains very empirical and there is always the risk of failure, but as with the preceding test, it deals with compost generally and provides information rapidly. Also, putting it into practice requires neither sophisticated equipment, nor qualified personnel.

4. General phytotoxicity tests (JUSTE et al. 1980)

The techniques hitherto described may prove to be quite insufficient to determine the presence of toxic substances responsible for phytotoxic damage to crops. Pot experiments may be necessary in order to identify such problems, which are impossible to forecast through simple analysis or respirometry.

The proposed test allows phytotoxic properties in urban compost to be revealed by growing two test plants (maize and beans) in the compost and comparing growth with those planted in non-enriched peat whose pH has been raised by addition of calcium carbonate. The growth of corn and bean seedlings, sown directly in the same container containing either pure compost or compost diluted with 10-25 % of non-enruched peat is measured for 18 days.

This test was used to compare 6 composts of different origins and degrees of maturity (6, 60, 120, 240 days). Dilution of the compost with peat produced a beneficial effect in the same way as ageing of the compost may do (Tables I and II).

This test allows the systematic diagnosis of some causes of the observed depressive effects when using massive doses of compost (e.g. excess salinity, nitrogen immobilization, lack of oxygen. We have shown that these parameters, considered in isolation do not explain completely the effect observed. In the case of immature composts (less than 120 days of maturation) risk remains high of the appearance of phytotoxic effects due to the depression of the oxygen level, nitrogen deficiency excess salinity and very probably, the presence of transitory phytotoxic organic substances subsequently biodegraded. These immature composts may be improved by the addition of small doses of nitrogen as organic fertilizer; mineral fertilizers in similar conditions are inefficient. In the same way, dilution of the compost with polystyrene chips or bark can reduce phytotoxicity in sufficiently mature compost.

The other remaining depressive effect for mature compost (180 to 240 days of maturation) used in a pure state, would seem to be excess salinity, which has a tendancy to increase during maturation. In this case a moderate leaching of the compost, improves its agronomic value considerably. The addition of increasing amounts of peat to mature compost reduces the effects of salinity considerably and permits a plant growth which is superior to that observed in the same volume of peat by between 55% and 90% on average. Thus, using the fertilizing potential of compost, importation of peat may be reduced to a large extent. A mixture of compost 75% fresh weight and peat 25% gives the best results.

This test does not require costly material (small plastic growing containers, weighing scales) and may be put into practice easily by non-qualified personnel. Its use may be considered suitable therefore in compost plants which makes it possible to separate lots of compost that seem doubtful or that are insufficiently matured, corn reacts by an intense reddening with immature compost.

CONCLUSION

This series of experiments on urban compost, which has been supported by the ministry responsible for the Environment, and carried out using samples identical for all the laboratories which have taken part in the experiments has led to a) the definition of parameters that describe the evolution of urban waste during decomposition and b) the development of numerous tests to determine maturity in compost.

Some proposed techniques demand well equipped and specialized laboratories (respirometry, analysis of biochemical parameters) and provide considerable potential advantages for carrying out occasional control analyses of compost production as well as proiding a study of new composting processes and new organic raw materials. On the other hand, other techniques are directly usable in composting plants and provide rapid information on the maturity of compost (simplified respirometry, C/N ratio, chromatography, colorimetry, phytotoxicity tests).

The large number of tests existing at present shows very clearly that a technique has yet to be found which allows overall control of composting with assessment of the stage of the maturity of the compost.

This depends on the final destination of the product (for use either fresh or mature) and can only be carried out validly if a sufficient number of parameters are taken into consideration (decomposability,

Table I - Influence of compost ageing and dilution by peat on growth of maïze and bean

Plant	Growth medium	Length of maturation (days)				Effect of mixing with peat
		6	60	120	180 or 240	
maize	pure compost	55	49	61	58	**56**
	90 % compost 10 % peat	53	68	96	109	**82**
	75 % compost 25 % peat	62	104	134	189	**122**
	mean	**57**	**74**	**97**	**114**	
bean	pure compost	62	67	64	53	**62**
	90 % compost 10 % peat	75	82	101	107	**89**
	75 % compost 25 % peat	78	100	127	155	**115**
	mean	**68**	**83**	**97**	**105**	

growth on pure peat = 100

JUSTE et al. 1980.

Table II - Influence of the origin of municipal refuse compost on maize and bean growth

Origin of the refuses (town)	Maize	Bean
Auxerre	65	73
Montbéliard	72	71
Ponten-les-Forges	125	102
La Loupe	105	109
Chalon-sur-Saône	69	81
Chateaubriant *	195	141
pure peat	100	100

* grinding refuses after a 2 year evolution period into a sanitary landfill.

JUSTE et al. 1980.

humification, ...).

These tests are complementary as are the classic parameter controls of temperature and humidity.

REFERENCES

1. BENISTANT, D. (1978). Caracterisation de la maturité des composts d'ordures ménagères. Mémoire ENITA Dijon-INRA Laboratoire de microbiologie des sols. Dijon. 40 p. + annexes.

2. CHROMETZKA, P. (1968). Determination de la consommation d'oxygène des composts en voie de maturation. Bull. Inf. GIROM, 33, 253-256.

3. COLIN, F. (1978). Recherche de paramètres caractéristiques de la maturité des composts. Etude de l'évolution de la biomasse au cours de la maturation. Compte-rendu de fin de contrat n° 78-142, Ministère de l'Environnement et du Cadre de Vie. Institut de Recherches Hydrologiques, Nancy, 71 p.

4. DUBOIS, M., GILLES, K.A., HAMILTON, J.K., REBERS, P.A., SMITH, F. (1956). Colorimetric method for determination of sugars and related substances. Annal. Chem., 28, 350-356.

5. GERMON, J.C., NICOLARDOT, B., CATROUX, G. (1980). Mise au point d'un test rapide de détermination de la maturité des composts. Compte-rendu de fin de contrat n° 79-509, Ministere de l'Environnement et du Cadre de Vie. Laboratoire de Microbiologie des Sols INRA Dijon. 11 p.

6. GUCKERT, A. (1973). Contribution a l'étude des polysaccharides dans les sols et leur rôle dans les mécanismes d'agrégation. These d'Etat, Univ. Nancy I, 124 p.

7. HERTELENDY, K. (1974). Paper chromatography, a quick method to determine the degree of humification of refuse compost. IRCWD News n° 7, 1-3.

8. JUSTE, C., POMMEL, B. (1977). La valorisation agricole de déchets. I - Le compost urbain. Ministère de la Culture et de l'Environnement, Ministère de l'Agriculture, 75 p.

9. JUSTE, C., SOLDA, P., DUREAU, P. (1980). Mise au point de tests agronomiques légers permettant de déterminer simultanément la phytotoxicité globale des composts d'ordures ménagères et leur degré de maturation. Compte-rendu de fin de contrat. Ministère de l'Environnement et du Cadre de Vie. INRA Bordeaux, 19 p. + annexes.

10. NICOLARDOT, B. (1979). Valorisation des composts d'ordures ménagères; caracterisation de leur maturité, accélération de leur évolution dans le sol. Mémoire ENITA Dijon-INRA Laboratoire de Microbiologie des Sols Dijon. 60 p. + annexes.

11. MOREL, J.L. (1982). L'évaluation de la maturité des composts urbains par une méthode colorimétrique. Compost Information 10, 4-8.

12. MOREL, J.L., JACQUIN, F., GUCKERT, A., BARTHEL, C. (1979). Contribution à la détermination de tests de la maturité des composts urbains. Compte-rendu de fin de contrat n° 75 124. Ministère de l'Environnement et du Cadre de Vie. ENSAIA, Nancy, 32 p. + annexes.

13. SPOHN, E., KNEER, F. (1968). Organischer Landbau, 4-5, 68-71.

DISCUSSION

EVANS: Did you measure or control the pH value of the sample under test. Why did you incubate at 20°C? Did you consider or study other temperatures?

MOREL: We did not measure pH but I think that the changes in pH of compost are negligible during incubation in the simple test. 20°C is the room temperature most commonly found so it is well suited to the conditions of a compost plant laboratory. Using electrolytic respirometers, temperatures such as 5, 15 and 28°C were also tested, each of them giving specific results but similar classification of compost samples in relation to their respiratory activity.

DE BERTOLDI: I have two questions:-

1. Why do most of the tests used evaluate the composting process and not the end product?

2. Why does ATP decrease during composting?

MOREL: 1. Some of the proposed techniques require well-equipped and specialised laboratories and are effective only for following the composting process. However, techniques such as simplified respirometry, colorimetry, rapid plant tests provide information about end-product quality because results obtained are directly related to the effect of compost on plant growth.

2. Decrease in ATP content during the phase of accelerated decomposition is mainly due to a reduction in biological activity probably, with reference to Finstein's paper, as a result of increased temperature.

VERDONCK: The method for the determination of compost maturity based on circular chromatography is difficult to interpret. Have you any comments?

MOREL: We also experienced problems of interpretation with this method. The main difficulty is the heterogeneity of urban compost but also this is a qualitative test. Considerable technical skill is necessary for correct interpretation.

BIDDLESTONE: In order to achieve reproducibility, it is important that one person should carry out all testing.

PHYTOTOXINS DURING THE STABILIZATION OF ORGANIC MATTER

F. ZUCCONI, A. MONACO and M. FORTE
Istituto di Coltivazioni Arboree, Portici, Napoli
M. de BERTOLDI
Istituto Microbiologia Agraria, Pisa

Summary

The control of composting in commercial plants appears to be empirical, and the product is seldom satisfactory. This situation reflects a lack of standard procedures (and of related methodologies) for evaluating process performance and product quality. To be of practical value, analytical methods should reflect the dynamic nature of the process, or some fundamental event related to it. A number of methods (physical, chemical and biological), with complementary analytical results, is necessary to satisfy all required specifications. The complementary value of analysing metabolic toxins is discussed in this paper. Toxicity analyses help to determine processes (aerobic/anaerobic, properly/poorly conducted) and phases (decomposition, stabilization, curing). Microbial production of toxins is high during the initial decomposition stage, and it subsides in the following stabilization (humification, mineralisation) stage. This evolution is altered (and toxicity persists) in anaerobic processes, or when dealing with unbalanced substrates. Also, if processes are not completed, the organic matter resumes metabolic activity (including production of toxins) related to the degree of stabilization already achieved. Based on this principle, laboratory analyses for "latent toxicity" were developed which help to reveal the degree of stabilisation of products of unknown origin. Finally, unstabilized products may endanger crops or plants when used as soil conditioners. A number of factors (physical and biological) interact with toxins to account for toxicity to plants, which explains the irregular nature of plant damage observed in these circumstances.

1. INTRODUCTION

Organic matter represents a highly polluting fraction of refuse, and the advantage of composting, in reducing environmental hazard and recovering residual energy, is generally recognised. Yet, the debate over the value of industrial composting is still unsettled; mostly because the quality of products from commercial plants is not always satisfactory or compatible with agricultural requirements. To be called compost, the organic matter must be stabilized to a humus-like product with an earthy odour, and it must be degraded to fine particles, having lost its original identity. A stable product can be stored without further treatment or applied to the land without damage to crops or trees. When stabilisation is not achieved, the decomposition continues, foul odours develop, and metabolites are produced which are toxic to plants (phytotoxins).

The problem is not that of the process itself (composting allows good results in terms of efficiency and product quality), but rather of a lack of understanding of process performance. Consequently, products vary greatly, most of them being unstabilized (decomposing) organic matter rather than mature compost. Increasing the time of processing guarantees stabilisation, but it becomes incompatible with industrial programmes due to the additional requirements for space, handling and equipment. Rapid efficient processes are necessary which in turn require definitions of standard quality and of suitable methods to monitor process performance (7, 9, 18).

The following discussion is mostly centered on the metabolic production of toxins which characterises organic matter decomposition (Chapt. 3). Metabolic toxicity is, indeed, strongly dynamic, and supplies a valuable guage for monitoring stabilization and assessing process performance and product quality (Chapt. 4 and 5). Also, toxicity to plants involves a number of determinants, other than toxins, which will be discussed as well (Chapt. 2).

2. THE CONCEPT OF TOXICITY

Injuries to growing plants or germinating seeds have often been reported following application of organic matter in agriculture. The literature often refers to the injurious effects of manuring, green manuring, or ploughing-in crop residues (Fig. 1), not to mention the allelopathic problems observed when continuous cropping is practised or replanting is done. Information on organic matter phytotoxicity has not always been systematically gathered, nor properly interpreted. Uncertainty thus persists as to the amount and frequency of damage. Yet, an explanation for the potential phytotoxicity of decomposing organic matter does exist, and it may be related both to the nature of the root system and to the origin of the organic matter. This chapter concentrates on the plant side of the problem, and on the damage and adaptability observed when toxins are present in the root environment.

2.1 Plant damage

Impairment of the root system in the presence of "toxins" is a complex phenomenon, strongly influenced by a number of interacting factors (physical and biological). It follows that the concept of phytotoxicity of decomposing organic matter needs to be understood in a broad and dynamic sense.

Broad because a large number of organic molecules appears to be toxic to the roots. Also, during decomposition of organic matter, we are confronted with a large number of toxic substances which may interact and are combined with different phases of organic substrates (Fig. 2). Finally, the effects of the concentration and duration of these materials, even for indentified toxic compounds, remains to be established.

The dynamic nature of plant response to toxins deserves attention, too. Firstly, because the production of toxins represents a transient condition throughout decomposition (see Chapt. 3). Secondly, because different plants are differently affected (specific sensitivity) by these products (Fig. 3). Thirdly, because sensitivity to toxins may be overcome by plant adaptation. When the first contact of roots with the organic matter is not lethal, plants show some capability to adjust to, and in some cases to thrive on, organic matter enriched substrates (Fig. 4).

2.2 Plant adaptability

Plants may grow in a range of different substrates varying in physical, chemical and biological characteristics. Far from implying

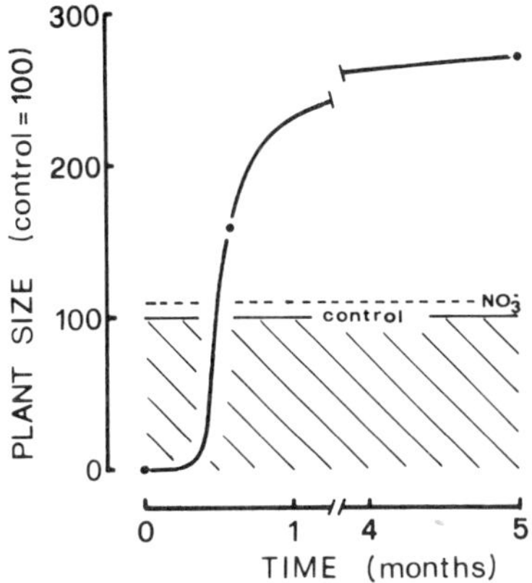

Fig. 1 - Growth of wheat plants sown at different time intervals after green manuring (lucerne). Growth is compared to control or to nitrate fertilization (data elaborated from: 10).

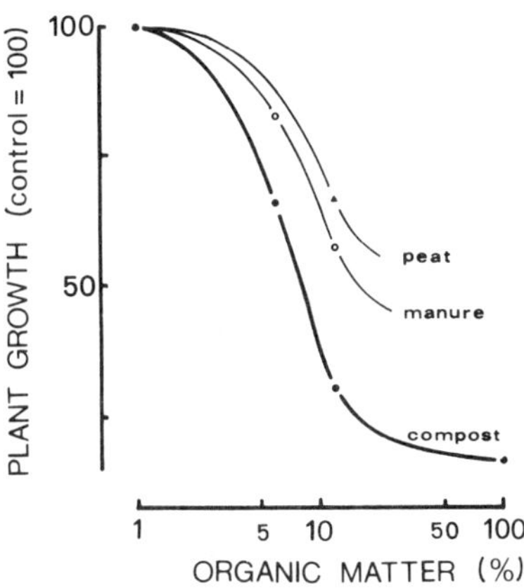

Fig. 2 - Olive plants stunting in organic matter enriched substrates. Specific dose response curves are associated with each organic product, suggesting the existence of different toxic groups (from 17).

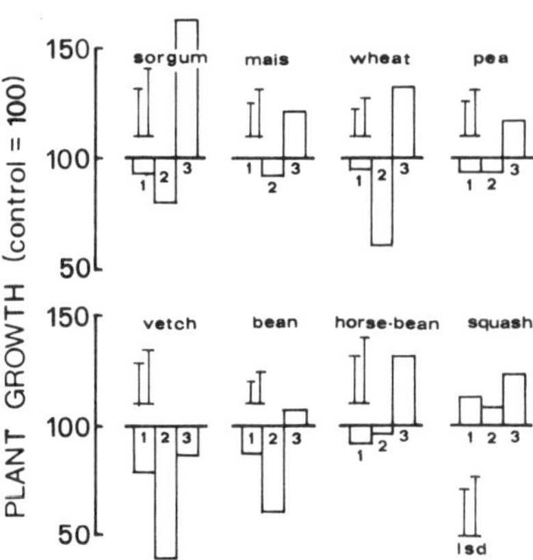

Fig. 3 - Response of a number of species to fresh (1), immature (2) and mature (3) compost underlying perlite seed-germination bed. 5% and 1% probability of least significant differences (lsd) are given for each specie (from: 17).

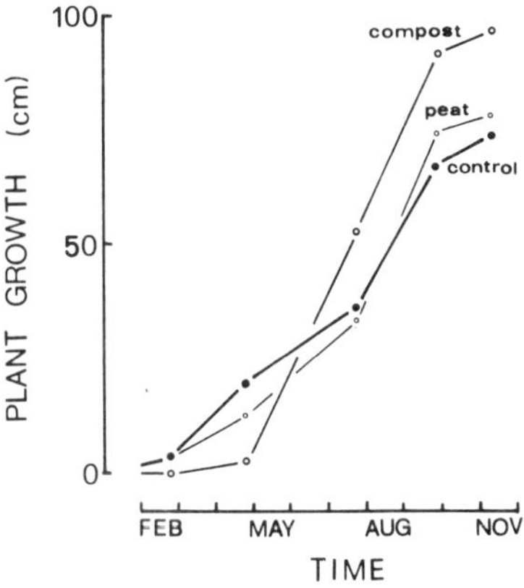

Fig. 4 - Growth of olive plants transplanted (2nd year of life in the nursery) in organic matter (12%) enriched substrates. The organic matter causes a period of stunting before the plant recovers and grows (from: 17).

indifference to the substrate, this flexibility appears to be associated with a highly-specific functional differentiation of roots. Plants develop specialized root systems, each adapted to a distinctive type of adsorption, and fit for the particular substrate in which the roots were formed (17, 19). Therefore, changes in physical, chemical or biological conditions cause a reduction (or arrest) in root absorption in proportion to the time required for adapting to the new environment. The plant responds to the stress by decreasing its metabolic rate (17) and concentrating its energies on generating a new absorbing apparatus (root hair, cortex,root tips, and entire roots or root systems may decay, at progressively higher levels of toxicity).

Plant damage then reflects the combined effects of the environmental change, together with the effect of specific root sensitivity and of metabolic conditions e xisting in the plant. This concept of phytotoxicity (PhT) may be represented by the following expression (I):

$$\text{I)} \qquad PhT = f\left(\frac{\Delta \cdot M \cdot S}{t_{\Delta}}\right)$$

where Δ represents the amount of environmental change, M the plant metabolic rate (mostly rate of absorbtion), S the specific sensitivity (when comparing different species, ages, etc.), and t_{Δ} the time lapse (or rate) of the change.

Consistent with expression (I), amount of change, metabolic rate or rate of change may have an individual impact on toxicity comparable to or greater than that associated with the nature of the change itself. Support for this conclusion comes from observing that plants may be adapted to conditions, otherwise lethal (High toxin concentrations), by lowering environmental temperature (as illustrated in figure 5). The same may be achieved by gradually increasing the concentration of toxic compounds (adaptation: total Δ results from summation of smaller changes), as illustrated in figure 6. The effect of the amount of change, compared to the nature of its causative factor, may also be observed in figures 6 and 7, where analogous toxic effects are induced by organic solutions (extracts from decomposing organic matter) or by inorganic ones (Hoagland solution).

2.3 Differential root absorption

Differential root absorption of organic molecules remains a rather obscure physiological event. From the information presented it appears, however, that roots are defenseless against "new" molecules which might enter cortical cells, upon first contact, causing cell poisoning and hampering absorbtion. This conclusion is consistent with the damage observed (root degeneration, wilting, generalised mineral deficiencies, reduced growth). These symptoms are non-specific, reminding us of similar responses to allelopathic factors, soil compaction, water-logging, or root-damaging diseases, all of which affect absorption.

Changes of soil environment require major root adaptations to meet the new absorption requirements. Growth in other parts of the plants, or high transpiration rates oppose root recovery. When a conspicuous change in the root environment occurs during active growth, the plant may experience a profound water stress due to a high water demand at a time of arrested root absorption. This explanation may account for the lethal and non-lethal responses which are equally observed in these circumstances.

A peculiarity of the lethal effect is that it is immediate (upon first contact with toxins or in conditions reproducing it), or it does not

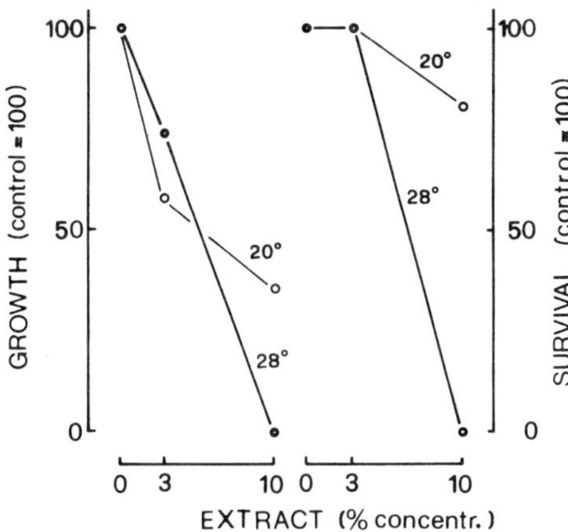

Fig. 5 - Growth (left) and survival (right) of 12 day old bean plants transferred from a Hoagland solution to a solution enriched with extracts from decomposing organic matter. Different environmental temperatures (20 or 28°C) affect plant response to toxins.

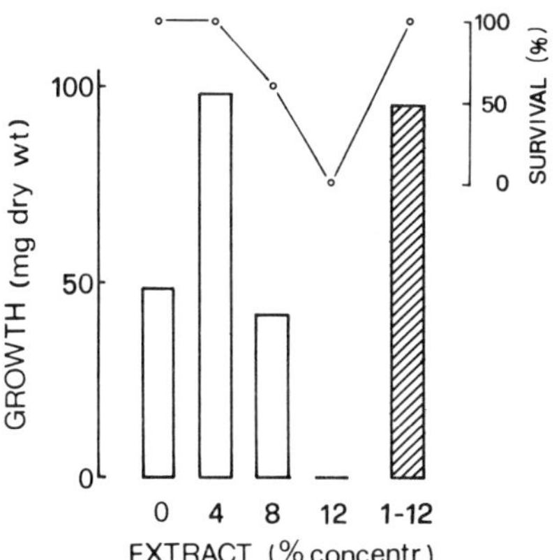

Fig. 6 - Survival and growth of bean plants (12 d old) transferred from Hoagland solution into solutions enriched with different concentrations of a toxic extract from decomposing organic matter. The shaded histogram represents the response when solution is gradually enriched by steps 1, 2, 4, 6, 8 and 12% concentration, each lasting two days.

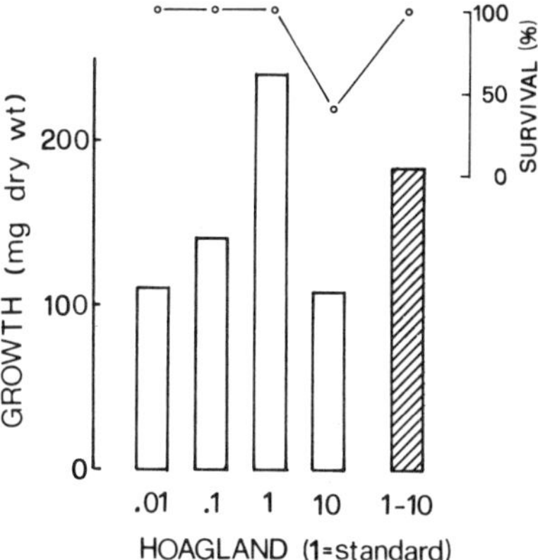

Fig. 7 - Survival and growth of bean plants (12 d old) transferred from water to Hoagland solutions of different strengths (1 = standard concentrations). The shaded column represents the response to progressive adaptation obtained by exposure to intermediate concentration steps (1 and 3%), each lasting two days.

come at all. When initial conditions are not lethal (no matter how close to the limit) the plant has the opportunity to become adapted and to resume its activity. However, the plant is in an unstable condition, where any additional factor may kill it.

This also explains the complex nature of plant tolerance to changes, which involves intrinsic (genetically bound) resistance to exogenous conditions, and adaptation. Adaptability in turn, may depend on extrinsic conditions (amount of changes) and intrinsic conditions (metabolism) affecting the plant.

3. TOXICITY DURING ORGANIC MATTER STABILIZATION

Though mostly centered on composting, this discussion includes information on other processes, and on the impact of some special substrates.

3.1 Production of toxins during industrial composting

A peculiarity of metabolic toxicity is its association with some, but not all stages of composting. Production of phytotoxins in biooxidative processes characterizes the initial stage of decomposition, its intensity and duration being dependent on a number of different factors. Figure 8 illustrates the dynamic changes in phytotoxicity occurring during composting. The impact of metabolic toxicity is here parallelled in the laboratory (cress bioassay) and in the nursery (organic substrates for potted plants).

The process of stabilization during composting appears to be more rapid with increasing oxygen availability. Consistently, the duration of toxin production was reduced in ventilated systems compared to turned windrows, as illustrated in figure 9. In ventilated systems toxicity may disappear in 2-3 weeks, preceeding the completion of the thermophilic stage (Fig. 10). In our experience, insufficient care in supplying oxygen appears to be the most common limiting factor preventing efficient composting in commercial plants. In these conditions we have observed lethal levels of toxicity lasting months, and even beyond a year.

3.2 Production of toxins as affected by different stabilization processes of substrates

Anaerobic processes are characterized by high levels of toxicity which persist in stabliized products. Sludges from water treatment plants (aerobic and anaerobic), sampled in different stages, show great differences when biologically tested for toxic response (Fig. 11).

Independently of aeration, production of phytotoxins may be affected by the nature of the substrate (see also Chapt. 3.3). This is the case with nutritionally unbalanced substrates, more frequently with agricultural or industrial wastes than with the organic fraction of urban waste. Animal wastes provide a good example. These are often characterized by a low, or very low, C/N ratio as well as by increased stabilization time and phytotoxicity. The process seems to undergo a prolonged stage of ammonia production, possibly to increase the C/N ratio as required to enhance the activity of microorganisms.

A peculiar case is presented by the residues from olive-oil extraction (meal and water fractions), which seem to undergo a process of prolonged toxic metabolism even when aerobic conditions are maintained (Fig. 12). The nature of this "persistent" production (as compared with the transient production observed in most substrates) remains unclear. It reminds us, however, of the metabolic routes of organic matter decomposition in soil. Here we may be seeing humification processes

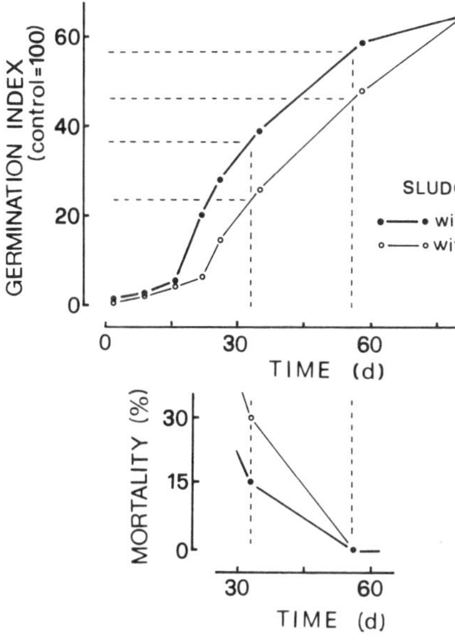

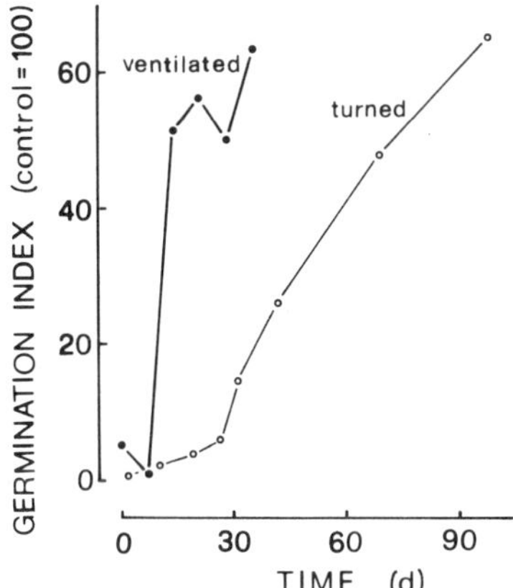

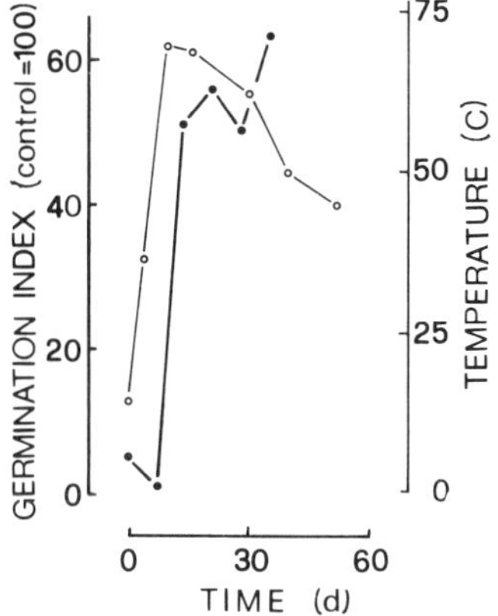

Fig. 8 - Toxicity during composting of the organic matter from urban waste. Toxicity is evaluated by cress germination test (above), or by mortality of olive trees (below) potted in substrates enriched with organic matter (12%) sampled at two different times during composting (from: 17).

Fig. 9 - Evolution of toxicity (cress bioassay) during composting of municipal wastes (organic solid fraction plus waste-water sludges). Turning and forced ventilation are compared (from: 18).

Fig. 10 - Dynamics of temperature and toxicity evolution during composting under conditions of forced ventilation (from: 18).

alongside anomalous routes responsible for "soil sickness" ("replant", or similar, diseases).

3.3 The origin of toxins

Plant damage, following contact with decomposing organic matter is often related to high C/N ratios, and to the presence of ammonia and of a number of different products (phenols, fatty acids, etc.). However, nitrogen deficiency, at times observed with high C/N ratios, must be considered as a problem of competition apart from phytotoxicity. Also, toxicity persists in extracts free of ammonia (solvent-extracted fractions concentrated under vacuum: 17) and in the absence of phenols and fatty acids.

Although toxic materials may be present originally, metabolic toxins produced by microorganisms become most relevant when dealing with industrial stabilization of organic matter. The impact of microbial synthesis is evident in figure 13, where metabolic toxicity evolves from a toxin-free, cell-free sucrose substrate (19). High rate of toxin production is related, even in aerobic conditions, to the accessibility of the substrate and to the rate of growth of microbial populations (possibly an allelopathic reaction in the competition for the available substrate). Figure 13 shows how toxicity can evolve differently with sucrose or cellulose as a carbon supply.

During composting, the initial microbial population, mostly bacteria (2), increases geometrically which leads to rapid exhaustion of the readily available substrate. It follows that the existing population decreases and a new population develops on the residuel of the original one. This new wave of growth is generally characterised by a "return" of toxicity which, however, may seldom be noticed in the composting pile, where the evolution appears less uniform, and stages develop out of phase. Many factors contribute to the disappearance of toxicity at more advanced stages of the process. These include changes in the microbial composition (actinomycetes and fungi prevail at the end), metabolic destruction of toxins, and synthesis of new polymers. In the final stage (curing), the presence of free molecules is limited to products of mineralisation and to a few microbial excretions.

3.4 Decomposition/stabilization

The term stabilization, though commonly attributed to the whole of composting, would be more appropriate for the second stage of the process (characterized by mineralisation and humification), following a decomposition stage, where instability and increasing entropy prevail.

Toxic metabolism differs greatly between the two stages. It appears maximally expressed during decomposition, tending to subside when the process progresses towards stabilization. Interruption of composting leads to a later completion of residual stages (when conditions become favourable again). *In vitro* analyses in the laboratory, monitored by toxicity evolution, indicate that residual metabolic processes (latent toxicity) are strictly complementary to the degree of decomposition or stabilization already achieved. In figure 14, latent toxicity appears more intense and of longer duration when metabolizing fresh organic matter. Progressively smaller activity characterizes samples taken at more advanced stages.

Routine analyses of latent toxicity on marketed soil conditioners have shown a range of stabilization stages. Few products appear sufficiently stabilized to deserve being called "mature compost". The majority is characterized by the development of an intense latent toxicity

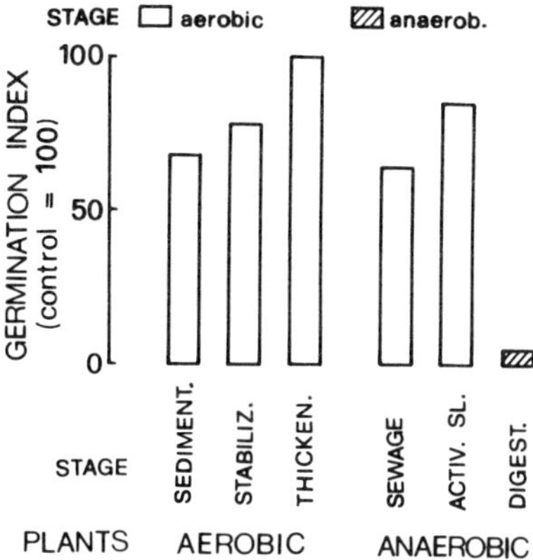

Fig. 11 - Toxicity (measured by cress bioassay) at different stages of waste-water treatments in aerobic (City of Viareggio) and anaerobic (City of Massa) plants (from: 16).

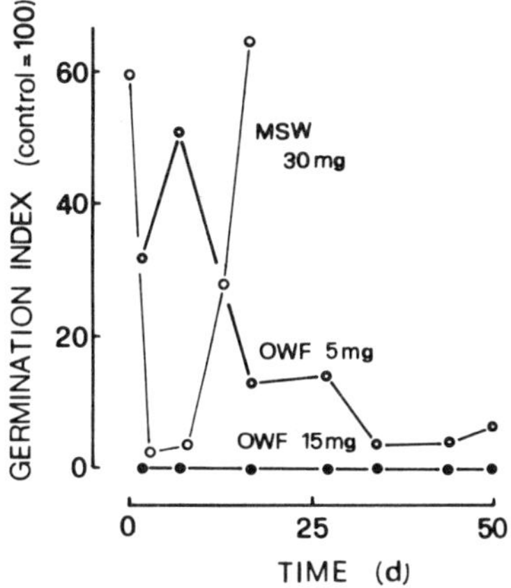

Fig. 12 - Toxicity of olive water fraction (OWF) aerobically metabolized. Cress bioassay reveals a toxic metabolism persisting far beyond that of the organic fraction of municipal wastes (MSW). Different amounts were tested (mg/Petri dish).

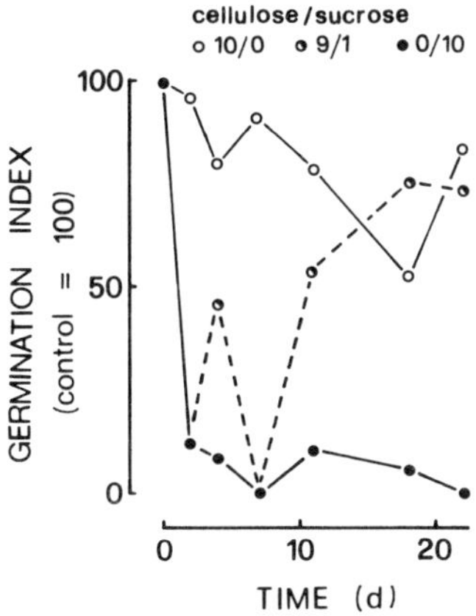

Fig. 13 - Evolution of toxicity (cress bioassay) in sucrose and/ or cellulose substrates aerobically metabolized in a liquid phase (10% dry wt) in the laboratory

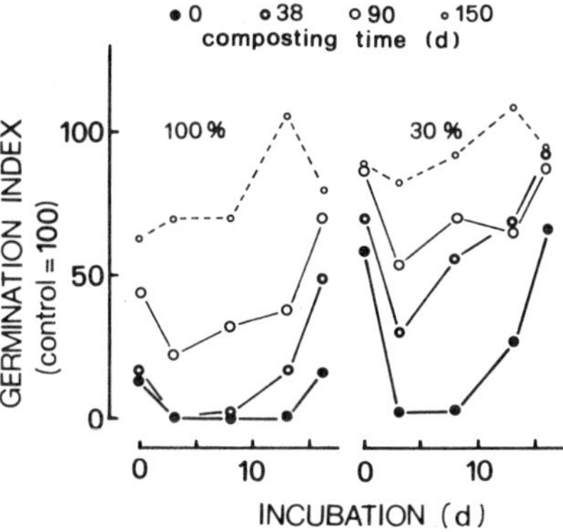

Fig. 14 - Residual toxic metabolism (latent toxicity) when using organic matter at different stages of composting (turned piles: Fig. 8). Toxicity is measured with cress at 100 and 30% concentration of the original solution (from: 18).

when analysed in the laboratory. For some products, the analyses indicate that no biological stabilization had occurred before they were dried, packed and marketed, because dehydration, screening, pelleting, chemical additives, etc. may reduce foul odours and provide the visual characteristics of stabilized matter. If used for soil conditioning, these products resume a production of toxins which is proportionally greater, and of longer duration, than in industrial composting due to the lower oxygen availability in the soil.

The introduction of massive amounts of poultry manure (dried, pelleted) in the Italian market, supported by "aggressive" advertising, has caused extensive damage (16) and concern among growers in the last few years.

4. THE IMPORTANCE OF MONITORING STABILIZATION PROCESSES

Control of composting in commercial plants today appears to be most empirical, while the product is seldom satisfactory. As stated in the premise, this mostly results from lack of standard procedures for evaluating process performance and product quality, which in turn reflects a lack of suitable methodologies (9, 19). A number of methods has been proposed (1-6, 11-15), none of which, however, has been pursued far enough to allow full appreciation of its potential value. Theproblem with these methods is that their capability of discrimination between processes or stages must be better clarified before they become applicable. It seems therefore appropriate to consider what are the necessary requirements for proposed analytical methods.

4.1 Requirements for methodologies

The value of analytical methods is related to the general nature of the parameter they refer to, when compared in different conditions and processes. Proposed methods should therefore show a number of characteristics, namely:

a) sensitivity to the dynamic changes occurring during the decomposition and stabilization stages. Parameters should reflect some fundamental aspects of the process analysed and be able to recognize relevant characteristics, critical phases or phase transitions;

b) ability to discriminate between processes (aerobic/anaerobic; properly/poorly conducted; etc.). Proposed methods should be selected by testing them in a number of situations with varying physical, chemical and biological conditions;

c) the potential for revealing abnormal processes (physical stabilization, chemical additions, etc.). in finished (marketed) organic products;

d) to be simple, reproducible and rapid. Rapidity is essential for biological tests. Long-lasting ("cultural") tests must be avoided as they would be affected by the continuing stabilization processes and by the adaptability of test plants.

It is safe to say that no single analytical method (physical, chemical or biological) may be expected to satisfy all the specified requirements, and that a number of parameters, with complementing spectra, will need to be used in parallel.

4.2 The value of analysing toxicity

Analyses of metabolic toxicity may supply a wide spectrum of information on stabilization processes, thus providing a useful instrument for industrial, agricultural and environmental analyses. The advantage of introducing toxicity analyses (parallel to other analyses) may be appreciated if we consider that:

1) evaluating phytotoxicity is consistent with the final use of composts, and with the need for assessing physiological compatability and best use of organic soil conditioners;

2) metabolic production of toxins is strongly dynamic, well reproduced in different processes, with a good distinction between the decomposition/stabilization stages. It would then offer: a) a common parameter for comparing different composting systems and b) a means for reducing composting to the minimum time required for a standard product (progress may be assessed well ahead of the temperature drop).

3) "latent toxicity" (ie, toxicity which remains to be expressed *in vitro*) provides a useful method for revealing the degree of stabilization previously achieved by products of unknown origin.

In conclusion, analyses of toxicity provide objective criteria for evaluating composting process performance and product quality, as well as the relation between the two. Further action is necessary, however, to increase knowledge of metabolic toxicity, and to improve the analytical value of methods based on toxin evolution. This includes:

a) undertaking in-depth studies on the nature of toxic metabolism and of factors which may affect it (through physiological, biochemical and microbial research);

b) checking the progress of toxic metabolism in parallel with that of other (and potentially useful) physical, chemical and microbial measurements;

c) analysing the potential for introducing analyses of "latent toxicity" among standard methods required for assessing the stability of marketed organic soil conditioners.

5. APPENDIX: CRESS BIOASSAY AND ANALYSES OF PHYTOTOXICITY

This chapter illustrates standard methods used in our laboratory for the analysis of toxicity. Experiments conducted using these analyses were illustrated in previous chapters (figures 8-14).

5.1 Extraction of toxins from solid organic fraction

Extraction is limited to free water, the fraction most responsible for immediate toxicity. Although methanol extraction may provide good results (17), we prefer pressure extraction procedures, which are faster and simpler. The extraction procedure is:

a) measure dry wt on a separate sample;

b) standardize moisture content (we use 60% content) by water addition (1/2 h incubation). If the original material exceeds 60%, adjustment may be made in the subsequent dilution steps;

c) extract 5 min at 2.5 atm^3cm^2;

d) clean the extract by millipore filtration (sterilizing);

e) the extract may be used immediately, or frozen (-20°C) for later use;

f) that a number of concentrations, (100, 30, 10, 3%) of the extract. Concentrations of 30% and 10% are the most indicative for evaluating toxicity.

5.2 The cress bioassay

In our work, requirements were for a simple and rapid bioassay to be quickly handled with simple instrumentation. We explored germination tests, since they require no special preparation. Also, we selected cress (Lepidium sativum L.) which from preliminary experiments attracted us because of its rapid response (24 h).

A number of variables were studied to make such a test more effective (8). According to this study, maximum sensitivity is at 27°C. We chose to limit incubation time to 24 h, consistent with the lower variability observed at this stage. Also, 24 h incubation gives the advantage of accelerating the test and reducing the time required for measurements (a 4-7 mm root is present at this stage).

An ad hoc "germination index" was obtained by multiplying germination and root growth, both expressed in % values (root growth as % of control). This index has proved to be a most sensitive parameter, able to account both for low toxicity (10^{-7} M, or 25 ppb, abscisic acid) which affects root growth, and toxicity, which affects germination.

In brief, the method of bioassay is as follows:

- prepare Petri dishes (4-5 cm diameter), lined with filter paper, containing 0.5 ml of the solution to be tested;
- use 6-8 seeds per dish with 10-15 replications;
- seeds are incubated at 27°C for 24 hours in the dark;
- at the end of incubation period germination is stopped by adding 1 ml of denatured ethanol (50%);
- the number of germinated seeds and the average root length per dish are expressed in percent of control (average of all control reps);
- results are expressed as germination index obtained from the product of % germination and % root length divided by 100.

5.3 Latent toxicity

The method consists of the aerobic digestion of organic matter in a liquid phase. The digestion lasts for the time required to stabilize the organic matter and to end toxin production. Organic matter is stabilized rapidly in this condition (10-15 d), thus allowing one to test "finished" products or suitability of organic substrates for composting. The method is as follows:

- organic matter (10% wt) is incubated at 27-36°C under aerobic conditions (air or oxygen must be forced through the solution);
- the supernatant liquid is sampled at regular intervals (3-7 d) filtered (millipore), and tested with cress at 100%, 30%, 10% and 3% concentrations. The toxic phase of organic matter may be considered terminated when the germination index rises above 60%. At this level the organic matter only inhibits growth temporarily if put in direct contact with a plant root system.
- secondary peaks of toxic metabolism may be observed possibly indicating successions of microbial populations. Waves of secondary metabolism are a common feature in the digestion of organic matter even during humification in the soil.

REFERENCES

1. ALLENSPACH, H. and OBRIST, W. (1969). Determination of the degree of maturity of refuse compost. International Research Group on Refuse Disposal. Information Bulletin; 35.

2. CHANYASAK, V., YOSHIDA, T. and KUBOTA, H. (1980). Chemical components in gel chromatographic fractionation of water extract from sewage sludge compost. J. Ferment. Technol; 59: 533-539.
3. CHANYASK, V. and IUBOTA, H. (1981). Carbon/organic nitrogen ratio in water extract as measure of composting degradation. J. Ferment. Technol.; 59: 215-219.
4. CHANYASAK, V., HIRAI, F.M. and KUBOTA, H. (1982). Changes of chemical components and nitrogen transformation in water extracts during composting of garbage. J. Ferment. Technol.; 60: 441-448.
5. de BERTOLDI, M., VALLINI, G. and PERA, A. (1983). The biology of composting: a review. Waste Management Research; 1: 157-176.
6. de VLEESCHAUWER, D., VERDONCK, O. and VAN ASSCHE, P. (1981). Phytotoxicity of refuse compost. BioCycle, January-February; 44-46.
7. FINSTEIN, M.S., CIRELLO, J., MAC GREGOR, S.T., MILLER, F.C., SULER, D.J. and STROM, P.F. (1980). Discussion of : Haug, R.T. "Engineering principles of sludge composting". Journal Water Pollut. Contr. Fed. Vol. 52 (7): 2037-2042.
8. FORTE, M. (1980). Messa a punto di un bioassaggio per lo studio della tossicita della sostanza organica in via di compostaggio. Thesis. University of Pisa.
9. GOLUEKE, C.G. (1977). Biological reclamation of solid wastes. Rodale Press Emmaus, PA. USA, 249.
10. KONONOVA, M.M. (1966). Soil organic matter. Pergamon Press. London 544.
11. LOSSIN, R.D. (1970). Compost studies. Compost science; Nov-Dic., 11: 16.
12. NIESE, G. (1963). Experiments to determine the degree of dceomposition of refuse by its self-heating capability. International Research Group on Refuse Disposal. Information Bulletin, 17.
13. ROLLE', G. and ORSANIC, E. (1964). A new method of determining decomposable and resistant organic matter in refuse and refuse compost. International Research Group on Refuse Disposal. Information Bulletin, 21.
14. SUGAHARA, R. and INOKO, A. (1981). Composition analyses of humus and characterization of humic acid obtained from city refuse compost. Soil Science Plant Nutrition, 27 (2): 213-224.
15. USUI, T., AKIKO, S. and YUSA, M. (1983). Ripeness index of wastewater sludge compost. BioCycle, Jan-Feb.; 25-27.
16. ZUCCONI, F. (1981). Utilizzazione in agricoltura dei residui solidi e liquidi urbani. Il problema della fitotossicita dei prodotti intermedi della biodegradazione. Trattamento e smaltimento delle acque residue, dei fanghi e dei rifiuti urbani ed industriali. Centro Scientifico INternazionale. MI: 448-459.
17. ZUCCONI, F., FORTE, M., MONACO, A. e de BERTOLDI, M. (1981 a). Biological evaluation of compost maturity. BioCycle, 22 (4): 27-29.
18. ZUCCONI, F. PERA, A., FORTE, M. e de BERTOLDI, M. (1981 b). Evaluating toxicity of immature compost. BioCycle, 22 (2): 54-57.
19. ZUCCONI, F. (1983). Processi di biostabilizzazione della sostanza organica durante il compostaggio (Biostabilization of organic matter during composting). International Symposium on Biological Reclamation and Land Utilization of Urban Wastes. La Buona Stampa, Naples; 379-406.

DISCUSSION

FINSTEIN: A test is required both for the process and the product. One possibility is to show the absence of toxicity on process dynamics. Another is the start of nitrification, the presence of NO_2 or NO_3 showing that the process is well advanced. Should we discuss this as a rough and ready test?

ZUCCONI: I have no idea whether this can be used.

LYNCH: You concentrate on the phytotoxicity aspects of composts as measured by the germination index. Have you considered measuring root extension as an index of beneficial effects of composts on crop growth?

ZUCCONI: I do not wish to give the impression of attributing a 'toxic' connotation to compost because the opposite is true. I am suggesting that measurements of toxicity may help in establishing the degree of compost maturity and its best use in agriculture. Beneficial effects are produced by mature products.

The bioassay we use interprets both germination and root growth thus improving the sensitivity of the test which may range from very high levels of toxins, affecting germination, to very low levels affecting root extension. Measured against a natural plant inhibitor, abscisic acid, this sensitivity ranges from concentrations of 5×10^{-3} to 5×10^{-7} molar.

WOOD: What is known about acceleration of maturation?

FINSTEIN: Table 3 of my paper demonstrates that nitrification is an index of maturation. This is achieved rapidly in the process.

LE ROUX: What is known about the effects of moisture content on particular microbes or particular wastes in terms of maximum activity?

FINSTEIN: There is little evidence for the importance of water content of materials. It is known that the matrix potential varies between materials.

SESSION II

MECHANISATION OF COMPOST MAKING AND EARTHWORM CULTURE

Machinery aspects of compost making

Practical experience with farm scale systems

MACHINERY ASPECTS OF COMPOST MAKING

A.D. MARTEGANI (°) and M. ZOGLIA (°°)

(°) Istituto di Fisica Tecnica e Tecnologie Industriali, Università di Udine - Italy

(°°) DANECO - Danieli Ecologia S.p.A. - Udine - Italy

Summary

The compost produced in plants for treating agricultural and other wastes usually suffers from the drawbacks of a lack of exact quality standards to make the product reliable for use in agriculture and in the presence of impurities which make the product less desirable although microbiologically acceptable.

This paper describes the machinery used for typical composting plants. It considers and examines the various kinds of machinery for the materials handling, their size reduction, screening and classification, as well as the machinery used for compost making.

Particular reference is made to the machinery which will be used for the actual composting process.

1. INTRODUCTION

The machinery used in waste treatment plants and in composting plants particularly, varies according to the kind and quality of the materials to be treated, the products required and, not least, the treatment cycle and/or composting method adopted (1),(2).

All plants have similar machines or machines working on the same principles, and treatment systems which include the composting process of the organic matter can have more differences and particular characteristics in the method used depending on which microbial process of mineralization and partial humification of organic substances takes place.

However three main areas can be considered:

a) selection and preparation of the materials to be composted;

b) composting;

c) handling and possible compost treatments (for example).

Some machines are used for operations of both areas a) and c), even if the composting process takes place differently. In contrast, the machines for area b) (composting) are normally typical for the process adopted.

Therefore the machines can be subdivided ignoring their position in the plant cycle and making particular reference to the kind of process they are used for.

The following subdivision has then been adopted:

- handling
- sorting
- size reduction
- composting

2. HANDLING MACHINERY

These are machines used for moving materials in the plant. The choice of the kind of machine mainly depends on the following two parameters:

i) Transporting profile, i.e. the distance the material must be conveyed horizontally (horizontal carry); the height the material must be moved vertically (vertical lift); the steepness of the slope up (or down) at which the material must be conveyed (angle of inclination); and the variety of the directions the material is to be conveyed, as compared to linear movement (complexity). Equipment with greater distances between centres requires stronger components and structures; the height of vertical lift is often critical also.

ii) Materials properties like material size and flow characteristics. When handling solid waste material, special consideration should be given to maximum size of items to be handled and the percentage of the large items in the total product handled. For the flow characteristics, the angle of repose is normally an approximate measure of this for most bulk materials; however the flow characteristics of wastes are not easily defined and may require a test programme for the true final determination.

The following properties of materials should also be considered: corrosiveness and abrasiveness (material in the range pH 1-7 may require special consideration).

Conveyors suitable for handling wastes are generally chosen from the following designs:

- cranes with grapple bucket;
- bucket elevators;
- belt conveyors;
- hinged metal belt conveyors;
- vibrating conveyors;
- drag chain conveyors;
- screw conveyors;
- pneumatic conveyors.

Table 1 is a quick reference guide for selecting a conveyor to meet the conditions as determined by the basic requirements.

Cranes with grapple bucket are generally employed for taking materials from a pit, or a heap, and placing it on the machinery or plant.

A bucket elevator is a conveyor for carrying bulk materials vertically or in an inclined straight path. It consists of an endless belt, chain or chains, to which buckets are attached, and it is usually more suitable for conveying with separate vertical or horizontal movement than doing both si-

Table 1. Selection of Solid Waste Handling Conveyors

SELECTION OF SOLID WASTE HANDLING CONVEYORS	Material Characteristics													Profile					Type of Action				Typical Material & Conveyor Selections								
	Size					Flowability				Abrasiveness																					
Conveyor Type	Very fine	Fine	Granular	Lumpy	Irregular, stringy	Very free flowing	Free flowing	Avg. flowing	Sluggish	Non-abrasive	Abrasive	Very abrasive	Very sharp	Horizontal	Inclined-declined	Vertical	Horizontal & inclined	Horizontal & vertical	Carries material	Pushes material	Drags material	Trash Type 0	Rubbish Type 1	Refuse Type 2	Wood shavings & sawdust	Glass cullet, bottles, etc.	Glass cullet, plate	Metal chips	Metal turnings bushy	Die cast scrap	Stamping scrap
Belt Conveyor	●	●	●	●	●	●	●	●	●	●	●	●		●	●		●		●			●	●	●	●	●	●				●
Hinged Metal Belt Conveyor		●	●	●	●		●	●	●	●	●	●	●	●	●		●		●			●	●	●				●	●	●	●
Vibrating Conveyor		●	●	●	●	●	●	●		●	●	●	●	●	●				●							●	●	●		●	●
Drag Chain Conveyor		●	●	●			●	●	●	●	●		●	●	●		●				●							●			
Screw Conveyor	●	●	●			●	●	●	●	●	●		●	●	●	●				●					●						
Pneumatic Conveyor	●	●				●	●			●				●	●	●	●	●		●					●						

multaneously with an inclined bucket elevator. However, the size of material must be limited and the weight of the moving equipment limits the height of elevation. The maximum elevation for special types is about 30 m.

A belt conveyor is an endless rubber or treated fabric belt which carries the solid waste material directly upon it (Fig. 1). It is used for horizontal movement or a maximum of 25-30° inclined straight path and can move from a few kilogrammes per minute to thousands of tonnes per hour of material, depending on belt width and belt speed (with free-flowing material, the maximum speed is about 2.5 m/sec).

In those cases where slopes are greater than 25-30° (depending on materials) special belts with cleats fastened to the carrying surface can be used up to angles of 60°.

Hinged metal belt conveyors consist of a series of overlapping metal pans mounted on chains running over terminal sprockets. The pans can be provided with side wings to form a metal trough. This unit can have a horizontal section for loading, with the discharge section inclined up to 45°. If the waste contains fine materials or wire particles, the material can work its way into the unit's moving parts, causing wear and conveyor stoppage. A safety clutch can be incorporated in the drive mechanism to prevent possible damage.

A vibrating conveyor is a simple trough, flexibly supported and vibrated at relatively high frequency and small amplitude to convey bulk material and objects. The vibrating conveyor is one of the simplest, most trouble-free types of conveyor but it is limited to linear moving. It will handle any material which is not sticky, and particle size and shape are no real problem. Conveyor widths are 0.25 to 1.5 m. Unfortunately, this is the most expensive form of conveyor and can seldom be used to elevate materials.

A screw conveyor consists of a steel helix mounted on a shaft suspended in bearings, usually in a U-trough (Fig. 2). As the shaft rotates, material is moved by the thrust of the lower part of the helix, and is discharged through openings in the trough bottom or at the end. Screw conveying can be done with the path inclined upwards, but the capacity decreases rapidly as the inclination increases: a standard pitch screw inclined at 25° loses 60% of its horizontal capacity (Fig. 3).

Material characteristics play an important role in speed, size and consequently in capacity of machinery. Light, free flowing, non abrasive materials fill the trough deeply and high rotation speeds are possible. However, this type of conveyor is not easily adaptable to handling solid waste material, except when the product size is less than 100 mm. It is suitable for sludge. Stringy material can wrap around the shaft and helix. If not properly selected, this type of conveyor can be most unreliable, and has a high operating cost. Standard units can be from 150 mm to 600 mm diameter: usually, for handling wastes, sizes between 300 and 500 mm are used.

The length of screw is limited by the maximum torque available at the shaft or couplings.

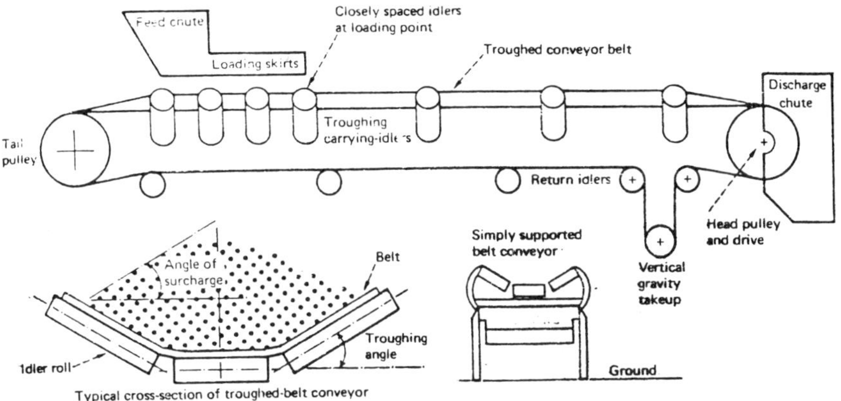

Fig. 1 - Schematic lay-out of belt conveyor system showing major components

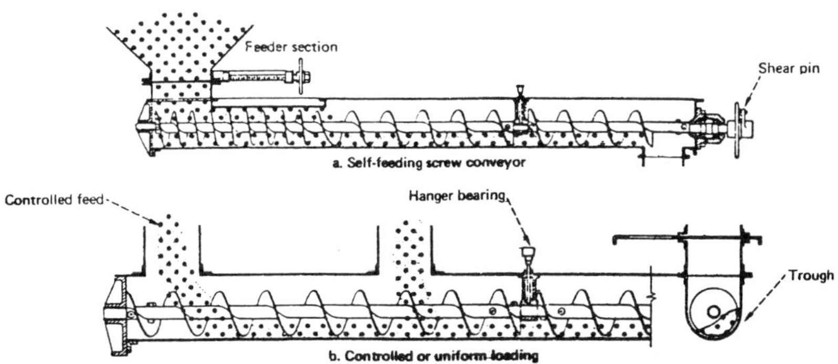

Fig. 2 - Two arrangements by which solids enter screw conveyors : self-feeding, controlled feeding.

A screw conveyor properly used works well with costs often less than half those of other conveying equipment.

The drag chain conveyor has an endless chain in a trough which drags materials by fixed bars or pans.

Drag conveyors combine the features of easy accessibility to the moving parts, minimum-sized trenches, total containment, no extra wear due to fine materials, and maximum flexibility of conveyor layout. Widths of these units can range from 0.3 m to 1.5 m.

A variety of pneumatic conveyor systems have been designed to handle solid wastes.

They consist of a pipe line in which material is fluidised and moved by an air stream from a fan or a blower. For dense material conveying the blower is connected at the start of a pressurized pipe. Often, volumetric blowers are used because they are able to overcome the effects of a pipe blocking. A rotary valve is necessary to introduce the material into the air stream and the line can be branched after inlet section.

With the blower connected at the end of pipe line the conveyor is suitable for lighter and more fluid materials and requires larger diameters of pipe.

In this case, the line can be designed with several inlet sections.

In both cases, a separating facility, like a cyclone, is necessary often followed by a bag filter in order to separate fine dust.

Pneumatic conveying requires small sized and homogeneous materials but it is very suitable and useful for complex paths. On the other hand it suffers the drawback of low energy efficiency.

This summarizes the units most commonly used for handling of waste materials.

Many others are available, like enclosed mass conveyors, harpoon conveyors, skip hoists, elevators, and can be adopted in special cases.

3. SORTING MACHINERY

The need to sort a mixture into components is often apparent in composting plants; for example separation of the organic fraction of refuse and recovery of wood chips used as bulking agent in composting sewage.

Three cases are considered:

a) subdivision of one material of different sizes into size classes;

b) separation of different materials with the same size;

c) separation of different materials with generally differing sizes.

In all three cases the machine characteristics depend on their particular application. They can be roughly divided into screens and separators.

The screens allow division of a material by particle size, i.e. splitting the material into different fractions, each with a defined upper and lower size limit. There are two types of screens:

i) rotating screens;

ii) vibrating screens.

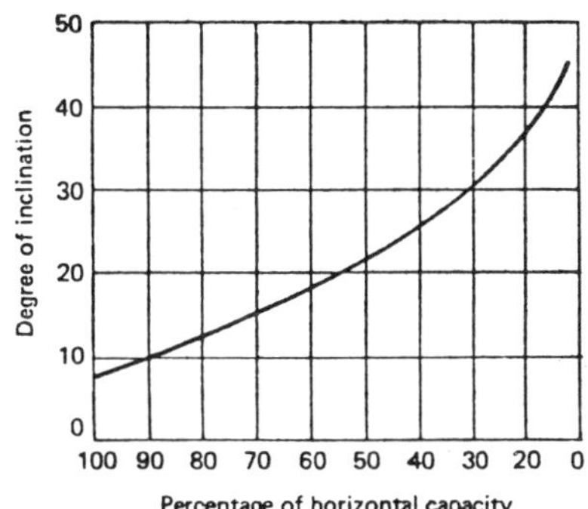

Fig. 3 - Decrease of screw conveyor capacity with inclination.

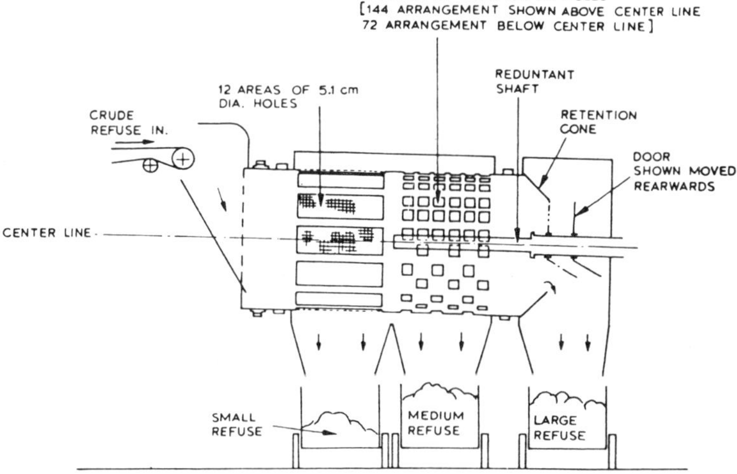

Fig. 4. Diagrammatic section through rotating screen.

i) Rotating screens, or so-called trommels, consist of a plate cylinder with holes in it, with its rotational axis slightly inclined to the horizontal (2-6°). The material to be treated is introduced at the upper end; the large size fraction comes out of the lower end, while the fraction smaller than the holes passes through and is collected under the screen.
The main important parameters of this kind of machines are:
- screen hole diameter;
- rotating cylinder diameter;
- rotating cylinder length;
- rotational speed;
- angle of inclination of the rotational axis.
The above parameters influence the size of the fine materials to be separated, the machine capacity, the screening efficiency.
Rotating screens are very simple and reliable machines but the degree of utilization of the screen surface is much less than for flat bed screen: this is one of the reasons for their normally large dimensions but, on the other hand, they are relatively lighter than the flat bed ones (Fig. 4).
ii) Flat bed vibratory screens require smaller screen surfaces than revolving cylindrical screens but, at the same time, because of their particular vibratory mode(s) (either in a more or less vertical direction, horizontal direction, or a combination of the two), do not destratify particles with large surface area to thickness ratio (e.g. leaves and plastics) which entrain the finer material (e.g. dirt, stone fines, etc.). Hence, in a flat bed screen most of the fines remain sandwiched and attached to the larger particles.
This type of machine can be usefully adopted for sorting final products, i.e. compost.

Separators are machines which, by utilizing various operating principles, allow separation of materials which cannot be separated with the simple screens: for example materials having equal dimensions but different densities.

The main types of separators are:

i) ballistic separators: the material is thrown and splits up; various portions are collected in different hoppers depending on the trajectory;
ii) bounce separators: they exploit the differing elasticity of materials and the resulting different bounce trajectories;
iii) sloping belt conveyors: less dense materials with irregular form are conveyed to the top, while the dense ones with spherical form go to the bottom;
iv) magnetic separators: they exploit the various ferro-magnetic characteristics of the materials, particularly for ferrous metals separation;
v) aeroballistic separator: this is a patented machine with separating based on throwing the mixed material to be separated, against a counter current of air (3) (Fig. 5);

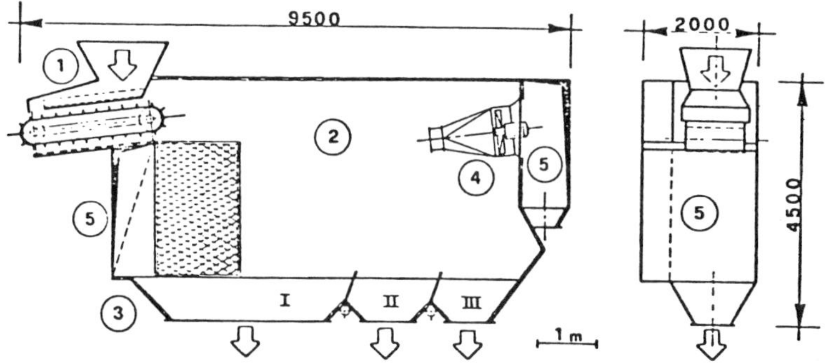

Fig. 5. Aeroballistic separator.
1. Feed hopper, 2. Throwing chamber, 3. Hoppers, 4. Jet fan, 5. Air ducts.

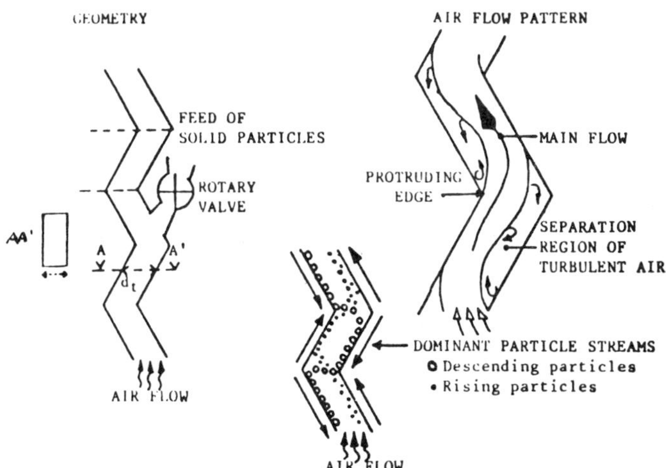

Fig.6. Geometry of the zig-zag air classifier, the air flow pattern and the dominant particle streams.

vi) air flow separators: they usually exploit the different aerodynamic characteristics of the particles present in the mixtures and are often called classifiers, as the separation of various size fractions of the same material can be obtained.

There are three typical classes of air flow classifiers, referring to the direction of flow:

- vertically;
- horizontally;
- with a rotating cylinder.

The limits of such subdivision can be often confused.

In the vertical path separators the material is allowed to fall from the top against a counter current of air.

One of the most efficient machines of this type is the zig-zag air classifier (Fig. 6) consisting of a number of compartments with rectangular cross section connected to each other at a fixed angle to create a zig-zag channel.

Material is fed to the column via a rotary valve. The particles are classified by their falling behaviour in the channel against a continuous, growing stream of air.

The geometry of the channel and the air flow patterns induced by it cause two distinct particle streams:

- a stream of "light" particles carried upwards by the upflowing air current;
- a stream of "heavy" particles moving downwards along the lower wall of each section.

At each junction of two sections, i.e. at each stage, the particles of both streams are subjected to further classification.

The separation performance of the zig-zag classifier is determined both by the particle behaviour at a single stage and by the interaction between the stages.

In the horizontal air-flow classifiers (Fig. 7) the materials are allowed to fall into a chamber through a slow speed horizontal air stream (5 m/s). The heavy particles are little influenced by the air flow, while the lighter ones are subject to more movement horizontally during their fall.

This differing behaviour is utilized for collecting the separated materials into two or more hoppers.

The authors have tuned up one of this kind of patented separator having a particular configuration, called "Tapyro" (4).

In the air-flow separators with a rotating cylinder the air flows through the cylinder with a sloping axis into a vertical plenum which also receives the material to be classified. The light fractions are conveyed to the top and the heavy ones fall down to the bottom (Fig. 7 bis).

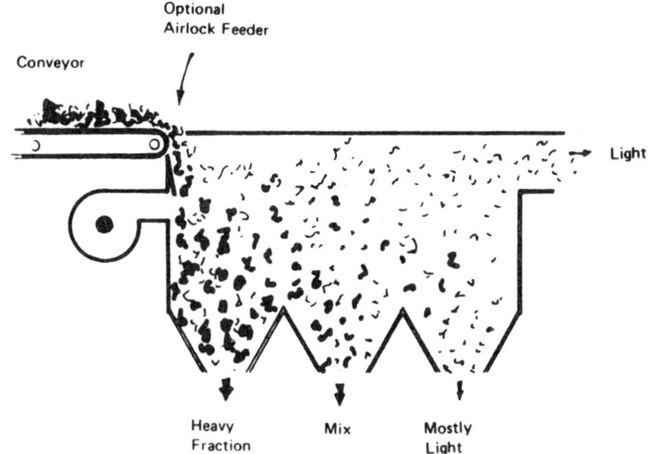

Fig.7. Horizontal air classifier.

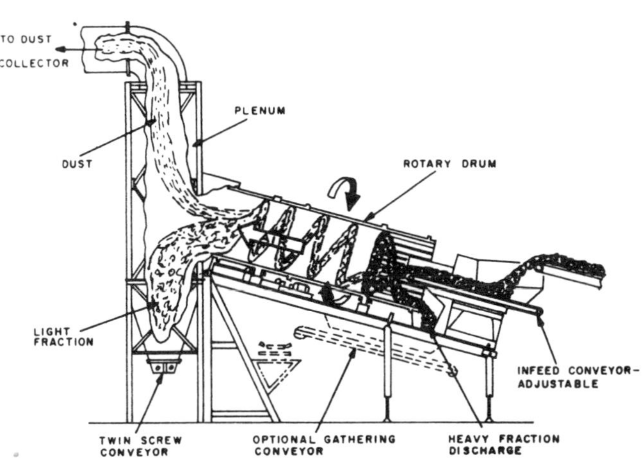

Fig. 7bis. Aeraulic separator with rotating cylinder.

4. SIZE REDUCTION

In composting plants, material preparation may require a size reduction. The machines to be used depend on the initial sizes, volume and type of material.

Some kinds of machines are:

- hammermills: these are a broad category of size reduction equipment utilizing pivoted or fixed hammers with a tip velocity from 40 to 70 m/s;
- bale breaker: this is basically a hammermill with special provision to control the rate and direction of the feed of a bale into the tip circle;
- disc mill: this has a high-speed single or contra-rotating disc between which pulpable material is processed; maximum input size is generally 125 mm; the output is fine grain or pulp;
- chipper (knife hog): this is a size reduction device having sharp blades or teeth, attached to a rotating shaft, which shave or chip off pieces of wood such as tree branches or brush wood;
- cutter: this is a size reduction machine in which wide, rotating blades shear thin sections of flexible material past a fixed, sharp edge;
- shredder: a machine reducing the size of solid objects primarily by cutting and tearing;
- shear: a size reduction machine that cuts material between two large blades or between a blade and stationary edge.

5. COMPOSTING MACHINERY

The machinery used for the composting process should allow the development of microbial mineralization of organic matter and partial humification to satisfy the following three fundamental points (5):

- short processing time and low energy consumption;
- to guarantee a standard end product which is safe for agricultural use and also of satisfactory fertilizing value;
- hygenic safety of the plant and end-products.

In order to reach these objectives several kinds of composting systems have been devised, as reported in Table 2 (5).

The main factor, on which the various designs have substantially developed is the availability of oxygen to the composting mass. The systems for oxygen supply vary from very simple to very complex depending on the solution adopted: in this regard it must be pointed out, however, that, from the industrial point of view, the system giving the best results in terms of cost/benefit analysis is to be preferred.

5.1. Closed systems

Some constructors of closed composting systems claim an "accelerated composting process" of a few days (even 2 or 3) residence of the organic matter in the system. Other authors (5) have shown that this is practically

Table 2. Summary of Composting System for Waste

CLOSED SYSTEMS	
Vertical reactors	- continuous
	- discontinuous
Horizontal reactors	- static
	- with movement of material
OPEN SYSTEMS	
Turned pile	
Static pile	- air suction
	- air blowing
	- alternating ventilation (blowing and suction)
	- air blowing in conjunction with temperature control.

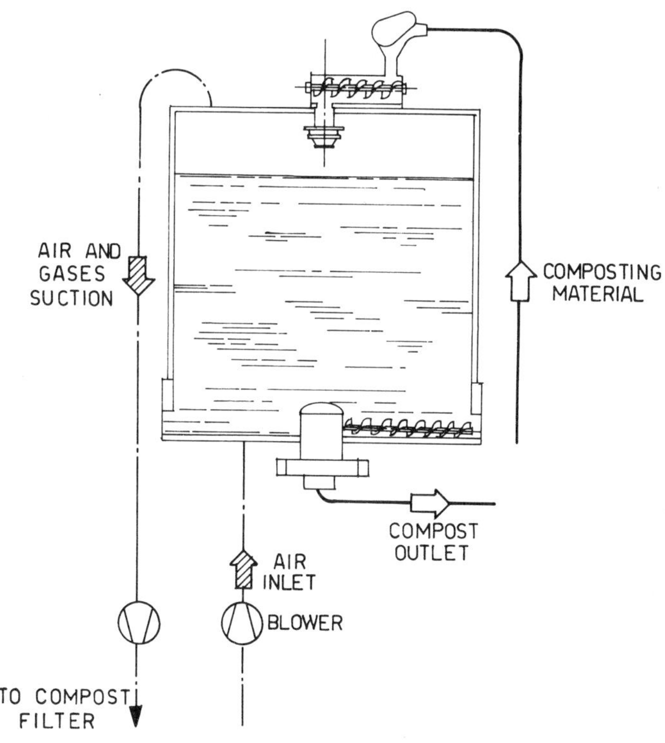

Fig. 8. Vertical continuous closed reactor.

impossible and that the product obtained, if it is not subject to a further curing process, cannot be considered "compost".

The closed systems consist of vertical or horizontal reactors. The vertical ones can be either continuous or discontinuous in operation. In the first case the composting material is present in one large mass, in the second case the mass is arranged on different floors.

In the continuous closed systems cylindrical reactors 4-10 m high are used, with total volumes of 1000-2000 m^3 (Fig. 8).

A continuous closed system is briefly described.

The bio-reactor consists of a thermally insulated air-sealed closed cylinder. The material is introduced at the top and passes down by gravity for a period of about two weeks. The entire mass is aerated by a current of air from bottom to top, with positive pressure at the bottom and removal by suction at the top. There is no mechanical agitation in order not to disturb the biological processes.

The material is introduced into the cylinder through a screw feed and spread by some underlying discs (situated below), which rotate and provide for uniform distribution of the material. The speed of rotation of the discs depends on how full the reactor is.

The air flow is introduced into the reactor from the bottom through four pipes connected to a blower providing constant flow and pressure.

The reactor bottom is divided into four sectors by means of four radial gates slightly higher than the aeration pipes.

The air distribution is made easier by using ballast of size 4-5 cm. Each aeration pipe is controlled by a special valve linked to a humidity probe in the outgoing gases.

The air and the carbon dioxide produced by the biological transformation are then sucked from the top by means of another fan and discharged through a cured compost filter into the air.

After two-weeks residence, the material is discharged from the centre of the reactor bottom. The discharge device consists of a screw mounted on a central cylinder, which can rotate on its axis and through which the material is discharged. The screw conveys the material from the periphery to the centre, while the central cylinder allows the screw to traverse all the base surface. The two movements are operated by hydraulic motors.

Downstream of the discharge device a motorized air-sealing sluice-gate is installed.

Other work (5) describes some microbiological limits to the system and some maintenance problems can occur with the charge and discharge systems, particularly if the discharge system breaks down.

Better ventilation can be obtained with vertical discontinuous reactors, i.e. with material arranged in layers not higher than 3 m on several floors.

Such a reactor consists of a vertical cylindrical tower containing up to six process stages built up from identical modules (Fig. 9). In operation shredded refuse, or classified refuse together with sewage sludge enters the unit at the top and remains there for a period of one day. It then

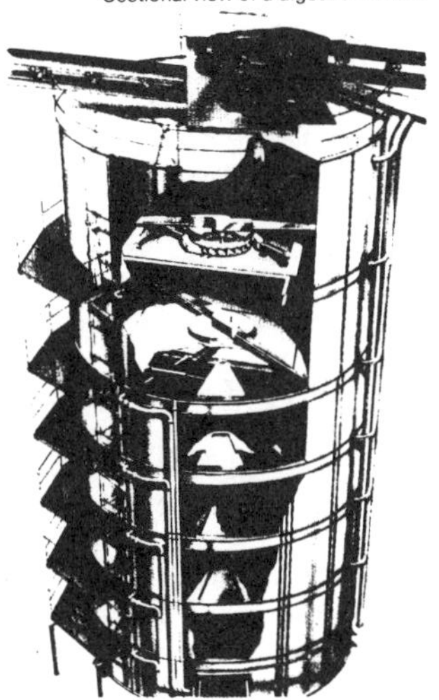

Fig.9. Vertical discontinuous closed reactor.

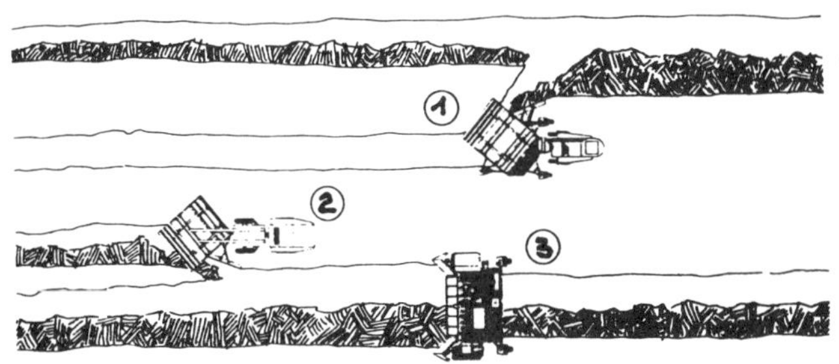

Fig. 10. Pile turning machines: 1. lateral turning, right hand machine, 2. lateral turning, left hand machine, 3. forward turning, windrow machine.

passes to the second and subsequent stages where it remains for a further day in each and leaves the unit after a total process time of one week.

Each stage has a slowly rotating arm which aerates the composting material by providing periodic agitation and spreading. In the initial stages, moisture can be added if required. The rotating arms can be controlled independently to a pre-set programme, in order to allow the operating conditions on each floor of the digestion tower to be varied to provide a better combination of air flow, temperature and agitation.

Reactors are used in parallel to reach the required plant capacities.

This kind of reactor can also give some of the same troubles as the continuous vertical reactor.

A variant of the discontinuous closed system is the system studied by Krupp, which consists of six fixed planes placed within three rectangular-section overlapping chambers, provided with forced aeration system.

On the planes the shredded material, classified by size, is continuously turned over by means of "ploughs" installed on a continuous belt (one for each chamber) whose velocity and direction of movement are adjustable.

Each of the three belts can be separately adjusted.

The temperature of the material is controlled through the turning, the addition of hot water by means of spray nozzles to regulate humidity, and by hot air circulation.

In the horizontal closed systems the process takes 20-30 days and the material is periodically turned by means of special devices (for example rotating screws installed on bridge cranes) and ventilation of the mass, which must not be over 3 m high, is often effected by blowing air from the bottom.

The containers for composting material can be made from reinforced concrete. For industrial use, these systems must also be as simple and cheap as possible.

Rotating cylinders (for example DANO) cannot be considered to be composting reactors, although they allow the differentiation of the various components of waste, through biological and physico-mechanical means. This simplifies subsequent separation of inert material from biodegradable organic matter (5).

On some of its plants DANECO utilizes a rotating cylinder in which air is continuously blown along all the material mass. After a residence time of about 72 hours in the biothermic cylinder, the classified and selected material is conveyed to a covered forced aeration area by alternate air suction and blowing. In a recent 80 t/day plant the cylinder length is about 26 m and its diameter is about 3.8 m.

The above systems often use ventilation systems; fans, blowers and air distribution devices will be described briefly at the end of the next paragraph.

5.2. Open systems

With composting in heaps the process is accelerated by turning or by using forced aeration.

Turning the heaps has some biological and partial-sterilization limits (6); furthermore a larger surface is needed, both when the heaps are moved sideways (Fig. 10) and when machines are used which turn the material and leave it in the same place. In the latter case the heap height is limited by the machine height (Fig. 11).

Systems which include forced aeration of static heaps are more effective and do not cause problems.

Table 2 shows that the aeration can be effected by suction or by blowing.

The suction system was studied at Beltsville (7) and is widely adopted in the U.S.A..

The process based on temperature control by blowing was devised at Rutgers University, New Jersey (8). Some remarks and comments on this method can be found in (5).

A process devised by DANECO includes continuous aeration by alternate suction and blowing; the times are controlled by temperature and CO_2 and O_2 percentage in heap (Fig. 12).

However all forced aeration systems include some common elements such as:

- possible machines for the mechanized production of heaps in the open or in a covered area;
- fans or blowers;
- piping for covering and air blowing/suction;
- possible condensate separators;
- possible filters of cured compost for treating the extracted air in the suction or alternate suction/blowing systems.

In large plants (100-300 t/day) one of two ventilation systems must be choosen:

- centralized facilities with air distribution by pipes;
- subdivision of the ventilation system into groups, each serving a limited number of heaps, or a limited zone in the case of a continuous heap.

The first solution allows for reduction in costs of the machines with a relative short increase in the costs of the air distribution piping; but at least one group must be kept in reserve (which increases the cost of technical investments), to ensure aeration in case of breakdown. The air flow regulation system in the various zones of the processing area is thus more complex.

The second solution allows a complete ventilation group to be kept as spare and the installation costs of the air distribution system are reduced, control is simplified; on the other hand the fans are smaller, so that the maximum efficiency does not reach that of the centralized groups.

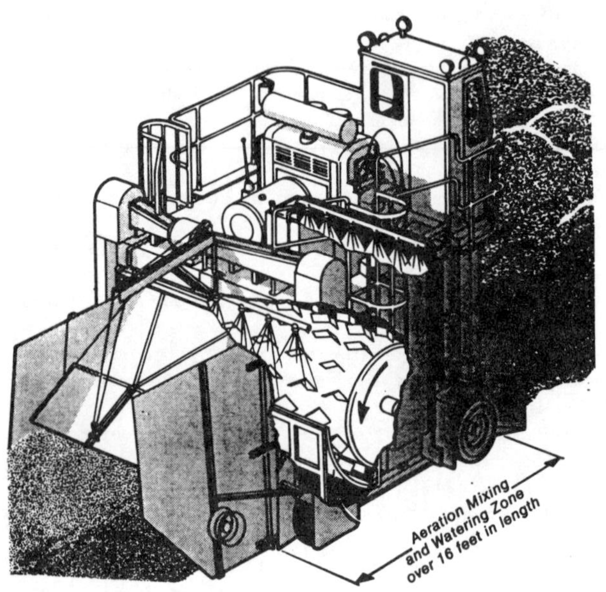

Fig. 11. Pile turning machine.

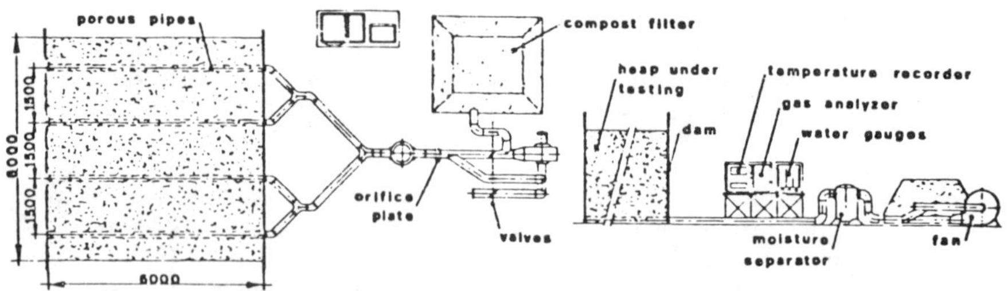

Fig.12. Forced ventilation of static pile. Alternating ventilation (blowing and suction) according DANECO system.

Centrifugal fans are installed with radial blades, possibly with low rotation speed in order to control noise and with anticorrosive surface treatments.

From our experience, air flows of 15 m^3/h for the equivalent of each ton of dry material (Fig. 13) and maximum pressures of 3,000-5,000 Pa must be considered depending on the height of the heaps, the type of material and the complexity of the air distribution piping.

We have carried out research into optimizing composting using both the suction and blowing methods (some machine characteristics of which have been patented) and for the location of the air distribution piping under the heaps, with both trapezoidal section and continuous heaps. Other published work gives (4) the test description and we want here simply to remember the system configuration as far as:

- heap height and sections
- distance between centres of the porous pipes for air
- air-flow rate

are concerned. The configuration has been tested, for fluid-dynamic characteristics by developing a mathematical model and determining the isovelocity curves by means of a computer (Fig. 14). The theoretical results were verified during experimental tests (4) and we found that the theoretical results corresponded to the experimental ones. For a trapezoidal section heap of height 2.5 - 2.8 m and with base length, B, the optimal distance between centres for the distribution pipes is 0.3B. On the other hand for continuous heaps the best velocity and temperature distribution has been found for distances between centres of between 1.5 and 1.6 m with heights up to 2.8 m.

The processes testing continuous aeration using alternating blowing and suction cycles allow:

- production of material with pre-set physical and microbiological features;
- curing time reduced to 3 weeks;
- elimination of the stage of pre-composting in a closed reactor (very expensive as a rule), which does not allow treatment plants to increase their working capacity for incoming refuse with the increase in the running time without any additional equipment.

A brief note on piping: cast iron and plastic piping located in special trench ducts under the heaps have been tested with very good results, with particular reference to corrosion. As regards the piping, the layout must allow head losses to be limited to within the power available.

Finally, the machines for making the heaps and compost collection are included in the ones indicated under point 2.

6. CONCLUSIONS

For composting plants some machines are common to the various possible designs and others are typical of the composting method adopted.

A brief review of the machinery used has been made of both types.

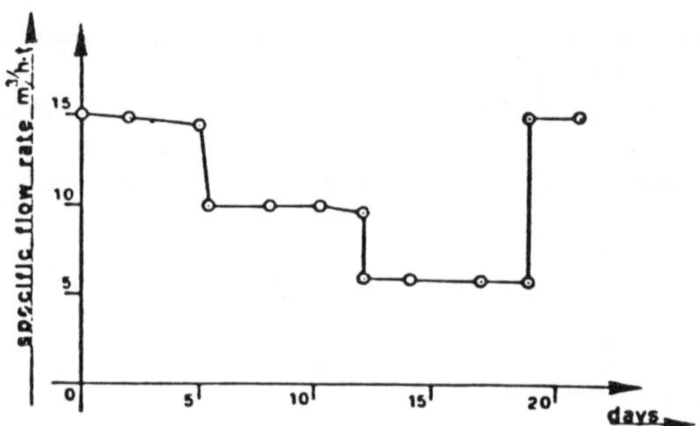

Fig. 13. Specific air flow rate.

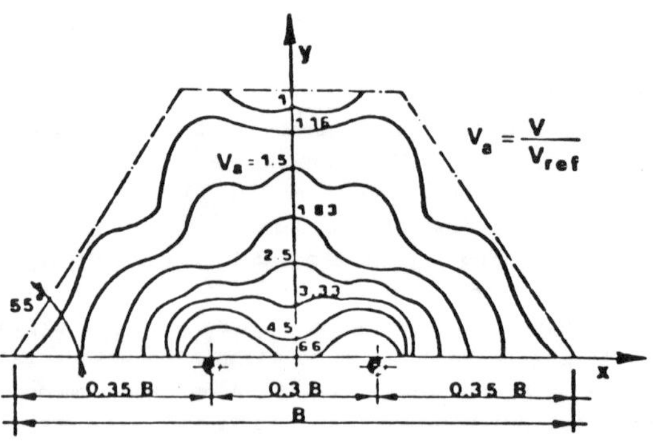

Fig. 14. Trapezoidal heap section: isovelocity curves.

It has been shown how the machines for the open composting processes are simpler and more reliable, as well as generally cheaper than the ones used for closed systems.

Futher action can be taken in order to reduce the plant and operating costs and the authors are convinced that the work will be more effective through direct cooperation between biologists and engineers.

REFERENCES

1. MARTEGANI, A.D. and BABOS, L. (1979). Municipal solid waste and disposal of it. Conference held in Rome at the Foreign Office. May 14th, 1979. DANECO Report.
2. BABOS, L. and MARTEGANI, A.D. (1980). Sistemi alternativi nello smaltimento dei rifiuti urbani e dei residui industriali in riferimento all'utilizzo in agricoltura. Proceedings of the Symposium on "I rifiuti solidi urbani ed i residui industriali". C.I.S.E.S. - Pordenone, Italy.
3. BABOS, L. and MARTEGANI, A.D. (1981). Esperienze di esercizio e possibilità di separazione del vetro in impianti di compostaggio. Proceedings of Symposium on "Ambiente e risorse". Bressanone, 1-5 September 1981. Italy.
4. BABOS, L. and MARTEGANI, A.D. (1981). Advanced methods of composting: research of an industrial method of accelerated composting and compost refining. Proceedings of 2nd International Symposium on "Materials and Energy from Refuse". Antwerp, 20-22 October 1981. Belgium.
5. DE BERTOLDI, M., PERA, A. and VALLINI, G. (1983). Principi del compostaggio. Proceedings of Symposium on "Biological reclamation and land utilisation of urbane waste". Napoli, 11-14 October 1983, ed. by F. Zucconi, M. De Bertoldi and S. Coppola.
6. DE BERTOLDI, M., VALLINI, G. PERA, A. and ZUCCONI, F. (1982). Comparison of three windrow compost systems. Biocycle, 23(2), 45, 50.
7. WILLISON, G.B. et Alii (1980). Manual for composting sewage sludge by Beltsville aerated-pile method. USDA, EPA 600/8-80 002, Cincinnati, Ohio, U.S.A..
8. FINSTEIN, M.S., CIRELLO, J., MCGREGOR, S.T., MILLER, F.C. and PSARIANOS, K.M. (1980). Sludge composting and utilization: national approach to process control. Final Rep. to USEPA, NJDEP, CCMUA, Rutgers Univ., New Brunswick, U.S.A..

DISCUSSION

HOVSENIUS: What is your experience of the use of belt-conveyors with an inclination of 30 - 60°?

ZOGLIA: The angle of the belt seems to have presented no difficulties. The problems of holding material on the belt varies with its size.

STENTIFORD: Do you impose any limit on moisture content for the final screening of compost derived from domestic refuse?

ZOGLIA: For composts, moisture content should not be more than 25%, and when this is achieved a vibratory screen can be used to separate plastics.

PRACTICAL EXPERIENCE WITH FARM SCALE SYSTEMS

A.J.BIDDLESTONE and K.R.GRAY
Compost Studies Group, Department of Chemical Engineering,
University of Birmingham, Birmingham B15 2TT, England.

Summary

Work at the University of Birmingham has sought a reliable method for the farm scale composting of animal manure slurries with straw. Much difficulty has been experienced in intimately mixing the 2 materials and forming the mixture into windrow heaps for composting. The paper describes a small composting unit of 4 cubicles under forced aeration, with recycle of drainage liquor to total absorption/evaporation. This unit worked well in handling the slurry from 60 fattening pigs but involved much hand labour. In the scale-up and mechanisation of this unit to handle the slurry from 1000 pigs considerable problems were encountered. Mechanisms tried included a large walled midden, a forage box with elevator and cross shuttle, and a side-emptying muck spreader with slurry injection. None of these devices proved acceptable. A processor has now been developed which is reliable; it is based on 2 rotors. A vertically rotating impeller opens up straw bales and mixes in slurry; then a horizontally rotating impeller projects the mixture into windrows. The machine has potential for handling a wide variety of organic materials - manures, sewage sludges and vegetable wastes.

1. INTRODUCTION

Composting is the decomposition of heterogeneous organic waste material by a mixed microbial population in a moist, warm aerobic environment. The mature end product is 'humus' consisting largely of the humic acids. The microbiology, biochemistry, process factors and applications of composting have been dealt with extensively by the authors elsewhere (1-9).

Composting is relevant to agriculture/horticulture because they are the sources of many of the organic wastes and because the soil is the destination of much of the compost produced. Table 1 lists some of the organic wastes generated in the UK. Where possible these wastes are re-used; for instance sugar beet tops, brewers' grains and some chopped straw are used for animal feed. Some material is used to prepare substrates for mushroom production and a relatively small amount for high grade compost and potting material for the horticultural trade. After this, the remainder is largely regarded as a low grade waste to be disposed of by the cheapest route possible which is usually a compromise between financial and environmental considerations. Landfill, incineration, dumping and incorporation into the soil are all used in the UK. Current major issues in UK agriculture include the field burning of cereal straw at harvest, smell complaints from disposal of manure slurries on land close to housing, and the forthcoming EEC Directive on disposal of sewage sludge to land. In all these problem areas, composting has a part to play, both as a production process for an end product of value and as a waste treatment process to render lowgrade organic wastes less obnoxious to the environment.

This paper describes the problems encountered in the research and development of a process for composting animal manure slurry with straw. The project commenced in 1969 and is still in progress.

2. SMALL SCALE COMPOST UNIT

Following laboratory experimentation a small compost unit was built and tested on a nearby farm at Clent.

2.1 Design basis

The aim was a simple, rugged unit with a capital cost below £15 per pig place or £150 per cow place. Low energy input was desirable plus operability by one man taking about ½ h/d.

Pig manure slurry is not ideal for efficient composting. The carbon to nitrogen (C/N) ratio is only 6:1 so that high nitrogen losses result from composting under forced aeration at about 60°C. Secondly, high power input is required for good aeration due to the hydraulic head and the low oxygen transfer efficiency achieved by rapidly rising air bubbles. Manure slurry needs to be opened out in an absorbent solid matrix which provides extra carbon. In the UK agricultural context the obvious choice is cereal straw; in the USA wood chips are used for the Beltsville process.

Little information was available about the straw requirement. When composting sewage sludge with 7.4% solids content, Giffard (10) used 2.33 kg straw/ kg dry sludge solids and claimed that little drainage occurred. In our laboratory experiments Melcer (11) achieved little drainage when using 4 kg straw/ kg dry solids from pig manure slurry of 6-7% solids content. Melcer's straw usage would be unacceptably high in full-scale farm practice so Giffard's figure was taken. The 85% moisture content of the resulting slurry-straw mixture was extremely high compared with the 55% optimum value when composting municipal refuse. However, straw forms a stiff matrix which persists for several days, even when wet; this should allow adequate permeation of air during the initial period of high oxygen demand by the slurry.

Composting in cubicles was envisaged. Cubicle size was based on the expectation that the slurry-straw mixture would drain to 75% moisture content and 642 kg/m^3 density. The resulting cubicle volume was increased by 50% to allow for the straw being in a loosely-packed state initially.

The design air flow rate of $1.8m^3\ d^{-1}\ kg^{-1}$ of volatile (compostable) solids was taken from the results of Wiley and Pearce (12). Melcer (11) had shown that passing air at this rate through a 3.4m high bed of slurry-straw gave a pressure drop less than 150mm W.G. For the pilot plant, however, a rotary blower with a head of 83 kN/m^2 gauge gave an ample margin of safety.

2.2 Plant construction

The unit was constructed to treat the effluent from 60 fattening pigs raised in batches from weaners to 90kg weight; the output of slurry varied widely, averaging $0.4m^3/d$, Best (13). The process flowsheet is shown in Fig.1.

Four open-topped cubicles were built from railway sleepers 1.5m wide by 1.8m deep to permit emptying by front-end loader; their height was 2.1m. Perforated plate was mounted 0.3m above the base to form a plenum chamber for air distribution and drainage liquor collection. The walls were lined with hardboard and 500 gauge polythene. To prevent air short-circuiting up the walls, a strip of polythene 0.2m wide was laid round the edge of the perforated plate. The front of each cubicle was closed by two removable half-doors of marine plywood. The cubicles were covered with a roof of plastic sheeting.

Air for the cubicles was supplied initially by a rotary blower of

$0.05m^3/s$ and 83 kN/m^2 gauge head. As the blower proved noisy and the high head unnecessary, it was replaced by a fan of similar capacity and 400 mm W.G.head. The air was metered to each plenum chamber by a globe valve and rotameter.

Manure slurry was scraped daily into a pit of $3.6m^3$ capacity. The slurry was agitated then pumped round a ringmain which was carefully sloped to ensure self-draining when pumping ceased. From the ringmain a valved outlet led to each cubicle.

Effluent liquor drained from the plenum chambers through U-bends into a pit of $0.64m^3$ capacity. A small centrifugal pump of 7.6×10^{-5} m^3/s capacity returned the drainings to the cubicles where it was jetted onto splash plates to aid distribution.

2.3 Plant operation

The daily input to the unit was approximately $0.4m^3$ of pig slurry at about 4% total solids content and 2 standard sized bales of straw weighing 15.9 kg each.

To start up a cubicle the 2 front doors were bolted in place. Two straw bales were spread over the perforated grid to form the initial filtration bed. Air flow to the plenum chamber was adjusted to 9.4×10^{-3} m^3/s. The slurry pit was agitated and the slurry pumped round the ringmain; the day's supply was tapped off and hosed over the straw, taking about 10 minutes. The mixture in the cubicle was then covered with the normal daily requirement of straw, 2 bales. At first this was spread but experience proved this unnecessary; bale fragments were then simply tossed in. Daily addition of slurry and straw were made until the cubicle was full; this required about 34 bales. The cubicle was then left to compost under forced aeration.

Straw absorbs liquor fairly slowly so approximately one third of the input liquor drained through the straw into the plenum chamber and thence to the drainage pit.

2.4 Experimental work

Once the unit was working satisfactorily the experimental objectives were two-fold--

(a) to reduce to a minimum the straw requirement, and

(b) to dispose of the drainage liquor.

Over the course of 2 years, 19 runs were made and the objectives were largely achieved. Table II lists the operating conditions for these runs.

In runs 1-3 drainage liquor from the cubicle returned to the slurry pit, the recycle pit not having been built. This was only an interim measure as the slurry feed was thereby diluted. Straw input of 4 bales per day proved excessive and was reduced to 2 bales per day in runs 4 to 16 while methods of treating the drainage liquor were investigated. Straw input was increased in runs 17-19 which were aimed at producing compost for mushroom spawning trials.

Runs 4-8 were aimed at reducing the level of Biochemical Oxygen Demand (BOD) and Suspended Solids (SS) so that the drainage was fit for discharge to a water course. The drainage was recycled over an aerated cubicle filled with straw fragments hoping that this would act as a 'biological filter'. This aim was partly achieved but after 24 hours recycle the liquor was generally removed by absorption and evaporation, leaving little for final discharge.

It was then realised that an aerated, composting matrix could act as a 'biological filter'. Consequently in runs 9-10 and 16-19 the drainage liquor from a cubicle being filled was collected in the recycle pit; it was then sprayed slowly over the compost cubicle which had been filled previously and was then at a temperature of about 60°C. Within 24h the liquor was evaporated due to the heat of fermentation released in the matrix.

In runs 11-15 the case of weak slurries was considered. It was considered uneconomic to use vast quantities of straw to absorb and evaporate the drainage liquor entirely. With reasonable straw input a substantial discharge of effluent would arise. Hence a small commercial high-rate filter was used to try to reduce the BOD and SS of the liquor to Royal Commission standard. The raw slurry was deliberately diluted to about 3% total solids to provide ample drainage liquor from the cubicles. This liquor was recycled for 24 hours over a compost box before being pumped over the high-rate filter.

2.5 Results and discussion

Considerable experimental data were obtained on this Compost Unit, including flow rates and analyses for total solids, BOD , COD ,N, P, K and C. They are reported fully by Best (13).

Table III shows the performance of the initial compost cubicle in removing liquid and total solids from the input slurry on its passage through the straw. It indicates that some 65% of the liquor was retained by the straw and 84% of the total solids were removed. The drainage contained about 1.75% total solids the majority of which, about 1.5%, represented dissolved solids. Hence the straw bed had proved an effective filter for suspended solids. There was little evidence of silt-size particles accumulating in the recycle pit so the suspended solids remaining in the effluent would be mainly colloidal matter. At the slurry application rate of 230cm^3 s^{-1} m^{-2} there was no evidence of 'ponding' on the bed. The average BOD in the slurry was 9670 mg/l which dropped to 7460 mg/l in the drainage.

The temperature in the compost matrix rose quickly to over 60°C, remaining above this level for some 3 weeks before falling slowly. The average analysis of product compost over the 19 runs was 81.9% moisture on a fresh weight basis; then on a dry weight basis 21.1% ash, 2.98%N, 1.70%P, 1.84%K, 42.7%C. The C/N ratio averaged 14.9 : 1 and the density 507 kg/m^3.

Summarised briefly the work on the drainage liquor from the compost cubicles showed that recycling for 24h over the straw bed percolating filter in Runs 4-8 reduced the BOD from an average of 8340 mg/l to 960 mg/l at an application rate of 27 cm^3 s^{-1} m^{-2}. Total evaporation/absorption was achieved in 4 out of the 5 runs within 24h. Recirculation over a hot compost box in Runs 9-19 reduced the BOD from 6170 mg/l to 1680 mg/l and lowered the effluent volume from 3.9m^3 average to 1.3m^3. Complete evaporation was achieved in 3 of the 10 runs. The application rate of 27 cm^3 s^{-1} m^{-2} was set by the equipment available; it could have been reduced with benefit to about 1/10th of this rate, just keeping pace with the rate of evaporation into the air stream passing up through the cubicle. The commercial high-rate filter used in Runs 10-14 took the liquor which remained after recycling over a hot compost box. It reduced the BOD from 1880 mg/l average to 580 mg/l but achieved virtually no reduction in effluent volume.

The 2 year test period indicated the possibilities of a composting process for organic slurries. Slurry with a total solids content of 5% or more could be layered daily with straw in the proportion of 0.14m^3 slurry per standard bale; this process would continue for 7 days. The drainage liquor, about 1/3rd of the input slurry volume,could be recycled very slowly over the compost box filled previously which had reached about 60°C in temperature. Complete evaporation within 24 hours should take place. Where weaker slurries are handled, some final effluent would arise but this would be much reduced in volume and have a BOD below 2000 mg/l. During weeks 3 and 4 the composting matrix would be allowed to decompose further under forced aeration, without liquor recycle, and then be discharged and stacked. The process, termed 'ARCUB', was patented and assigned to the National Research Development Corporation (N.R.D.C.) for commercial exploitation (14).

3. LARGE WALLED MIDDEN

Following the small-scale experiments at Clent, scaling-up the process to handle the effluent from 1000 pigs was carried out in conjunction with A.R.M.Ltd, agricultural engineers, of Rugeley, Staffs.

3.1. Design basis

Straw consumption at Clent was undoubtedly excessive. Using 1 standard bale for $0.14m^3$ slurry for complete absorption/evaporation would involve 200 t/yr of straw on a 1000 pig unit. By reducing considerably the straw input the ARCUB technique would have much greater applicability. The new unit was designed, therefore, for a straw input of 50 t/yr for 1000 pigs; total evaporation could not be expected, especially in winter months. However, the effluent liquor should have virtually all its suspended solids removed, its BOD reduced to below 2000 mg/l and its volume greatly decreased. The excess liquid could be pumped to a lagoon and used for farm irrigation or be disposed of by evapo-transpiration through a tree belt.

To reduce operator's time the process was designed with 2 middens, each holding 3 months supply of straw, 12½ tonnes or 750 standard bales. The slurry would be pumped over the midden daily or weekly, the drainage returning to the slurry pit for recycling. After the day or week the liquor remaining in the pit would be discharged to the lagoon. After 3 months, slurry pumping would be switched to the second midden, allowing the first one to compost gently under natural aeration prior to removal and replacement by new straw bales. With this approach it was hoped that solids handling would be confined to a few days once every 3 months while slurry pumping would be controlled intermittently by timeclock; forced aeration would be continuous through the active midden.

3.2 Plant construction

The plant was built at Rugeley in 1978 on a farm having nearly 1000 fattening pigs producing some $32m^3$ of slurry per week at 8% solids. A slurry pit was constructed with 2 compartments, $36m^3$ capacity for slurry and $9m^3$ for treated liquor. The 2 middens were each 7.2m square with walls of grain walling 2.4m high. The demountable front wall was of Tufboard within angle iron framing. The concrete floor sloped to the rear so that drainage liquor flowed out to the slurry pit via a U-bend seal. To provide a plenum chamber the floor was covered with 36 wooden pallets; Netlon with 20mm mesh was spread on top. A 50mm diameter air pipe entered under the pallets from the rear wall; the air flow was at $0.08m^3/s$ and 400mm W.G. head.

Slurry was pumped from the pit with a double-disc diaphragm pump designed for 2.3×10^{-3} m^3/s and 15m head. The piping was arranged so that some passed to a distributor driven round over the midden; the remaining slurry returned to the pit via a plunging jet aerator.

At first sedimentation in the slurry pit gave varying slurry solids concentration. This was overcome using an air sparge pipe. Initially a sliding-vane blower of $0.037m^3/s$ capacity at 17 kN/m^2 was used; this was later replaced by a blower with $0.015m^3/s$ throughput which still gave excellent agitation. Secondly, the slurry pump was unable to meet its design flows; modifications to the inlet piping, an increase in speed and a change in diaphragm discs were needed to achieve reliable performance.

3.3 Experimental

The first midden was stacked with 750 standard bales of wheat straw; each measured approximately 0.9m x 0.6m x 0.45m, weighed about 16kg and was tied with sisal twine. The bales were placed with their largest faces horizontal and with strings uncut. Six courses were laid, each being swung through 90° from the course below. The 3 fixed walls and front wall of the midden were covered with polythene sheet which overlapped the edge of the base pallets to prevent air leakage up the walls. The outer rows of bales in each course were jammed tight to the walls; gaps 75mm wide were left between the inner rows.

To prevent liquid from the rotating distributor being flung outside the midden, four further courses of bales were arranged on top of the midden to form an amphitheatre shape. The distributor revolved within this arena.

The unit was started up with forced aeration, and with slurry distribution for one hour in every six actuated by a time clock. Within 3 days composting was under way with vapour issuing from several points around the midden. However, it soon became obvious that the slurry was not spreading out to the midden edges. The corners remained dry while the amphitheatre started to 'pond' with slurry. After a week of operation 'ponding' was severe at the midden centre while the straw at the midden edges was tilting inwards, leaving a wide gap up the walls. Slurry pumping was stopped to allow the midden centre to drain. A survey with a 2m long temperature probe revealed temperatures of 60-70°C in several places; the water-logged central zone was much cooler. The bales forming the amphitheatre were re-arranged to try to overcome the ponding problem but this was not very successful and the first run was stopped prematurely. On emptying the midden it was found that in many cases little liquor penetration into the bales had taken place. Similar behaviour, but with liquor from a mechanical slurry separator, was reported by Soper at the Oxford University Farm (15).

In the second trial the straw bales were placed on their sides, the strings removed and the bales sprung open to 1.2m length; 0.15m gaps were left between the bale rows. This gave a slightly better performance but 'ponding' still brought the run to a premature end although liquor penetration of many bales was often incomplete. These two runs showed that loading the middens for 3 months operation was unsatisfactory and indicated that straw and slurry needed to be loaded on a weekly basis.

In Run 3 a midden was subdivided into two by a cross-wall of bales. One half was further subdivided into two sections by another bale wall. Approximately 25m^3 of slurry were layered with straw bale segments alternately into the two front quarter sections; the slurry was hosed on by hand. The drainings were sprayed from a pierced pipe onto the rear half of the midden which had been stacked with bales to a height of 2m. Of the slurry put on, some 50.8% was absorbed in the 2 areas for slurry application; this amounted to 0.101m^3 of slurry per bale, agreeing closely with the initial absorption obtained on the Clent unit. The two areas for slurry application did not compost satisfactorily. However, heating up and vapour emission occurred where drainage liquor had been applied. It was obvious that the problem of straw 'blinding' by slurry solids was still occurring so that the increased bed pressure drop in the slurry areas caused the air to take the easier path through the areas where drainage liquor had been applied.

Later, on the same midden in Run 4, 32m^3 of slurry were layered with segments of 144 bales of straw; 42.8% absorption was achieved, corresponding to 0.096m^3 per bale. As in the previous run, composting did not proceed satisfactorily in the slurry application areas; vapour was being vented close to the midden walls, indicating a maldistribution of air through the matrix.

In an attempt to overcome the straw 'blinding' problem the quantity of straw was increased to the 0.14m^3 per bale usage at Clent. In Run 5, using a new midden, 33m^3 of slurry were layered with 248 broken bales of straw using the full midden area. This gave an absorption of 61.8%, equal to 0.084m^3 of slurry per bale. To try to prevent air taking a short circuit up the midden walls, a 0.6m wide area around the sides was filled with straw bales. After a week for composting to commence, some temperature readings and exhaust gas analyses were taken. Very high levels of carbon dioxide, up to 32%, were found in the central area; this indicated that virtually no air movement was taking place, even though air was being blown under the pallets. In Run 6 a further 23m^3 of slurry were hosed onto 160 broken bales of straw in layers, giving 62.2% absorption and 0.088m^3 slurry per bale, Table IV.

In Runs 4,5 and 6 the midden was loaded in layers of 5.5m^3 of slurry and 40 bales of straw. Although the latter was spread from the bale and the slurry hosed on by hand, this could only be done using a person working on the top of the midden within the walls; evidently this compressed the matrix and helped lead to the inevitable 'blinding'.

These last 3 runs showed that even with an increased input of straw and the use of a layering technique, results could not be achieved which were comparable with those obtained on the small plant at Clent. It was impossible to load the 7.2m square midden evenly over the walls with straw and slurry. A central mass was being formed which was virtually impenetrable to air; the latter took the easy route up the inside of the walls. On the Clent plant, by contrast, even if much air had moved up the walls it could readily diffuse 0.6 - 0.9m sideways so that the entire mass could be permeated and composting take place. On the large midden, a sideways diffusion of 0.6 - 0.9m still left a large central volume unaerated.

Uneven settling was very apparent in the midden, the central area being well depressed; this tended to drag material away from the walls so that a gap up to 0.6m wide opened up at the top. (This effect, on a smaller scale, has often been noticed in our 1m^3 compost boxes when composting batches. It was not apparent on the Clent plant where material was added daily). Another observation with the large midden was that the drainage liquor carried a substantial portion of silt-size particles which sedimented out as the drainage moved slowly across a horizontal concrete area; this effect was not noticed on the Clent plant.

It was now apparent that straw and slurry needed to be far more intimately mixed and the mixture placed in the midden without compression. To this end a slurry distribution box was mounted over the beater of a flat bed, rear-unloading muck spreader. Straw bales with strings removed were placed in the muck spreader; as they were fluffed out by the beater slurry was mixed in. The mixture formed a very open pile some 0.9m high at the rear of the muckspreader; excess liquor was allowed to drain away to the slurry pit. The intimacy of mix was very good. The muckspreader was then raised 0.9m above ground on railway sleepers so that a deeper pile would be formed. The mixture was picked up by a manure fork on a front-end loader and carried into the midden where it was stacked 2.4m high on the rear row of pallets. As more material was brought into the midden further rows of pallets were laid down. To improve movement of air through the mass a number of perforated plastic drain pipes 0.1m diameter by 2.4m long were mounted vertically on 1.2m centres as the mass was assembled.

The mass in the midden warmed up quickly, evolving vapour and within a few days had sunk to a height of less than 0.9m. Unfortunately there was no way of overlaying this material with further mixture. Moreover the whole operation had proved a difficult operation with liquor dripping from the mass on the manure fork on its passage to the midden.

The large-scale work had now shown

a) that a walled enclosure was impracticable due to problems of loading materials and obtaining even air distribution,
b) that the straw needed to have its strings removed and be opened out mechanically for mixing in the slurry intimately,
c) the resulting mass warmed up quickly but sank to about one third of its height in 3 days and needed fresh material added on top to make the best use of space,and
d) the matrix did not flow as does grain but took up an angle of repose of 90°.

Accordingly, on a large scale we were faced with far more difficult engineering problems than had been apparent at the pilot plant stage. Our conclusions were later substantiated by scientists from the Ministry of Agriculture, Fisheries and Food who erected an elegant midden at Askham Bryan College in Yorkshire (16).

4. OVERHEAD PLACEMENT

Although the large walled midden at Rugeley had failed, work continued in conjunction with A.R.M.Ltd because of the obvious necessity to solve the slurry disposal problem.

4.1 Design basis

It was now apparent that good mixing of straw and slurry could be achieved while straw bales were being spread out by the beaters of a muck spreader. The mixture sank quickly in height and needed overlaying periodically with fresh material. This could not be achieved easily by a machine straddling the windrow similar to those used for low heaps in the USA. A mobile processor was needed which moved alongside and parallel to the windrow heap, mixing straw and slurry and elevating the mixture to about 3m height, moving it sideways and discharging it onto the heap. In this way a heap 3m high could be built in the first instance and overlaid with new mixture as it sank.

4.2 Processor construction

An old forage box was rebuilt, twisting it 180° round on its chassis for rear discharge. An elevator was attached to the rear and a slurry distribution box constructed over the apex. Straw bales in the forage box were opened out in the beaters, dropping onto the elevator and being lifted to a height of 3m where slurry fell on it immediately prior to discharge. The whole device was towed and driven by a tractor and its p.t.o.

Once the combination was operable the attachment for sideways movement and discharge was built. This was based on a rubber belting moving between side plates; it could be reciprocated through a distance of 2.4m. The slurry-straw mixture dropped from the head of the elevator onto this belting which could then move it sideways and discharge it over a width of 2.4m. Combined with the forward movement of the whole device, a heap could now be laid out initially and overlaid back to 3m height subsequently.

When the basic machine had been tested an improved version was constructed based on a new forage box. Initially an auger was used for sideways movement of the mixture and discharge but it did not function well and was replaced by an improved reciprocating cross-shuttle, Fig.2.

4.3 Experimental

The mobile processor was tried out in several modes. Initially it was used with a slurry-straw ratio of 0.091m^3 per bale; a layer of straw was put down first without any slurry to prevent run-off from the first application of mixture. Uptake of slurry amounted to 0.082m^3 per bale without run-off when the heap was formed on a concrete base. Temperatures were in the range 51°-60°C.

The processor was then used to turn compost from previous runs. It handled the material well although a reduced load in the forage box was necessary because of the higher density. The appearance of the compost was greatly improved with long straw being broken up and lumps of dense material opened out. Extra slurry was added in some of these runs giving a final uptake approaching 0.14m^3 per bale as at Clent.

In further trials where straw input was reduced to 0.14m^3 per bale the heaps were formed over drainage/aeration channels. The drainage was collected in a separate pit and recycled over a previously formed heap. This was only partially successful. The initial uptake was very good, 0.11 - 0.13m^3 per bale, and recirculation brought this to 0.14m^3 per bale. However, liquor recycling through small pumps, lines and nozzles was made difficult due to fine silt-size particles brought down with the drainage liquor. Some method of fine solids separation was desirable as line filters clogged very quickly.

Windrows were constructed with several methods of air supply -

a) on the concrete base with no forced aeration,
b) over a drainage/aeration channel but with no forced aeration,
c) over a channel with forced aeration,
d) over a channel with air sucked down through the heap, and
e) over an aeration 'tunnel' within the heap, with and without forced aeration.

In each case heap temperatures in the range 55°-65°C were reached and maintained for 4-6 weeks. Use of metal aeration 'tunnels' required more labour and their recovery before moving the heap was difficult.

Although the mobile processor enabled good mixing of straw and slurry with placement into a heap which could be overlaid, in practice the machine proved unwieldy and liable to mechanical trouble. A simpler method of heap assembly was needed. Moreover the mixture tended to build up in tall columns which toppled over in the path of the wheels of the processor.

5. STRAW CHOPPER

The idea of chopping straw bales and blowing the fragments from a pipe to impinge with a jet of slurry was investigated. The concept had merit in that the two streams could be readily conveyed to the point of placement. An imported Danish straw chopper was hired for several weeks in 1980 but the trials were not successful. Considerable input of labour was needed to feed the bales singly into the chopper. Moreover the resulting absorption of slurry by the straw fragments was not as great as in the experiments on overhead placement.

6. THE BALLISTIC APPROACH

A simpler method was required of mixing straw with slurry and placing the mixture into windrow heaps than the complicated arrangement of beaters, auger, elevator and cross-shuttle as used in overhead placement.

6.1 Design basis

A re-appraisal of current muck-spreading equipment was made in Summer 1980. The rear-unloading machines were unable to produce tall heaps or overlay ones which had subsided. The rotating ones with chain flingers could not easily mix straw and slurry and the resulting side discharge was uncontrollable and could give rise to slurry droplet atomization. However, an Italian muckspreader was found which used a ballistic approach. In this machine, farmyard manure (FYM) moves forward on a chain and slat conveyor and meets a vertical rotating disc with several angle iron attachments. The material is swept up, constrained by an overhead shroud, and projected over the side of the machine at a slight angle to the horizontal. It was reported that the machine had been used with some degree of success with straw bales in mulching an orchard.

6.2 Experimental

A machine was purchased and tested in a static position with partially matured pig-slurry/straw compost. The result was a well-shaped windrow some 9m long, 2.7m high and 2.4m wide.

Tests were then carried out using straw bales with the strings removed. The bales were quite well broken up but, being less dense than manure, the material was not thrown as far. A lot of straw built up on the side of the machine at the exit and fell off into the path of the spreader's off-side wheel. This build-up was greatly reduced by mounting on the side of the machine a flicking device driven from the tractor p.t.o.

Slurry was then pumped in close to the rotating impeller, being mixed with the straw as the latter was torn from the bale. After trials with different slurry injection points and modifications to the overhead shroud an operable machine resulted. Straw bales were brought to the back of the machine, the rear door of which had been removed, and hand-loaded onto the conveyor. The machine could be towed along, trailing its slurry pipe, stopping to build new heaps or putting further layers on older heaps which had sunk.

Early work with the machine showed that the double-disc pump could not supply slurry fast enough to match the straw throughput of the spreader. It was replaced with a centrifugal pump which passed up to 4.5×10^{-3} m^3/s.

Initial trials showed that a slurry uptake of $0.082m^3$ per standard bale could be obtained with no obvious run-off. The partially composted material could then be put back through the machine and further slurry added to give a total uptake of $0.109m^3$ per bale without run-off or $0.127m^3$ per bale if run-off was permitted. The appearance of the compost was much improved.

The machine was given an extended trial during early Summer 1981; for a period of 3 months it handled the slurry output from a herd of nearly 1000 pigs. Some results are given in Table V. These show that to handle one month's supply of slurry some 1100 standard bales (18 tonnes) of straw were required. Absorption was $0.136m^3$ per bale, including recycled liquor. The machine could handle 90 bales per hour, requiring 14 hours running time per month. Two operators were required, one to load the bales, the other to bring up the bales, move the machine and handle the slurry supply.

This extended trial revealed the drawbacks to the system. Considerable labour was required, especially in lifting the bales onto the conveyor. To allow the chain and slat movement, the base of the spreader was not liquid-tight; consequently leakage of slurry took place. The operator loading the machine was totally exposed to the weather and on very windy days could suffer much inconvenience from slurry spray and straw dust. The trailing slurry pipe was also an inconvenience. Hence although the machine was mechanically reliable in operation it was not acceptable to the farmers who saw it in operation.

During this work on the muck spreader some experiments were made with big round straw bales. This was partially successful. Much of the bale could be unwound satisfactorily from a central spindle. However, the soft centre of the bale invariably collapsed, making an even straw flow impossible.

7. ARMIX PROCESSOR

The composting process needed a centrally sited ballistic device, with the operators under shelter if possible, and with the material being thrown out tangentially in a series of heaps or in a continuous pile over a sector.

7.1 Design and construction

During late Summer 1981 a horizontal disc, driven from underneath by a hydraulic motor, was set up on a framework at a height of about 1.8m. The disc, with angle-iron spokes attached, was surrounded by a shield with several exit ports which could be opened or closed. The projection of slurry and straw bale fragments was studied.

From this concept the ARMIX twin disc processor was evolved. Straw bales with strings removed passed up a chain and slat elevator to a height of about 3m and tumbled into a hopper where a spoked disc some 1.2m in diameter was rotating at several hundred rev/min. Slurry was injected to maintain a level in the hopper bottom. The first disc broke up the bales of straw, mixed in the slurry and discharged the mixture onto the second wheel which rotated in the horizontal plane. This wheel, surrounded by a shield with several exit ports, could project the mixture over an arc of 180°. The elevator and the two discs were driven by hydraulic motors.

7.2 Experimental

Trials were carried out to assess suitable speeds of rotation of each wheel, the best position for slurry injection, the throughput of the processor and the pattern of discharge into heaps. Following this work the machine was patented (17) and exhibited at several agricultural shows. Interest was shown by pig farmers in parts of the UK where slurry spreading gives rise to smell complaints and action by environmental health officers. Certain Water Authorities were interested in the ARMIX for treating partially-dewatered sewage sludge prior to land-spreading. Further interest came from firms involved in the washing and packing of vegetables.

In recent months an improved version of the ARMIX has been used by Cooper (18) for blending sewage sludge and straw, celery waste and straw, and leek waste and straw to form compost heaps in the 3-10 tonne range. The straw content has been varied between 5 and 12% by weight and the heaps laid on plain concrete or over 100mm diameter perforated drainage pipe under both natural aeration and forced aeration conditions. Temperatures and CO_2 concentrations have been measured, as well as the volume reduction and moisture release in the heaps.

In two trials with 5 tonnes of sewage sludge which had been dewatered to about 80% moisture content, 5 wt% of straw was added and the mixture projected into a heap on the concrete pad. Temperatures rose quickly to over $60^{\circ}C$. CO_2 concentrations in the lower regions of the two heaps built up to 20% in one heap and 50% in the other. The rate of composting was obviously very slow, being limited by poor oxygen penetration. Nevertheless, when the heaps were broken up after 2-3 weeks and spun out together through the ARMIX machine again, there were no severe anaerobic odours and the material appeared to be suitable for land spreading without environmental concern. Once amalgamated into a new heap the mixture regained its high temperature and CO_2 levels. With a fairly dense material like sewage sludge forced aeration should increase reaction rates significantly.

In two trials with 3 tonne heaps of celery waste with 7.5 wt% straw laid on the concrete pad the behaviour was very different. Even though the moisture content of celery is very high, 93.5%, the water is bound up in plant tissue and is only slowly released. The result was a stiff porous matrix so that air penetration was good and CO_2 levels were well below those with sewage sludge. The two heaps of celery-straw mixture were later amalgamated; then a 10 tonne heap was projected from the ARMIX machine. Considerable reduction in volume was obtained with this material and the resulting compost, although rather moist, had broken down well to a fibrous material. Because of its low energy content, 377 J/kg, forced aeration of celery will need to be carefully arranged, balancing composting air requirements and moisture evaporation against heat release so that the heap is not cooled too severely.

In recent trials with leek waste, two heaps with 5 and 10 wt% straw were projected over perforated drainage pipe which was not forced aerated initially. Leeks also have a high moisture content, about 90%. The mass soon sank to a fairly dense matrix. With 10% straw there was little moisture release but at the 5% straw level a lot of drainage liquor accumulated. CO_2 levels built up quickly in the heap with 5% straw. Close to the aeration pipe the CO_2 was 8.3%; at a distance of only 0.15m above the pipe the concentration had risen to 19.4%. At 10% straw, CO_2 levels were about 2.5%. Later in the trial a fan was connected to the piping of the 5% straw heap and forced aeration commenced, but by this time temperatures had fallen close to ambient values. The pressure drop required was about 325mm W.G.

8. PROCESS IMPLICATIONS

This paper describes work on mechanizing the blending of manure slurry with straw and placing the mixture into windrow heaps for composting. Although a few papers were published in the 1940's on composting sewage sludge with straw these were labour-intensive processes with little mechanisation.

This project has shown that cereal straw is a very difficult material to open out from the bale, to mix with other materials and to place uniformly in heaps. However, after a succession of inadequate mechanisms a reliable processor has been developed which has applicability for blending straw with slurries, sludges and solid organic wastes.

This work on composting slurry and straw mixtures has been carried out at moisture contents far in excess of the 55% normally regarded as optimum when handling municipal wastes. This is because cereal straw can maintain an open matrix for many days, even at moisture contents over 80%; by contrast the paper content of municipal wastes soon collapses at moisture levels much over 60%. Straw appears to have about 4 layers in its wall; the inner layers are attacked fairly readily while the outer, more resistant, layer retains its form considerably longer. Straw does eventually collapse with loss of matrix porosity. Hence it is most important when composting that the bulk of the high-rate oxygen demand is met within about 3 weeks, before collapse occurs; this is particularly important when working with high moisture contents and tall heaps where compressive forces in the lower region will tend to squeeze moisture into the interstitial spaces.

The use of straw as a bulking agent is very pertinent in the UK where a large excess of cereal straw is produced and is currently burnt in the field at harvest. Environmental pressures are already forcing major restrictions on straw burning and alternative ways of disposal are being sought urgently. In the composting process straw goes through to the final product; its presence does not impose any restraint as long as heap porosity is maintained up to the degree of maturity required in the compost.

This is in complete contrast to the use of wood chips as bulking agents in the USA, typified by the Beltsville process (19), and the work of Finstein et al(20) and Singley et al (21) at Rutgers University. The use of shredded rubber tyres as described, inter alia by Singley et al(21) is similar to that of wood chips. With these materials, their cost is very significant to the process economics and great effort is made to recover them by sieving from the product compost. To achieve an acceptable separation the moisture content of the compost must be below 50%. This immediately imposes a major restraint on the initial moisture content of the mixture prior to composting; as this is normally in the range 55-65% it means that only partially-dewatered sludges can be handled and not the slurries of 6% solids content that have been used in the present research with straw. A further drawback with wood chips, as shown by both Finstein (20) and Singley (21), is that in order to achieve an average moisture content suitable for sieving the compost, some parts of the windrow heaps are reduced to moisture contents below 30%; as shown by Snell(22), at these low levels the activity of the micro-organisms is severely depressed. The use of vast quantities of air to keep the compost temperature below 60°C, as opposed to controlling air supply to achieve a desired O_2 level in the heap, makes this dehydration aspect more acute. By contrast, with the much wetter mixtures used in the present work with straw, dehydration cannot easily progress to a level where microbial activity is impaired.

The work of Singley (21) shows that forced aeration upwards and outwards through the heap leads to less pressure drop through the compost mass and aeration piping than does vacuum-induced aeration down through the mass. It also overcomes the problem of accumulation of liquor in the aeration piping. Cereal straw in the UK is fairly expensive, about £20 per tonne. Hence any straw-based composting process is almost certain to be operated with minimum straw and therefore be close to the maximum moisture-holding capacity of the straw, about 85%. Forced aeration outwards is certain to be necessary to oppose the gravitational movement of moisture, especially in tall heaps where compression forces could lead to moisture accumulation at the base.

A significant observation in the present work arises from batch composting in walled vessels or enclosures. The higher temperatures and generally greater microbial activity in the centre of the mass causes the middle to sink faster than the edge. If the material is long-fibred, such as straw, then this action causes the outer material to be pulled away from the walls, leaving gaps up which air can permeate. This means that uniform aeration is impossible. The effect is less likely to occur where the process is continuous, with fresh material being added at the top and old withdrawn from underneath, or when agitation is employed.

In conclusion, it appears that the essentials of a straw-based composting system applicable to a wide range of organic wastes in solid, sludge and slurry forms have now been delineated. Its applicability as a reasonably priced, reliable and readily operable farm process has been demonstrated.

9. ACKNOWLEDGEMENTS

The work described in this paper represents the efforts of a team of over 12 people spread over some 14 years. Their contributions, often under exceedingly difficult farmyard conditions, are gratefully acknowledged.

Financial support by the Agricultural Research Council, the National Research Development Corporation, the Wolfson Foundation and the Department of Industry has made the work possible.

Mr.D.Clement allowed us to erect the small-scale ARCUB unit on his farm at Clent while subsequent experimentation was done on the farm of Mr.G.Cooper at Rugeley.

Finally, the many inputs - financial, technical and commercial - of Mr.D.Cooper and A.R.M.Ltd of Rugeley, Staffs has enabled the work to be brought to the point of commercial take-off.

REFERENCES

1. GRAY, K.R., SHERMAN, K,,and BIDDLESTONE, A.J. (1971). A review of composting - Part 1, Microbiology and biochemistry. Process Biochemistry. 6,(6), 32-36
2. GRAY, K.R., SHERMAN, K., and BIDDLESTONE, A.J. (1971). A review of composting - Part 2, The practical process. Process Biochemistry. 6, (10), 22-28
3. GRAY, K.R., BIDDLESTONE, A.J., and CLARK, R. (1973). A review of composting - Part 3, Processes and products. Process Biochemistry. 8, (10), 11-15 & 30
4. GRAY, K.R. and BIDDLESTONE A.J. (1974). Decomposition of urban waste. Chapter 24 in 'Biology of plant litter decomposition',Vol.2. Eds. Dickinson, C.H.and Pugh, C.J.F. Academic Press, London. 743-775
5. GRAY, K.R.and BIDDLESTONE, A.J. (1976). The garden compost heap - Part 1. Journal of the Royal Horticultural Society. 101, (11), 540-544
6. GRAY, K.R. and BIDDLESTONE, A.J. (1976). The garden compost heap - Part 2. Journal of the Royal Horticultural Society. 101, (12), 594-598
7. GRAY, K.R. and BIDDLESTONE, A.J. (1980). Agricultural use of composted town refuse. In 'Inorganic pollution and agriculture'. M.A.F.F. Reference Book 326. H.M.S.O., London.
8. DALZELL, H.W., GRAY, K.R., and BIDDLESTONE, A.J. (1979). Composting in tropical agriculture. Review Paper Series 2. International Institute of Biological Husbandry, Stowmarket.
9. GRAY, K.R. and BIDDLESTONE, A.J. (1981). The composting of agricultural wastes. Chapter 6 in 'Biological Husbandry'. Ed.Stonehouse,B. Butterworths, London. 99-111
10. GIFFARD, W.H. and GRAY, K.R. (1972). Conjoint composting of sewage sludge and straw. Journal of the Soil Association. 17, (1), 27-31
11. MELCER, H. (1972). Process investigations into a treatment system for farm animal effluents. Ph.D thesis. University of Birmingham
12. WILEY, J.S. and PEARCE, G.W. (1955). Progress report on high rate composting studies. Proceedings of the American Society of Civil Engineers, Journal of the Sanitary Engineering Division. 81,Paper 846
13. BEST, P.R. (1975). Farm-scale trials of the conjoint composting of pig slurry with straw. Ph.D thesis, Part 2. University of Birmingham
14. GRAY, K.R. and BIDDLESTONE, A.J. (1978). Improvements in and relating to composting. British Patent No.1498938
15. CROFTS, A. (1978). Two practical ways of handling slurry. N.A.C. News, (3), 10
16. ROBINSON, I. and EDMUNDS, J.R. (1983). Composting pig manure by modified ARCUB technique. Proceedings of International Conference on Composting of Solid Wastes and Slurries. University of Leeds
17. GRAY, K.R, BIDDLESTONE, A.J., BALL, D., PATTERSON, W.T. and COOPER, D. (1983). Spreader. British Patent No.8305622.

18. COOPER, J.D. (1984/85). Studies on air flow in composting processes. Ph.D thesis (in preparation). University of Birmingham
19. WILLSON, G.B., PARR, J.F., EPSTEIN, E., MARSH, P.B., CHANEY, R.L., COLACICCO, D., BURGE, W.D., SIKORA, L.J., TESTER, C.F. and HORNWICK, S.B. (1980). Manual for composting sewage sludge by the Beltsville aerated-pile method. USEPA-USDA Publication No.EPA 600/8-80-022
20. FINSTEIN,M.S., MILLER, F.C., STROM, P.F., MACGREGOR, S.T.and PSARIANOS, K.M. (1983). Composting ecosystem management for waste treatment. Biotechnology. 1, (4), 347-353
21. SINGLEY, M.E., HIGGINS, A.J. and FRUMKIN-ROSENGAUS, M. (1982). Sludge composting and utilization: a design and operating manual. New Jersey Agricultural Experimental Station, Rutgers University,USA
22. SNELL, J.R. (1957). Some engineering aspects of high-rate composting. Proceedings of the American Society of Civil Engineers, Journal of the Sanitary Engineering Division. 83, Paper 1178

Table 1 Organic wastes in the UK

Waste	10^6 Tonnes per year (fresh wt)
Wood-shavings, sawdust and bark	> 1
Food processing wastes, brewers grains	> 1
Potato haulms, sugar beet tops	1.6
Garden and nursery wastes	Possibly 5
Cereal straws, surplus	5-7
Municipal refuse (about 50% organic matter)	18
Sewage sludge, (dry solids)	35, (1.2)
Farm manures	120

Table II General operating conditions on small scale unit at Clent, 1972-74

Run No.	Straw	Slurry	Straw/slurry applications	Days in cubicle	Air flow-rate, m^3/s		Remarks
1	B.S.	wk	10	62	0.013	Blower 12 hrs./day	High straw. Drainage to slurry pit.
2	B.S.	wk	17	50	0.013		
3	B.S.	wk	20	100	0.013		
4	B.S.	st	11	80	0.013		Low straw. Drainage recycled over straw in separate box.
5	B.S.	st	13	66	0.012		
6	B.S.	st	17	100	0.012		
7	B.P.	wk	20	50	0.010	Fan 24 hrs./day.	
8	B.P.	wk	15	48	0.010		
9	B.P.	wk	14	62	0.010		Low straw. Drainage over compost box.
10	B.P.	wk	16	98	0.007		
11	B.P.	wk	15	98	0.007		Low straw. Diluted slurry. Drainage over compost box then over high-rate filter.
12	B.P.	wk	14	101	0.007		
13	W.P.	wk	10	104	0.007		
14	W.P.	wk	16	103	0.007		
15	W.P.	wk	10	73	0.007		
16	W.P.	wk	6	95	0.007		Low straw. Drainage over compost box.
17	W.P.	wk	6	69	0.008		High straw for mushroom compost. Drainage over compost box.
18	W.P.	wk	7	42	0.007		
19	W.P.	wk	10	42	0.007		

Key:- B = Barley straw W = Wheat straw S = bales "scuffed out" P = bales broken into pieces
wk = weak slurry below 5% total solids st = strong slurry over 6% total solids

Table III Liquid and Total Solids (TS) removal by straw bed - small scale unit, 1972-74

Run	Straw added kg	Slurry added kg	Slurry added %TS	Drainage kg	Drainage %TS	Slurry removed %	Dry slurry solids added kg	Dry slurry solids effluent kg	TS removed %	kg slurry/kg straw: slurry added	kg slurry/kg straw: slurry removed
	a	b	c	d	e	100(b-d)/b	f=bc/100	g=de/100	100(f-g)/f	b/a	(b-d)/a
1	477	2918	4.64	1273	2.11	56	135	27	80	6.1	3.4
2	605	5632	4.56	1432	2.06	75	257	29	88	9.3	6.9
3	509	12850	3.41	4186	1.87	68	438	78	82	25.2	17.0
4	382	3618	6.49	1027	2.57	72	235	26	89	9.5	6.9
5	445	5000	6.69	1700	2.81	66	335	48	86	11.2	7.4
6	573	6727	6.28	2168	2.33	68	422	51	88	11.7	8.4
7	668	12464	3.34	4109	1.27	67	416	52	88	18.7	12.5
8	509	10545	4.10	2918	2.18	72	432	64	85	20.7	15.0
9	477	9545	4.44	3136	2.00	67	424	63	85	20.0	13.4
10	541	10909	2.60	3705	0.81	66	284	30	89	20.2	13.3
11	509	10227	3.03	3373	1.31	67	310	44	86	20.1	13.5
12	573	10086	2.62	5273	1.17	48	264	62	77	17.6	8.4
13	509	10318	3.32	4832	1.08	53	343	52	85	20.3	10.8
14	636	11936	2.38	4518	1.13	62	284	51	82	18.8	11.7
15	636	12132	2.75	4059	1.20	67	334	49	85	19.1	12.7
16	573	8486	4.02	2495	1.96	71	341	49	86	14.8	10.5
17	684	6468	4.58	2227	1.93	66	296	43	85	9.5	6.2
18	573	5355	3.68	1795	1.93	67	197	35	82	9.3	6.2
19	605	8114	1.90	3895	1.49	52	154	58	62	13.4	7.0
Total	10484	163332		58121			5901	911			
Range			1.90-6.69		0.81-2.81	48-75			62-89	6.1-25.2	3.4-17.0
Mean			3.94		1.75	65			84	15.6	10.1

Table IV Trials on large walled midden at Rugeley, 1978

Run	Straw added kg	Slurry added kg	Slurry removed %	kg slurry/kg straw	
				slurry added	slurry removed
3	2545	31968	50.8	12.6	6.4
4	2291	31818	42.8	13.9	5.9
5	3945	32727	61.8	8.3	5.1
6	2545	22727	62.2	8.9	5.5

Table V Some results on ballistic processor at Rugeley, 1981

Date	Straw added kg	Slurry applied kg			Drainage kg	Slurry removed %	kg slurry/kg straw	
		by machine	by hose	total			slurry applied	slurry removed
12.5.81	3245	36509	-	36509	8877	75.7	11.3	8.5
13.5.81	2068	24155	-	24155	7523	68.9	11.7	8.0
14.5.81	2545	24041	2591	26632	4277	83.9	10.5	8.8
15.5.81	700	6445	8045	14490	6623	54.3	20.7	11.2
18.5.81	223	3464	5455	8919	5245	41.2	40.0	16.5
19.5.81	3818	40132	-	40132	9918	75.3	10.5	7.9
20.5.81	1273	14618	-	14618	10673	23.0	11.5	3.1
21.5.81	1591	17218	5818	23036	10341	55.1	14.5	8.0
22.5.81	1352	15977	-	15977	2577	83.9	11.8	9.9
26.5.81	2020	21809	2727	24536	7909	67.8	12.1	8.2
29.5.81	-	-	1727	1727	-	100	∞	∞
3.6.81	-	-	2045	2045	-	100	∞	∞
5.6.81	-	-	2318	2318	-	100	∞	∞
Total	18835	204368	30726	235094	73963			
Average						68.5	12.5	8.6

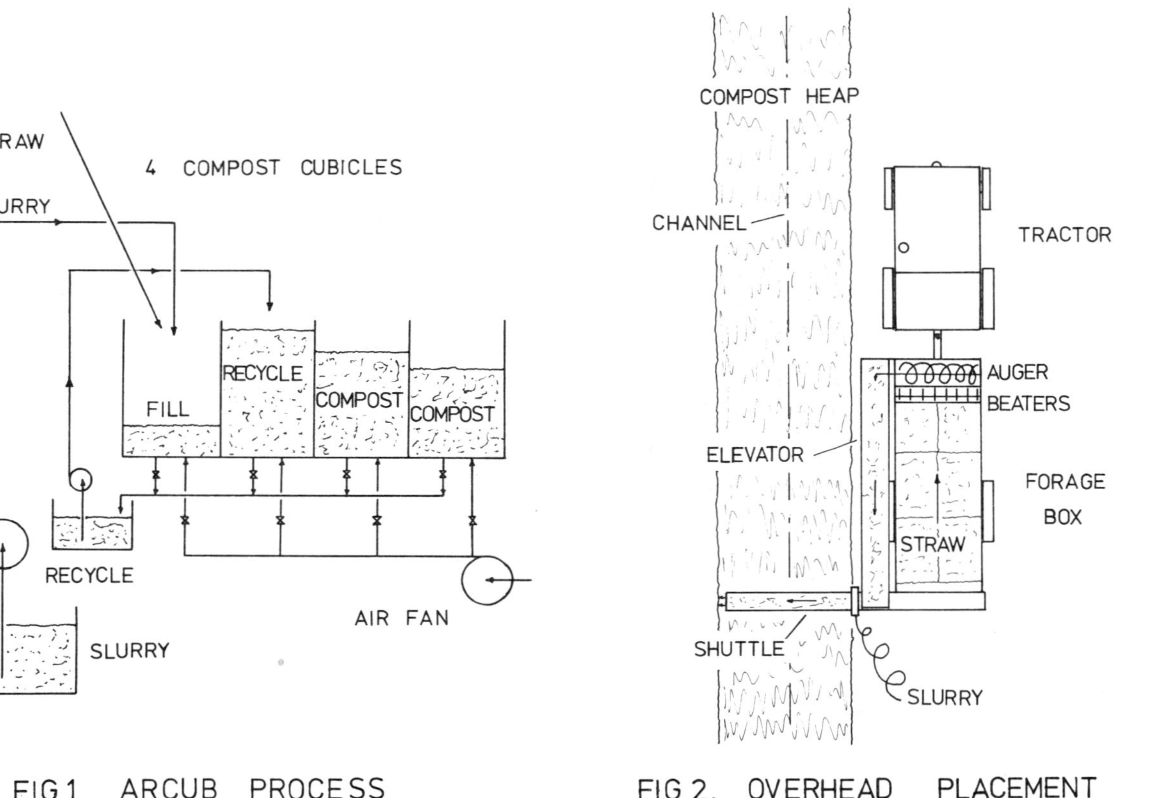

FIG 1. ARCUB PROCESS

FIG 2. OVERHEAD PLACEMENT

DISCUSSION

SVOBODA: Would it be better to use the straw as bedding and let the pigs mix it with slurry or to treat the slurry on its own as a liquid?

BIDDLESTONE: The work was commenced specifically to deal with the problem of slurry from pig housed on slatted floors and using slurry systems. This was the practice at the time and many of these 'intensive' systems did not have land for disposal. There is little interest in a return to straw as bedding in pig housing. The process and system described is appropriate to other animal wastes, food wastes and sewage sludge.

DE BERTOLDI: Did you find any problem in composting pig slurry when antibiotics and heavy metals were present?

BIDDLESTONE: We had no problem in composting the pig slurries available to us. I have no experience with antibiotics. Copper is likely to be present in significant amounts because of copper additives in pig feed and would only be decreased by reducing the level of copper in the feed.

VERDONCK: In all your trials straw was used as the carbon source. Is this too expensive for use and can you also use other carbon sources such as bark or sawdust?

BIDDLESTONE: We have used straw not only as a carbon source but also because it retains its structure for some time and provides a suitable matrix for air diffusion as well as having a high moisture content whilst retaining structure. With regard to cost and availability of straw this may be a problem although it may be possible to arrange exchange of straw for compost. I would add that some 5 - 6 million tonnes of straw are burnt annually in the field in the UK. This practice is likely to be stopped thus affecting availability. The use of other carbon sources is possible.

EDWARDS: Could you say what your machine costs?

BIDDLESTONE: The cost of processing machine is in the range of £12,000 to £15,000 depending upon specification. The manufacture of the machine is the responsibility of our industrial partners, ARM Ltd (Colton Road, Rugeley, Staffs, UK), who would provide detailed costs.

LOPEZ-REAL: You mentioned that straw was used with the pig slurry for physical improvements and not to adjust the C:N ratio. The carbon straw is largely unavailable and in terms of compost product quality considerable losses of nitrogen presumably occur?

BIDDLESTONE: We have concluded that the amount of straw required in terms of C:N balance would be prohibitive. The straw usage therefore has been determined by the requirement of a suitable matrix for gaseous diffusion and filtering of solids. I accept that inevitably nitrogen will be lost as ammonia from the heaps.

LE ROUX: For a 1,000 pig farm would you envisage running the process continuously?

BIDDLESTONE: No. The processor has a capability of $2\frac{1}{2}$ standard straw bales per minute and 50 gallons of slurry per minute. For a 1,000 pig unit this would require no more than 1 hour operation daily. You could operate on a weekly basis or whatever time interval is appropriate for the particular farm situation.

HOVSENIUS: Have there been any odour problems?

BIDDLESTONE: There have been problems with heaps settling but none with odour. This is mainly a problem with sludge.

BRUCE: I wish to raise two points i) was the sewage sludge/straw composting system you described force ventilated or was it aerated by natural aeration?

ii) With a mixture of straw and sludge, would it be advantageous or disadvantageous to add the straw to the liquid sludge before dewatering?

BIDDLESTONE: i) The results I described were for natural aeration although we formed a new heap by passing it through the machine again. We have now introduced forced aeration.

ii) We hope to investigate this question in the near future.

SESSION III

HEAT FROM COMPOSTING

Heat recovery from composting and comparison with energy from anaerobic digestion

Heat from aerobic treatment of liquid animal wastes

Heat recovery from composting solid manure

HEAT RECOVERY FROM COMPOSTING AND COMPARISON WITH ENERGY FROM ANAEROBIC DIGESTION

A. VEROUGSTRAETE, E.-J. NYNS and H.P. NAVEAU
Unit of Bioengineering, Catholic University of Louvain
B-1348 Louvain-La-Neuve, Belgium

Summary

Composting and anaerobic digestion are two processes for treating wastes that allow recovery of energy, in the form of warm water from composting and methane from anaerobic digestion.
Both processes are briefly presented from the point of view of energy recovery and their operational parameters are discussed as they relate to energy production.
Results are available from laboratory or pilot size composting experiments and from farm-scale anaerobic digesters. Both processes allow energy to be extracted from manure with a yield of approximately 4-8 MJ per kg dry matter, depending upon the system and parameters used. The power produced per m^3 fermenter is 5 to 10 times less for methane digesters than for composting plants and this should influence investment costs.
Utilisation of the methane in a boiler gives an efficiency of 65-75 % depending on the boiler to produce hot water or steam. Biogas can also be used in an engine to provide mechanical power or to generate electricity. On the other hand, warm water from composting will not exceed 60 °C. Composting shows new promises as a process for recovering energy from bedding manure or straw and research and development should continue. On the other hand, net energy yields from farm waste digesters need to be improved through better design and operation, especially for the treatment of bedding manure.

1. INTRODUCTION

The production of energy from biomass residues has taken an increasing importance in recent years, particularly in the agricultural world. Amongst the biological processes, biomethanation or anaerobic digestion and its energy product, biogas, have been known for many years, although real values of net energy production in full-scale digesters are not commonly found in the literature. Composting, being an oxidative process, also produces energy in the form of heat, recoverable as warm water. This heat-recovery from composting is still at the investigative stage but encouraging results have been published. It appears worthwhile at this stage of development of both processes to try to evaluate their relative potential, benefits and problems in relation to the recovery of energy from agricultural residues.

2. ENERGY RECOVERY THROUGH COMPOSTING

2.1. The problem

The temperature during composting readily increases to 50-70 °C in the composting pile, due to the oxidative reactions taking place. When composting is done in heaps or windrows without forced aeration, the heat produced is evacuated through losses at the surface of the heaps to air and ground, or when turning the heaps. The composting material being a bad heat conductor, the loss of heat is low and the temperature in the heap may reach 70 °C or more. When composting is controlled with forced aeration, the heat is mainly lost through the aeration air : the air and its water content are warmer at the exit than at the entrance of the composting unit; there are also some losses through the walls of the unit.

The theoretical amount of energy that could be produced is equivalent, at the maximum, to the calorific value of the substrate (about 18.5 MJ/kg TS). Although only partial oxidation is possible, the biodegradability of the substrate also influences the amount of recoverable energy, which can be estimated as 50-60 % of the calorific value (9-11 MJ/kg TS).

2.1.1. Recovery with heat exchangers inside the pile

The idea has been used to remove this excess heat through exchange with water circulating in heat exchangers placed inside the piles. If warm water has been obtained by this method, the power or the amount of energy recovered daily is low. Typical results are given in table (1).

The main reasons for the low energy recovery are the bad heat conduction properties of the cellulosic substrates, the difficulty of installing the system and repairing it inside the piles, and the necessity to keep high temperatures inside the heap for stable operation. Even if this system can give satisfactory results to a limited number of interested people, it should not be considered for widespread use on farms or industrial applications.

2.1.2. Recovery with heat exchangers working on aeration air

The idea has been developed of using the aeration air to extract the heat from the composting unit and to recover this heat using an air-water heat exchanger. The warm water coming out of the exchanger is then utilized for space heating or as washing water (2,3).
Only this type of process and its results will be further presented and discussed.

2.2. Parameters and processes using air-water heat exchangers

Composting systems for manure with air-water heat exchangers for heat recovery have been studied amongst others at the CEMAGREF (Groupement de Clermont-Ferrand, France) (2) and at the Institute of Agricultural Engineering of the Royal Veterinary and Agricultural University at Tastrup (Denmark) (3) where Berthelsen has developed preliminary designs (4).

2.2.1. Parameters

The air is used as the carrier of the heat out of the composting system.
This exhaust warm air is humid and, to avoid drying out of the compost and

losses of heat, this humidity must be condensed and returned to the composting material, eventually after heat exchange (2). Since a content of 10 % oxygen in the exhaust air is compatible with good composting, part of the exhaust air can be recycled (3). Oxygen consumption being low, a high air flowrate seems necessary to ensure an homogeneous aeration. The optimum temperature for a high composting rate has been found to be between 50-60 °C or even 65 °C (2,3).
Moisture content should be between 70 and 80 % (2,3). Bulk density is recommended at 300 kg/m^3 (3) or 400-600 kg/m^3 (2), but this depends on the proportion of straw and the water content. A free air space of 70 % is shown to be optimal (3). The runs are of 5-6 days duration or even shorter when an inoculum is used.

2.2.2. Processes

CEMAGREF has studied heat recovery through aeration at laboratory stage (30 l) and with one fermenter of 350 l. Recoverable heat production is calculated from measurement of air moisture, temperature and flowrate at the entrance and the exit of the fermenter (2).

At the Danish Royal Veterinary and Agricultural University, a pilot plant (10 m^3) with mixing has been run, based on extraction of heat with aeration air and its recovery through heat exchange in a so-called "double cooling tower heat exchanger/scrubber" (3). This heat exchanger, shown in figure 1, is characterized by :

A) an air-water exchanger : the warm exhaust air from the composting plant is circulated upwards in a packed bed (n° 2) against water sprayed on this packed bed.
B) a water-water exchanger (n° 2) where the warmed water from A gives its heat content to the central-heating water system or any warm water end-use system.
C) an air-water exchanger (n° 9) where the residual heat content of the water from A is transfered to the incoming air of the composting unit in a water-air exchanger similar to the one in A : the luke-warm water is sprayed downward on a packed bed (n° 9) where it exchanges its heat against fresh air cirulating upward before entering the composting plant.
D) the water is used again for spraying in A (n° 12, 13, 1) so that it circulates in a closed circuit.

This "double cooling tower heat exchanger/scrubber" allows the attainment of less than 5°C difference between temperature in the composting pile and the usable warm water. The compost does not dry out because only exhaust air is going through the exchanger independently of inside recycling.

2.3. Energy production from composting

Results from different experiments are summarized in table 1 in terms of power and usable energy output per kg dry matter.

It is seen that the power output of systems using tubing exchangers is low. The final energy yield as MJ/kg TS seems in some cases quite high if compared with the average calorific value of 18.5 MJ/kg dry biomass. However, the data on which these yields have been calculated by us are not well documented and these values should be treated with caution. Results obtained with the two air-water exchangers systems show a much higher power capacity of about 1 kW/m^3 or 170-270 MJ/ton fresh material and per day. Experimental energy yields are about 4 to 8 MJ/kg TS or about 22 to 44 % of the energy contained in the substrate (18.5 MJ/kg TS).

Figure 1. Schematic representation of a double cooling tower heat exchanger/scrubber for recovery of energy from composting (3).

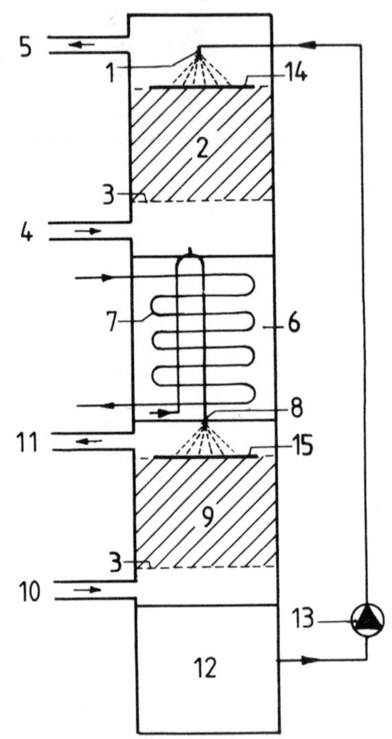

1. Water sprayer
2. Packing material for exhaust air cooling tower
3. Perforated plate
4. Air from composting plant
5. Exhaust air (cooled down)
6. Water
7. Exchange pipe to central heating system
8. Overflow pipe
9. Packing material for in-coming air
10. In-coming air
11. Heated air to composting plant
12. Water sump (cold water)
13. Pump
14. and 15 : Distributing plates

CEMAGREF estimates that using an optimized continuous system and a pre-fermentation phase or an inoculum, the gross power could reach over 1.4 kW/m^3 during 4 days. With a 10 m^3 fermenter, calculated losses through walls and air plus process energy would then leave a net power of 1.15 kW or 250 MJ/t fresh substrate and per day with an energy yield of 27 %.

The Danish double-cooling tower (DCT) heat exchanger used with 5 m^3 manure in a 10 m^3 pilot fermenter for three weeks also shows a good power capacity of about 1 kW/m^3 fermenter but, the bulk density of the solid manure being lower (300 kg/m^3), the energy yield is higher and reaches 8.1 MJ/kg TS (dry matter). Operation of the fermenter at full capacity would allow, with a yield of 83 % of the produced energy as output energy, to reach a power of 1.25 kW/m^3 (360 MW/ton fresh material and per day) and an energy yield of 10.9 MJ/kg TS (dry matter), representing a recovery of 59 % of the energy contained in the substrate. Taking into account the 83 % yield of the fermenter, the complete energy production would then be 0.59/0.83 = 0.71 or 71 % of the energy in the starting material. Such a process yield seems very high and before accepting it, complete control of all parameters, including the bulk density of the solid manure, would need to be made.

At this time, a power of 1 kW/m^3 fermenter (86.4 MJ/m^3.d) and 200-270 MJ/ton fresh material and per day together with an energy yield of 4-8 MJ/kg dry matter (TS) should be acceptable for comparison with anaerobic digestion.

Table 1. Energy recovery from composting processes

Substrate	Process	Duration	Energy recovered				Ref.
			W/m^3	MJ/m^3d	MJ/t d	MJ/kg TS	
Forestry residues (a)	T	18 m	10-17	0.86-1.5	2.6-4.3	(3.5-5.8)	(1)
Milled Forestry residues (a)	T	5 m	16	1.38	4.1	(1.5)	(1)
Milled Forestry residues (a)	T	12 m	35	3.0	9.2	(8.3)	(1)
Municipal wastes (a)	T	3 m	115	9.9	30	(6.8)	(1)
Manure (e)	AW	5 d	800	69	173	4.3	(2)
Manure (e)	AW	4 D	940	81	203	4.1	(2)
Manure, continuous, optimized (b,e)	AW	4 d	1150	99	250	5.0	(2)
Manure (c,b)	DCT	6 d	940	81	270	8.1	(3)
Manure (100 % capacity used) (d,f)	DCT	6 d	1260	109	363	10.9	(3)

Process (heat exchanger) : T = tubing in the mass, AW = air-water, DCT = double cooling tower; t = ton of fresh substrate, m^3 = m^3 of fresh substrate in the fermenter; TS = total solids or dry matter content; duration in months (m) or days (d).

(a) assuming bulk density of 335 kg/m^3 and 60 % moisture content, and power stable during length of experiment.
(b) based on calculations of lab-results scaled up to a 10 m^3 fermenter.
(c) 10 m^3 pilot fermenter used at 50 % capacity (5 m^3).
(d) calculated for use of 10 m^3 pilot fermenter (c) at 100 % capacity (10 m^3).
(e) bulk density of 400 kg/m^3 and 80 % moisture content.
(f) bulk density of 300 kg/m^3 and 80 % moisture content.

3. ENERGY RECOVERY THROUGH BIOMETHANATION

3.1. The biomethanation process

The biomethanation process is an anaerobic process in which several bacterial communities work to decompose organic matter to produce a gas called biogas, which is composed of methane (CH_4) and carbon dioxide (CO_2). The process takes place in a closed vessel called the digester whose characteristics and accessories are described in Figure 2.

Figure 2. Schematic representation of a completely mixed digester for sludges or slurries

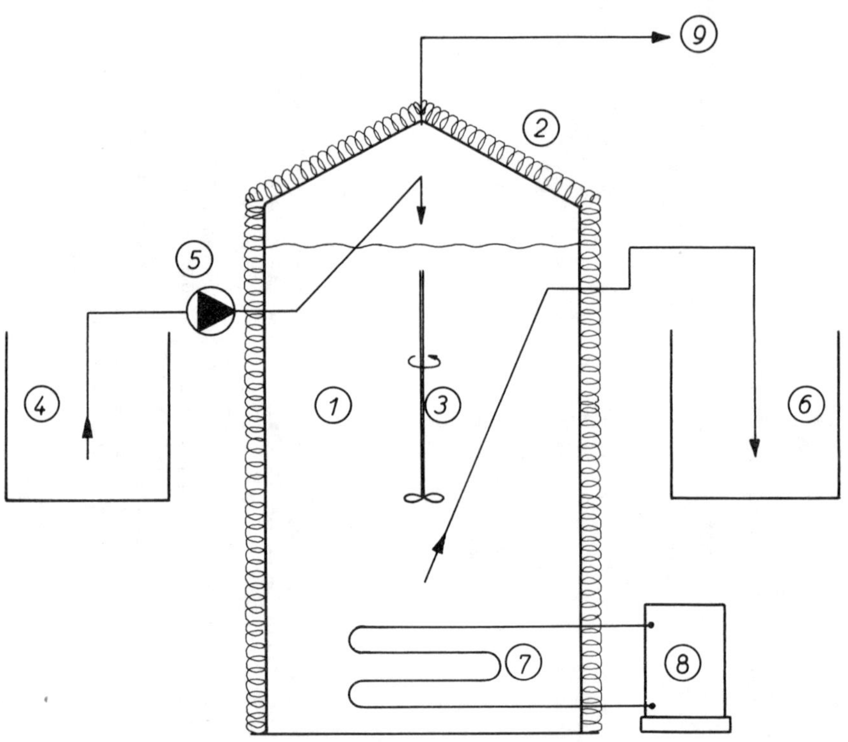

1. Digester
2. Insulation
3. Mixing system
4. Influent storage
5. Loading pump
6. Effluent storage
7. Heating system
8. Boiler
9. Gas storage or use

Biomethanation or methane digestion can occur in the range 4 °C to 60-70 °C, but it has been shown that 30-40 °C is the most favorable temperature range to use with agricultural residues. Lower temperatures give a much too slow gas production rate, while higher temperatures show unstable operation and need too much energy for maintainance, at least in average European conditions. A heating system is needed which will maintain the desired temperature inside the digester. To reduce the heat losses and thus to optimize the heating efficiency, the tank walls will need to be well insulated. Mixing is recommended to achieve a uniform temperature inside the methane digester and to prevent scum formation and settlement inside the digestion tank. Mixing may be mechanical or by gas or liquid recirculation.

The biogas produced by a methane digester has a methane content which varies between 55 and 75 % ($vol.vol^{-1}$). Its calorific value ranges between 20 and 26 MJ/m^3. In comparison, gas oil has a calorific value of 36 MJ/l and methane of 35,5 MJ/m^3.
Heating and mixing consume energy and this must be accounted for in the energy balance of a digester.

Three categories of wastes treatable by biomethanation may be distinguished :
- solid wastes with more than 15 % dry matter ($wt.wt^{-1}$) (manure)
- semi-solid wastes with more than 2 % and less than 15 % dry matter eg. all types of slurry produced on farms.
- liquid wastes with low concentration of suspended solids (2 % wt wt^{-1}).

Different biomethanation systems can be applied to these three categories of wastes, with different productivities and yields. In this paper, only agricultural solid or semi-solid wastes will be considered; agricultural digesters may be either batch, plug-flow or completely mixed digesters.

3.2. Parameters of the biomethanation process

The running of continuous digesters (plug-flow or completely-mixed) is characterized by several parameters (5) which influence the amount of methane produced (see also table 3) :

Table 2. Process parameters for biomethanation (5)

Symbol	Definition	Unit
θ	retention time	d
B_V	volumetric loading rate	kg (VS)/m^3_{ML}.d
$r_{V.CH4}$	methane production rate	m^3 (CH_4)/m^3_{ML}.d
T	temperature	°C
$Y_{CH4/VS}$	methane yield	m^3 (CH_4)/kg VS
S_o	feed concentration	kg VS/m^3_{ML}

Abbreviations : VS = volatile solids or organic matter, ML = mixed liquor or digester working volume

- the temperature of the digester, most often 35 °C,
- the mean retention time (θ) which, together with the temperature influences the extent to which the organic substrate will be decomposed.
- the volumetric loading rate (B_V) or the amount of organic matter (kg VS) fed into the methane digester per m^3 of digester working volume (m^3_{ML}) and per day (d),
- the feed concentration (S_o) is related to the volumetric loading rate and the mean retention time by the relation : $S_o = B_V \times d$.

The biodegradability of the feed expresses the fraction of the feed convertible to biogas and can be determined experimentally, whilst the chemical oxygen demand (COD) determines the theoretical amount of methane that could be produced from each kg of organic matter (VS). The biodegradability is dependent upon the feed of the animals and upon any pretreatment such as a long storage period at ambient temperature, that decreases the methane production potential.

3.3. Energy production potential

Each kg of substrate organic matter (VS) is equivalent to a given amount of chemical oxygen demand (1.0 to 1.6 kg COD/kg VS) depending upon its more or less oxidized state.

In a methane digester, no net oxidation takes place due to the anaerobic conditions; the chemical oxygen demand of the feed must be found in the products, sludge and gas. Hence, the difference in chemical oxygen demand between feed and sludge should correspond to the chemical oxygen demand of the gas phase, that is of methane. From the COD of methane, it can be shown that 1 kg COD = 0.395 m^3 CH_4 at 35 °C or 14 MJ.

If all the organic matter were transformed into methane, the maximum yield would be per kg organic matter fed :

$$0.395 \ m^3 \ CH_4/\text{kg COD} \times (1.0\text{-}1.6) \ \text{kg COD/kg VS} = 0.40\text{-}0.63 \ m^3 \ CH_4/\text{kg VS}$$

Experience shows average values of anaerobic biodegradability of 50 % for cattle slurry (6), 30-80 % for pig slurry (7) and 40 % for manures (8). With a mean value of COD/VS of 1.3, one can then expect a maximum methane yield of 0.25 m^3 CH_4/kg VS from cattle slurries, up to 0.40 m^3 CH_4/kg VS from pig slurries and 0.21 m^3 CH_4/kg VS from manure. These numbers can be compared with practical figures observed from full-scale digesters (table 3)

3.4. Expressions of the energy production

The energy production can be characterised by the volumetric energy production rate ($r_{V.energy}$) in MJ per m^3 digester working volume and per day, gross and net, and the energy yield of the substrate ($Y_{MJ/VS}$) or the MJ content of the methane produced for each kg of substrate, which is the product of the methane yield ($Y_{CH4/VS}$ in m^3 CH_4/kgVS) by the calorific value of the methane (35.5 MJ/m^3 CH_4).

3.5. Energy needs of the digester

The net energy production of a methane digester may be defined as the fraction of the gross methane production really available for household or farm uses, such as water heating, house heating, cooking, conversion into electricity...

In fact, as seen before, the biomethanation process itself consumes energy. Energy is required to heat the influent, load the digester, maintain the required temperature and mix the contents.

Table 3. Energy production by biomethanation of various manures in different systems.

Type of manure	Ref.	Type of digester(a)	B_V	S_o	θ	T	$r_{V.CH4}$	$Y_{CH4/VS}$	$Y_{energy/TS}$	$r_{V.energy}$		
			$(\frac{kg\ VS}{m^3_{ML}\cdot d})$	$(\frac{kg\ VS}{m^3})$	(d)	(°C)	$(\frac{m^3\ CH_4}{m^3_{ML}\cdot d})$	$\frac{m^3\ CH_4}{kg\ VS}$	$\frac{MJ}{kg\ TS}$	$\frac{MJ}{m^3_{ML}\cdot d}$		$\frac{net}{gross}$
									(net)	(gross)	(net)	(%)
Dairy	9	CM	4.3	64.5	15	35	0.9	0.21	6.1	32.0	22.4	70
Dairy	9	PF	2.9	72	25	34	0.41	0.14	4.4	14.6	10.8	74
Dairy + bedding	9	UF	4	80	20	37	0.38	0.10	2.2	13.7	7.2	53
Bovine	10	CM	3.5	70	20	33	0.59	0.17	4.9	21.2	14.8	70
Bovine	9	CM	2.8	42	15	35	0.83	0.30	8.7	29.5	20.7	70
Bovine	9	PF	1.1	63	57	33	0.37	0.33	9.6(c)	13.0	9.1	70(c)
Bovine + bedding	9	B	3	168	56	34	0.27	0.09	2.6(c)	9.6	6.7	70(c)
Bovine + bedding	9	CM	2.7	108	60	35	0.54	0.20	3.3	19.2	12.3	64
Poultry	9	PF	1.8	45	25	35	0.48	0.27	7.9	17.1	12.0	70
Pig	11	CM	1.7	68	40	31	0.28	0.16	2.8	9.9	4.3	43
Pig	9	CM	2.5	58	23	30	0.50	0.20	6.7	17.7	14.3	81
Pig	9	CM	3.9	49	13	35	1.01	0.26	8.1	36.1	27.1	75
Pig	9	CM	2.5	30	12	35	0.75	0.30	9.9	26.6	21.0	79

(a) : B = Batch, CM = completely mixed, PF = plug flow, UF = upflow.

(b) : Considering that % VS = 0.85 % TS.

(c) : Not given. Based on 70 % net/gross.

In their survey of the European biogas plants, Demuynck and Nyns (9) report that the biogas fraction used for the digester maintenance ranges from a low 20-30 % up to higher values of 60-80 %, this being the case when the methane digester is poorly built or operated. Of that energy consumption, 90 % serves to heat the influent and the digester.

In fact, the percentage of the energy produced which is necessary to heat the influent from ambient to digester temperature is a function of (1) the relative volume of the daily feed, inversely proportional to the retention time ($V_{feed} = V_{ML}/\theta$ where V_{ML} = working digester volume), (2) of the temperature difference between feed and digester and (3) of the concentration of biodegradable matter in the feed.

The volumetric energy production rate ($r_{V.energy}$) is given by the following relations :

$$\underset{(MJ/m^3_{ML}.d)}{r_{V.energy}} = \underset{(m^3_{CH4}/m^3_{ML}.d)}{r_{V.CH4}} \times \underset{(MJ/m^3CH_4)}{35.5}$$

with

$$\underset{(m^3CH_4/m^3_{ML}.d)}{r_{V.CH4}} = \underset{(kg\ VS/m^3_{ML}.d)}{B_V} \times \underset{(m^3CH_4/kg\ VS)}{Y_{CH4/VS}}$$

and

$$\underset{(kg\ VS/m^3_{ML}.d)}{B_V} = \underset{(kgVS/m^3_{ML})}{S_o} \times \underset{(1/d)}{1/\theta}$$

The amount of energy needed to heat the influent is, per day and m^3 of digester :

$$Q = 4.18 \ . \ \Delta T/\theta \quad (MJ/m^3_{ML}.d)$$

In order to produce at least enough methane to heat the digester, it is necessary to observe the following relation, considering an efficiency of 0.75 for the boiler :

$$r_{V.energy} \times 0.75 > Q$$

which, assuming a methane yield ($Y_{CH4/VS}$) of 0.20 (see § 3.3) can be transformed to :

$$35.5 \times 0.2 \times S_o \times 1/\theta \times 0.75 > 4.18 \times \Delta T \times 1/\theta \quad \text{or}$$

$$1.27\ S_o > \Delta T$$

3.6. Energy produced in biomethanation processes

Table 3 gives the gas production rates ($r_{V.CH4}$) and the energy production rate, gross and net, for a number of farm digesters utilising various manures, together with the running parameters (B_V, S_o, θ , T, and $Y_{CH4/VS}$). These figures allow the calculation of the net energy yield ($Y_{energy/TS}$ in MJ/kg TS) or the energy available from one kg of dry substrate in the form of methane, assuming that the dry matter (TS) contains 85 % organic matter (VS). Only digesters working satisfactorily have been considered.

4. DISCUSSION

When comparing the results, one sees that the amount of energy produced per kg dry matter is of the same order of magnitude for composting and biomethanation processes (4 to 6 and up to 9 MJ/kg TS). Energy yields from cattle manure are sometimes lower with full-scale biogas systems than with the pilot composting units, but similar yields are reached when biomethanation process parameters are well chosen. The power, or the amount of energy produced daily per unit working volume of fermenter/digester is 5 to 10 times higher for composting, due to the lower concentration of substrate (liquid phase) in anaerobic digesters. This apparently should constitute a clear advantage in investment costs for the composting process although the economics of running full-scale energy producing composting plants are not yet known. Also, research is carried out to develop a digester that could digest manure without diluting it but without the handling problems encountered in batch systems.

The utilization of the energy also presents important differences between the two types of processes. Warm water from composting is produced regularly at a temperature of 50-55 °C, never exceeding 60 °C, and is convenient for space heating of for washing uses. No boiler is needed but it is difficult to store the energy for more than 24 hr.

Methane can be burned to produce hot water (or steam) for space heating and wash water, or it may be used for cooking or as fuel in engines for power or electricity generation. Multipurpose uses are possible and the flexibility of methane is an advantage. Efficiency of boilers gives a loss of energy of 25-35 % as compared to warm water from composting; storage of the methane for more than 24 hr is expensive.

5. CONCLUSIONS

Both composting and biomethanation can produce significant amounts of energy from agricultural residues (4-8 MJ/kg dry matter). This energy can be used on the farm.

Composting systems are still at pilot or demonstration scale. Very encouraging results have been obtained at that scale but more design development and more data are needed before a final assessment can be made.

Biomethanation is quite widespread in Europe (9). However there are still a number of poorly designed and/or run digesters with low efficiencies. Dissemination and use of rules for designing and operating digesters would allow great improvements in that situation. A new design of digester for solid manure would help to overcome the bottleneck concerning biomethanation of solid residues.

Both composting and biomethanation seem valuable for energy recovery and both have their advantages and disadvantages. The choice between them should be based on an integrated appraisal of a given situation, including type of residue, qualitative and quantitative energy needs, type of residue handling equipment.

REFERENCES

1. DAUDIN, D. La récupération d'énergie dans les processus de compostage, CEC, R & D Programme "Recycling of urban and industrial waste", DOc. XII/MPS/83/82-FR (1983).

2. Compostage et chaleur, Récupération de la chaleur produite par le compostage accéléré du fumier bovin, Etudes du CEMAGREF, n° 491, Septembre 1982, 58 pp (1982).

DISCUSSION

SVOBODA: Have you tried to express energy evolution during composting or anaerobic digestion as a unit of energy per kg of degraded total solids rather than overall total solids. The quantity of energy released is surely proportional to the percentage of degraded total solids?

NAVEAU: Values found in the literature on energy recovery from composts did not allow this calculation to be made as degraded total solids were not mentioned. We have expressed results in this way as it refers to COD degraded. In anaerobic digestion where no oxidation takes place, it is possible to base a balance on COD. However this value must be equivalent to the COD of the methane formed, 0.395 l CH_4/g COD degraded (35°C, atmospheric pressure). CH_4 produced per unit of total solids degraded is difficult to use because the oxidation state of total solids can change in the digester and there are errors in determining total solids due to some organic compounds being volatile at 105°C (volatile fatty acids). In practice the farmer only wants to know how much energy he can get from the material put into his composting plant or digester.

STENTIFORD: Have you carried out any work on the recovery of heat from the digester effluent and, if so, did the solids level pose any particular problem?

NAVEAU: We have not carried out any research on recovery of heat from the digester effluent. From our discussions and a survey on "Biogas Plants in the EEC and Switzerland" we know that many problems arise with effluent heat exchangers such as clogging, and deposit on the tube or exchanger walls, at least with effluents containing suspended solids.

HEAT FROM AEROBIC TREATMENT OF LIQUID ANIMAL WASTES

S.BAINES, I.F.SVOBODA and M.R.EVANS
Department of Microbiology
West of Scotland Agricultural College

Summary

Environmental pollution can be controlled by aerobic treatment of liquid animal wastes. Heat energy can be recovered continuously from a continuous culture system. The amount of heat released depends on the degree of biodegradation and is related to oxygen consumption, mean treatment time, temperature and level of dissolved oxygen. Recoverable heat is that which must be removed to maintain the operating temperature after heat losses. Depending on the treatment temperature, a heat pump may be required to upgrade the recovered energy. The level of dissolved oxygen during treatment affects mineral nitrogen contents of treated slurry and nitrogen may be lost or conserved. If nitrification occurs, the heat released is increased but the oxygen consumption is also increased. A computer programme is described, which incorporates a model developed from laboratory experiments, to calculate the oxygen requirement, total heat released, heat losses and net recoverable heat from the aerobic treatment of slurry from the excreta of pigs or cattle. Aeration and anaerobic digestion are compared in terms of characteristics of treated slurry in relation to odour, stability, nitrogen status and pathogen removal. Energy release and recovery are compared in relation to efficiency of utilisation.

1. INTRODUCTION

Biological treatment continues to provide the means of stabilising a majority of organic waste materials and permitting their safe return to the environment. The biological treatments are based on naturally occurring processes which can be controlled and rendered more efficient especially in terms of time and space.

Organic compounds, the substrate for microorganisms, can be degraded either by oxidation in aerobic processes or by fermentation in anaerobic processes. Representing organic material by the glucose molecule, the two processes are illustrated in the following equations.

Aerobic $$C_6H_{12}O_6 \xrightarrow[\text{Oxidation}]{6O_2} 6CO_2 + 6H_2O + \text{Heat Energy}$$

Anaerobic $$C_6H_{12}O_6 \xrightarrow[\text{Fermentation}]{} 3CH_3COOH \xrightarrow[\text{Methanogenesis}]{} 3CH_4 + 3CO_2$$

Combustion $$3CH_4 \xrightarrow{6O_2} 3CO_2 + 6H_2O + \text{Heat Energy}$$

During aerobic treatment oxidative metabolism releases a large proportion of the degraded substrate energy directly as heat. Because of dissipation from the reactor the slurry temperature may only increase slightly above ambient. With good insulation the temperature will rise but metabolic activity and hence further heat release will cease if the temperature reaches the thermal death point of the indigenous microbial population. Therefore aerobic treatment only provides a source of low grade energy.

In anaerobic treatment organic faecal residues are initially fermented to organic acids, in particular acetic acid. This is subsequently converted by methanogenic bacteria to methane and carbon dioxide. In this process less than 5% of the degraded substrate energy is released as heat (1). The remainder is conserved within the methane molecule. With regard to energy recycling, anaerobic digestion has an obvious advantage over aerobic treatment because methane provides a source of higher grade energy.

Because of the value of methane as a fuel to drive an internal combustion engine or, to fire a boiler, the objectives of anaerobic digestion normally include methane production in addition to those of waste stabilization and pollution control. Therefore in most installations a continuous culture system is essential in order to provide a continuous stream of gas of a constant $CH_4:CO_2$ ratio, or calorific value.

Aerobic treatment of organic materials is normally restricted to the composting of relatively solid organic wastes, or aeration of liquid wastes, to control atmospheric soil, and water pollution. The production and recovery of low grade heat energy released in the process is unlikely to be a primary objective. It should only be regarded as a means of recovering some of the costs of treatment. Aerobic processes use oxygen more efficiently in continuous culture systems than in batch treatment and more heat energy is recoverable on a continuous basis. However because of physical and mechanical limitations, composting of solid wastes is more commonly practised on a batch basis. Only liquid wastes are normally suited to a continuous culture system.

Until comparatively recently little attention has been given to the possibility of heat recovery from aerobic treatment of liquid organic wastes of which only those which are relatively concentrated are deserving of attention. Slurries of farm animal excreta are in this category and provide problems of odour, soil and water pollution. Aerobic treatment has been examined especially as a means of odour control by oxidising and preventing the further formation of odoriferous compounds which are formed in anaerobic fermentation. It is, therefore, useful to quantify the amount of heat evolved in achieving the various objectives of aerobic treatment and to assess the proportion of the evolved heat which can be recovered and utilised.

2. HEAT PRODUCTION

2.1 Heat from Metabolism

Heat produced during aerobic metabolism is proportional to oxygen consumption in both the oxidation of organic compounds by heterotrophic bacteria and the oxidation of ammonia by the autotrophic nitrifying bacteria. Cooney, Wang and Mateles (2) measured heat evolution by heterotrophs supplied with a variety of organic substrates. Heat production averaged 4.03 kWh/kg of oxygen consumed. Similar studies on nitrification are not available, but it can be estimated that approximately 25% of the energy released from the overall reaction is conserved as chemical energy in adenosine triphosphate (ATP) (3). The remaining 75% will be released as heat and is equivalent to about 1.19 kWh/kg of oxygen consumed. The oxygen consumed by heterotrophic respiration can be measured by the amount of chemical oxygen demand (COD) degraded. Provided that the amount of COD lost by volatilization is small, which is usually the case in aerobic treatment of animal slurries, the amount of COD removed is a close approximation of the heterotrophic oxygen requirement of the treatment system. As long as denitrification is prevented, by maintaining the level of dissolved oxygen within the aerated slurry above 10% saturation at all times, the oxygen consumed by nitrifying bacteria can be calculated from the quantity of nitrate nitrogen in the treated slurry. The total oxygen consumption, together with treatment time, level of dissolved oxygen in the aerated slurry and the reaction temperature, forms the basis of aerobic reactor design. These design and operating parameters have been examined in a series of continuous culture experiments (4).

A series of equations (4), based on a model derived from the Monod(5) description of the specific growth of microorganisms relate the quality of treated slurry to the mean treatment time and reaction temperature when growth is not limited by the dissolved oxygen level.

In the systems studied the mean treatment times were all greatly in excess of those necessary to maintain a maximum microbial growth rate. Microbial growth was always substrate limited and hence there was no proportional relationship between metabolic activity and reaction temperature. It was necessary to describe the heterotrophic oxygen consumption by three equations which correspond to the three temperature ranges psychrophilic, mesophilic and thermophilic.

Nitrification incurs an additional oxygen requirement. However an active nitrifying population could only maintain itself at reaction temperatures below 40°C and mean treatment times greater than 2 to 3 days. The proportion of total slurry nitrogen oxidised varies from 40% at 15°C to 70% at 25°C but falls again to 35% at 40°C. The additional oxygen requirement for nitrification can be calculated from equation (1).

Nitrification oxygen demand = 4.57pNin
(kgO_2/kg N oxidised) (1)

Where p is the proportion of total slurry nitrogen oxidised. Nin the total Nitrogen in slurry, kg.

Thus the total oxygen requirement for aerobic treatment, assuming a N:COD ratio of 0.049:1, of piggery slurry is described by equations 2 to 4 and is illustrated in figure 1.

Oxygen requirement (15°C, kgO_2/kg CODin) $= 0.621 - \frac{0.547}{1 + 0.14R} + 0.224p$ (2)

Oxygen requirement (25°-45°C, kgO_2/kg CODin) $= 0.465 - \frac{0.333}{1 + 0.4R} + 0.224p$ (3)

Oxygen requirement (50°C, kgO_2/kg CODin) $= 0.555 - \frac{0.429}{1 + 0.7R}$ (4)

Where R is the mean treatment time, days.

It is important to recognise that allowing nitrification increases the total oxygen demand of the treatment system by about 30% but only increases the metabolic heat output by about 10%.

The total heat released during metabolism can be described by a modification of equations 2, 3 and 4, with the addition of a factor for nitrification (derived from equation 1) to equations 2 and 3.

Metabolic heat (15°C, kWh/kg CODin) $= 2.50 - \frac{2.20}{1 + 0.14R} + 0.264p$ (5)

Metabolic heat (25°-45°C, kWh/kg CODin) $= 1.87 - \frac{3.42}{1 + 0.4R} + 0.264p$ (6)

Metabolic heat (50°C, kWh/kg CODin) $= 2.24 - \frac{1.73}{1 + 0.7R}$ (7)

2.2 Heat from Agitation and Aerator Motor

Mechanical energy from the aerator shaft will also be transformed to heat and absorbed into the slurry. This can account for up to 80% of the aerator power requirement for aerators without a gearbox. If a submerged aerator is used all the energy from the motor will be transferred to the slurry. This can be calculated as follows:-

Heat from aerator (kWh/d) $= \frac{X(Oc+On)}{E}$ (8)

Where Oc and On are the oxygen requirement for the heterotrophs and nitrifiers respectively. X is the proportion of aerator electrical energy consumption contributing to the slurry heat. E is the aerator oxygenation efficiency, which may be between 1 and 5 kgO_2/kWh.

3. HEAT LOSSES

Not all the heat evolved by heterotrophic and nitrifying microorganisms is available for extraction and utilisation. Heat is lost from the surfaces of the treatment tank and the liquid surfaces as well as in the exhaust air. Some heat is used to increase the temperature of the incoming slurry to the treatment temperature.

3.1 Heat Losses from Treatment Tank Surfaces

The treatment tanks can be of various shapes. The surface area should be as small as possible. To achieve it with a cylindrical tank, the diameter and height of the tank should be equal. In practice, however, this does not always suit the mixing characteristics of an aerator.

The insulation factor ('U' value) of the treatment tank surfaces depends on the tank construction and insulating material used. With continuous culture treatment of slurry an insulated lid may not be necessary because a constant depth of foam usually remains on the liquid surface. The foam layer then provides good insulation. Heat losses from surfaces of the tank and slurry can be calculated from the following equation.

Heat loss from surface $= 0.024\ (S_1U_1 + S_2\ (U_2 + U_3))\Delta t_1$ (9)
(kWh/d)

Where Δt_1 is the temperature difference between the inner and outer temperature of the tank (^oC). S_1, S_2, are the surface areas of the tank walls and the top or bottom of the tank respectively. U_1, U_2, U_3 are the insulation values of the walls, tank base and foam layer on the slurry respectively ($W/m^2\ .\ ^oC$).

From this equation it can be shown that the surface losses are relatively small. They range from 0.4 to 2.0% of the heat evolved by microorganisms in the treatment of piggery slurry (60g/1TS) for 3 to 14 days mean treatment times and a reaction temperature between 15°C and 50°C when the 'U' value of the surfaces is 0.5 $W/m^2.^oC$ and ambient temperature 0°C.

3.2 Losses of Heat in Slurry

Excreta leaves the animal body at about 37°C but will be expected to cool rapidly to the ambient temperature level. The rate of cooling will increase with addition of cold water from drinkers and rainwater from roofs and yards.

Thus the temperature of slurry entering the treatment tank must be increased to the treatment temperature. The heat is then lost in the treated slurry when it is discharged from the reactor. This is the largest heat loss and can exceed the heat evolved. The quantity of heat involved be calculated as follows:-

Heat lost in treated slurry $= 1.163qdc\Delta t_2$ (10)
(kWh/d)

Where q is the volume of slurry per day (m^3/d); d is slurry density (kg/m^3); c is slurry specific heat capacity ($Mcal/kg^oC$); Δt_2 is the difference between the temperature of the input and that of the output slurry from the treatment tank (^{o}C).

The effects of dilution, lowering the slurry TS from 60g/1 to 15g/1, on heat lost as a proportion of the metabolic heat released, for treatment temperatures between 15 and 50°C, assuming an input slurry temperature of 10°C, are illustrated in figure 2. It is clearly undesirable to lower the TS below the maximum which the aerator is capable of mixing and aerating. With currently available aerators this maximum is about 60g/1. It is also evident that the heat loss significantly increases as the treatment temperature is increased, and therefore, as the difference in temperature (Δt_2) between the input slurry and treatment temperature is increased.

Thus, in addition to minimising the slurry volume by restricting dilution, it is also worth considering methods of retaining the heat of fresh slurry and possibly introducing a heat exchanger between the input and output slurries.

3.3 Heat Losses by Exhaust Air

Large volumes of air, supplying oxygen for microorganisms are needed for aeration. The volume is determined by the oxygem demand of the

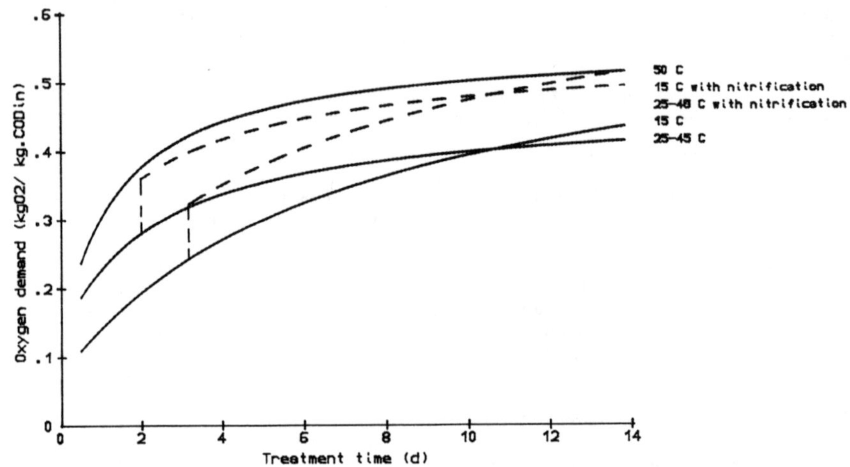

Fig. 1 - Effects of mean treatment time and temperature on oxygen demand.

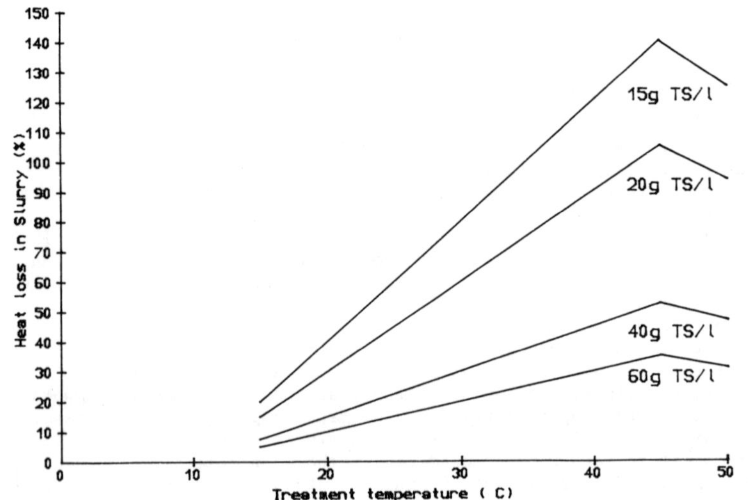

Fig. 2 - Effects of treatment temperature and TS of fresh slurry on the percentage of metabolic heat lost in treated slurry.

microorganisms, the type of aerator and characteristics of the slurry.

Woods et al (6) calculated that for a given volume of air the oxygen transfer coefficient (KLa) should be less than 175/h to achieve a treatment temperature of 55°C for piggery slurry of 45g/lTS with a mean treatment time of the slurry in the treatment tank of 2 days. The KLa value changes with the type of aerator, the treatment tank and the characteristics of the slurry. Only a few manufacturers provide the values of oxygen transfer rate for slurry treatment. Manufacturers do not normally provide KLa values for aerators, but some provide the percentage of oxygen transferred from air into liquid.

The following equation indicates the heat losses in the exhaust air from forced aeration.

$$\text{Heat loss in exhaust air (kWh/d)} = (Oc + On)\left[\frac{120}{Z}(k_2 - k_1) - \frac{k_2}{3.6}\right] \qquad (11)$$

Where Oc and On are the oxygen requirement of the heterotrophs and nitrifiers respectively (kg O_2/d). K_1 the enthalpy of input air (kJ/kg Air). K_2 the enthalpy of exhaust air (kJ/kg Air). Z the percentage of oxygen transferred from air into slurry.

The heat lost in the exhaust air from the treatment tank is related to the change in enthalpy between the input and exhaust air. The enthalpy of the input air is dependent on its temperature and himidity. Thus when submerged aerators are used, or the treatment tank is an intergral part of the animal house, ventilation air of the animal house can be used to conserve heat.

To minimize the air heat losses, the enthalpy of the input air and percentage of oxygen transferred from the air into the slurry should both be as high as possible.

Figure 3 illustrates the influence of the oxygen transfer on the amount of heat lost in the exhaust air as a percentage of the metabolic heat released by heterotrophs.

4. HEAT RECOVERABLE

In a system in which heat is not extracted, the operating temperature is determined by the treatment time, the occurence of nitrification and the amount of heat lost. When the operating temperature is controlled by the extraction of heat, the amount of heat recoverable is equal to the difference between the heat produced (calculated from equations 5, 6 or 7 plus equation 8) and the heat lost (calculated from the sum of the losses obtained from equations 9, 10 and 11).

These equations, together with similar equations for the oxygen consumed in the treatment of cattle waste (7) have been used to develop a computer programme. In response to a series of questions, about the quantity and quality of slurry, aerator performance, and required reaction temperature, the programme provides data about the oxygen required, heat released and net recoverable heat.

Results from this programme are illustrated in figure 4. In this example the extractable heat from slurry from 5000 pigs has been calculated for a range of mean treatment times between 0.5 and 14 days and a range of reaction temperatures between 15 and 50°C. It has been assumed that the pig excreta is diluted approximately 1.5 times to give a slurry containing 60g/l TS, 80g/l COD and 4g/l total N. The temperature of slurry entering the reactor is 10°C and ambient air temperature 0°C.

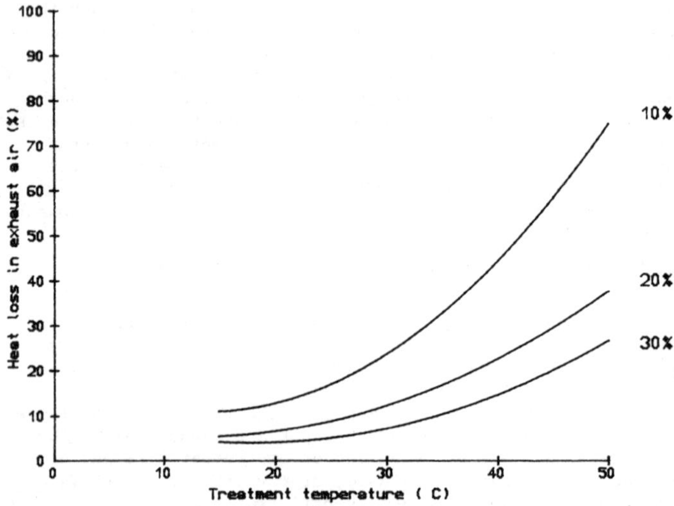

Fig. 3 - Effects of treatment temperature and oxygen transfer efficiency on the percentage of metabolic heat lost in effluent air.

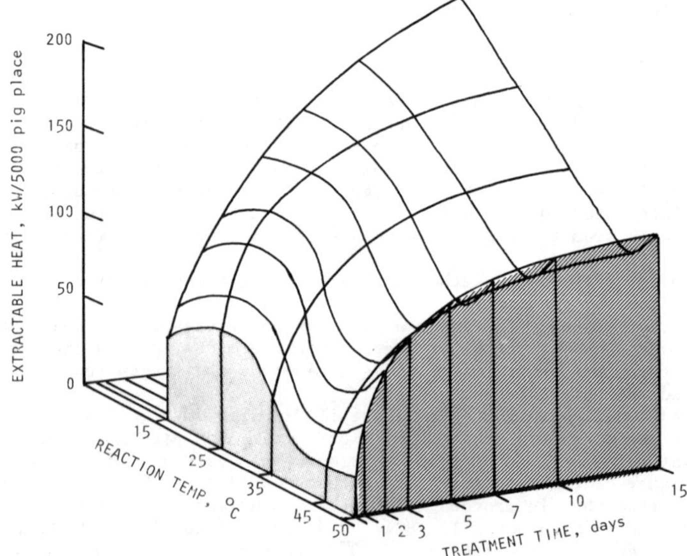

Fig. 4 - Effects of mean treatment time and temperature on extractable heat.

Influent air has a relative himidity of 50% but effluent air is water saturated and at the reaction temperature the aerator oxygenation efficiency is $2.5kgO_2/kWh$ and the oxygen transfer rate is 30%. Dissolved oxygen level is maintained just below 0.1% saturation in order to inhibit nitrification.

5. COMPARISON OF AERATION AND ANAEROBIC DIGESTION

5.1 Treatment Objectives

Anaerobic digestion will normally be used for pollution control in farm situations where methane production and utilisation is a major objective. The design and operating conditions are therefore restricted in terms of mean treatment time and temperature in continuous culture systems. Piggery slurry is usually digested at $35^{o}C$ for 7 to 10 days and cattle slurry at $35^{o}C$ for 14 to 20 days. Cold digestion over an extended period, although possibly satisfying some pollution control objectives does not permit the collection and utilisation of gas.

Aerobic treatment systems are primarily designed to meet a range of pollution control objectives. Recovery of low grade heat energy is secondary and variable, depending on the operating parameters dictated by the requirement to control pollution, especially odour. Continuous culture systems are most efficient in the use of energy for aeration and are essential for continuous heat recovery. However, the requirements for varying degrees of pollution control and stability of treated slurry can be met by a range of treatment times and operating temperatures which will result in different amounts of recoverable heat energy.

5.2 Odour

Most odour compounds contribute to the soluble 5 day biochemical oxygen demand (BOD_5) of the slurry and there is a correlation between soluble BOD_5 and odour offensiveness as measured by an odour panel (8).

The soluble BOD_5 behaves as a growth-limiting substrate in aerobic treatment and its residual concentration is inversly proportional to the mean treatment time (4). It is also, therefore, related to mean treatment time (figure 5) by the following equation:-

$$\text{Odour offensiveness} = 2.313 - 0.42 \log_e R \qquad (12)$$

Where offensiveness is assessed by an odour panel (9) on a scale of 0 to 5.

Odour is rendered faintly offensive (offensiveness rating 2) with a mean aerobic treatment time of only 2 days.

The soluble BOD_5 of anaerobically-digested slurry is usually 400 to 800 mg/l and the theoretical offensiveness remains between 2 and 3, a range supported by odour panel assessments of slurries which have been digested anaerobically for 7-14 days at $35^{o}C$ (10).

5.3 Stability

Anaerobic digestion provides an end product which remains stable during prolonged anaerobic storage. Further degradation and removal of odour can be achieved in aerobic conditions.

To prevent anaerobic degradation and regeneration of odour during storage of aerobically-treated slurry, the system should be designed to either encourage nitrification, or the treatment times should be extended beyond about 10 days. The value of nitrification is that nitrates provide a reservoir of oxygen to continue oxidative metabolism during storage, thereby preventing the regeneration of odour for several months.

5.4 Degradation of Organic Compounds

Anaerobic digestion of piggery slurry for 10 days at 35°C will remove about 40% of total solids (TS) 53% of COD and 83% of the total BOD_5 (11).

At psychrophilic and mesophilic temperatures aerobic cellulolytic activity is low so that removal of TS and COD is lower than that obtained by anaerobic digestion. COD removal of more than 50% requires mean treatment times of more than 15 days at temperatures below 45°C, but only 7 days at 50°C. TS removal is less than 15% at mesophilic temperatures, but exceeds 40% at 50°C, with mean treatment times of about 10 days (4).

However BOD_5 removal is greater aerobically than anaerobically. Total BOD_5 is reduced by more than 75% and 82% with mean treatment times in excess of only 5 days (figure 6) at mesophilic and thermophilic temperatures respectively. The residual soluble BOD_5, the major pollutant in slurry, is lowered to less than 100 mg/l after a mean treatment time of only 1.5 days and less than 10 mg/l after 7.5 days.

5.5 Nitrogen

The total quantity of nitrogen in slurry remains unchanged during anaerobic digestion. Much of the organic N is transformed to ammonia N and the fertiliser value of anaerobically-digested slurry may therefore be enhanced because of the increase in availability of the nitrogen.

The status of nitrogen in aerated slurries is dependent not only on treatment times and temperatures but also on the concentration of dissolved oxygen (DO) (12). At levels below 0.1% of saturation, ammonia N is retained in slurry irrespective of treatment times. At DO levels over 0.1% of saturation, and treatment times less than 3 days ammonia is released and total nitrogen is therefore reduced. When treatment is extended beyond 3 days nitrifying organisms are active at temperatures up to 40°C and nitrate N is conserved in the slurry at DO levels over 10% saturation. Between 0.1% and 10% saturation simultaneous nitrification is restricted or absent and nitrogen is again los t as ammonia gas (figure 7).

5.6 Pathogen Control

Pathogenic bacteria tend to die off in biological treatment systems of all kinds due to lack of competitiveness in the mixed culture. The proportion of survivors progressively reduces as the treatment time increases and as the temperature is raised.

In anaerobic digesters treating piggery slurry 2% of Salmonellae survive after 10 days treatment time at 35°C. Shorter treatment times and lower temperatures are necessary to achieve similar death rates in aerobic systems. At 15°C 90% of Salmonellae are killed in 1.6 days and 99% in 7.6 days treatment.

Anaerobic digestion and aerobic treatment at thermophilic temperatures above 50°C, will accelerate the death rate of bacteria. The fate of viruses and some parasites has not been fully investigated but survival is dependant on time and temperature in both aerobic and anaerobic treatment systems.

Much of the reported work on domestic sewage treatment has concentrated on the survival of pathogens and parasites in the final liquid effluent which is discharged.

Although it is known that many organisms are associated with particulate matter and survive longer in the separated solids.

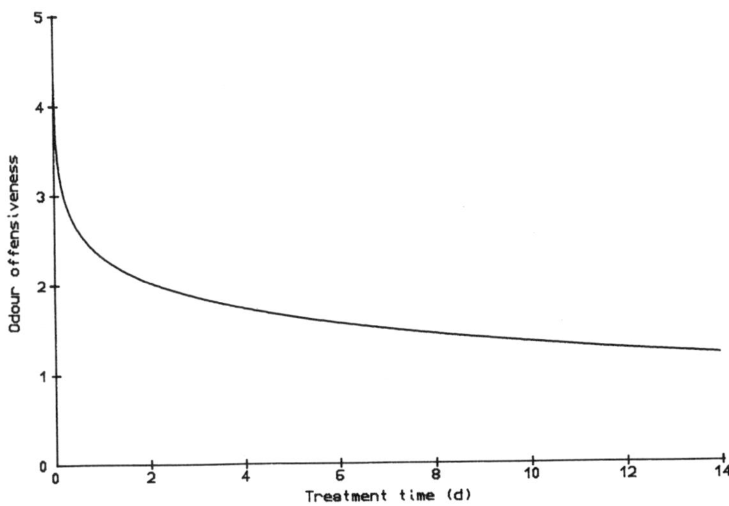

Fig. 5 - Effect of mean treatment time on odour offensiveness of treated slurry

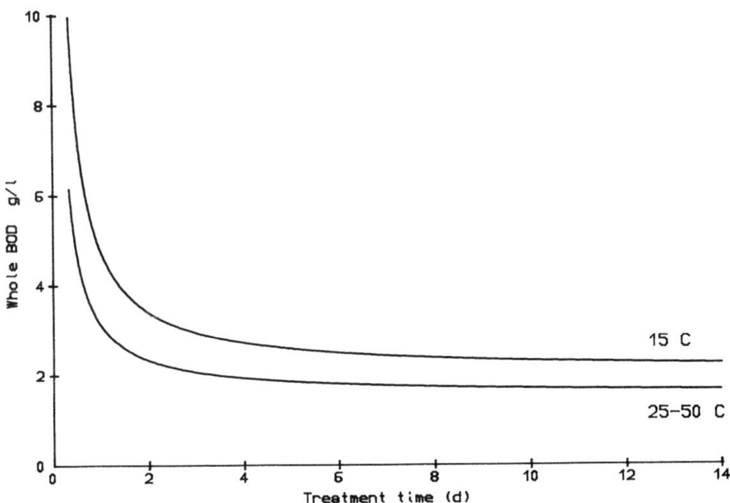

Fig. 6 - Effects of mean treatment time and temperature on residual whole BOD_5

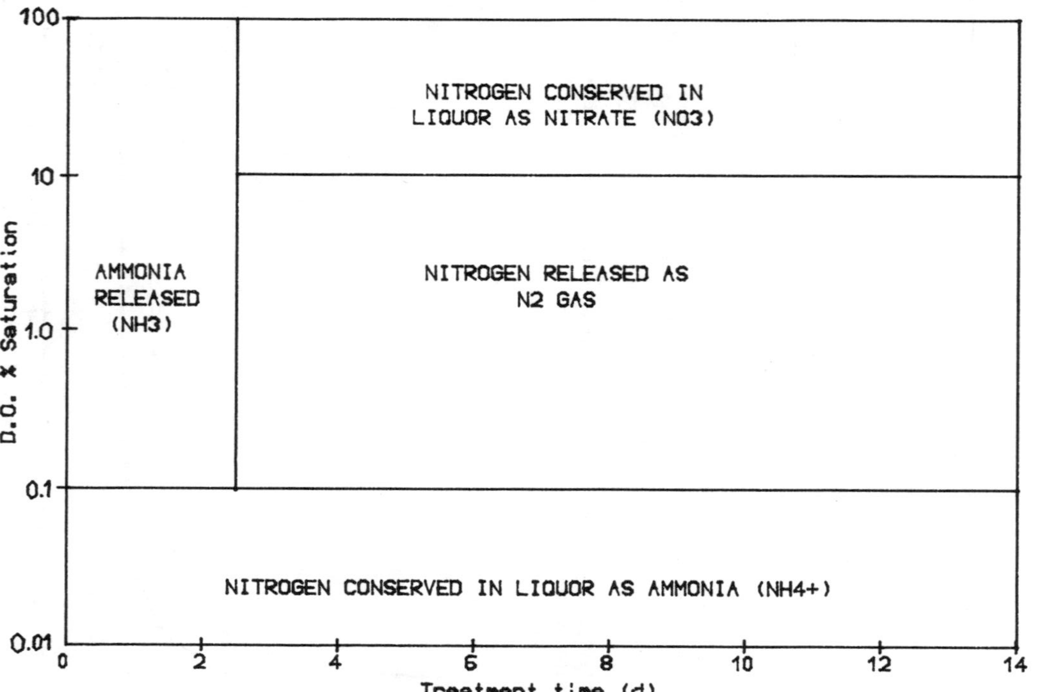

Fig. 7 - Effects of mean treatment time and dissolved oxygen on mineral nitrogen

5.7 Energy Production and Recovery

The quantity of gas produced during the anaerobic digestion of piggery slurry is about 0.3 m^3/kg TS (11).

The gas from anaerobic digestion can be used to heat a boiler, or drive a gas engine to generate electricity. In the UK about 30% of the heat from the boiler is needed to heat the digester. When generating electricity, waste heat from the gas engine can be used. Assuming a 75% conversion by the boiler then the gas will provide about 0.47 kWh per pig place each day of which 0.33 kWh are surplus. With a gas engine and a conversion rate of 30% then about 0.25 kWh electrical energy will be generated per pig place each day.

The heat energy released from aerobic metabolism during treatment can be more than 1.1 kWh per pig place each day (13). Depending on the aerator performance this may be as much as 20 times greater than the energy supplied for aeration. A proportion of this released heat is required to maintain the appropriate reaction temperature, after allowing for heating the incoming slurry and air, and losses from the reactor surfaces. However, up to 75%, 0.8 kWh of heat energy per pig place each day is recoverable.

It must be recognised that, in contrast to the high grade energy obtainable from anaerobic digestion, the heat recoverable from aerobic systems is only low grade energy and hence has more limited value.

REFERENCES

1. FIELDEN, N.E.H. (1983). The theory and practice of Anaerobic Digestion Reactor Design. Process Biochemistry 18 34-37
2. COONEY, C.L., WANG, D.I.C., MATELES, R.I. (1968). Measurement of heat evolution and correlation with oxygen consumption during microbial growth. Biotechnol. Bioengng 11 269-281
3. SHARMA, B., AHLERT, R.C. (1977). Nitrification and nitrogen removal. Wat. Res. 11 897-925
4. EVANS, M.R., DEANS, E.A., HISSETT, R., SMITH, M.P.W., SVOBODA, I.F., THACKER, F.E. (1983). The effect of temperature and residence time on aerobic treatment of piggery slurry - Degradation of Carbonaceous Compounds. Agric. Wastes 5 25-36
5. MONOD, J. (1949). The growth of bacterial cultures. Annual Review Microbiology 3 371-394
6. WOODS, J.L., GINNIVAN, M.J., O'CALLAGHAN, J.R. (1979). Thermophilic treatment of pig slurry. Engineering problems with effluents from livestock. Commision of the European Communities EUR 6249EN p 415-427
7. EVANS, M.R., HISSETT, R., SMITH, M.P.W., THACKER, F.E., WILLIAMS, A.G. (1980). Aerobic treatment of beef cattle and poultry waste compared with the treatment of piggery waste. Agric. Wastes 2 93-101
8. WILLIAMS, A.G. (1981). The biological control of odours emanating from piggery slurry. PhD Thesis University of Glasgow
9. SOBEL, A.T. (1972). Olfactory measurements of animal manure odours. Transactions of the Americal Society of Agricultural Engineers 15 696-699
10. MAFF. Unpublished.
11. HOBSON, P.N., BOUSFIELD, S., SUMMERS, R. (1979). Anaerobic Digestion of Farm Animal Wastes Engineering Problems with effluents from Livestock. Commision of the European Communities EUR 6249 EN p 492-504
12. EVANS, M.R.m SVOBODA, I.F. (1983). Recovery of heat from aerated liquid manure. Seminar - Composting of Organic Wastes - Jutland Tech. Institute Aarhus Denmark 133-135
13. EVANS, M.R., SVOBODA, I.F., BAINES, S. (1982). Heat from aerobic treatment of piggery slurry. J. Agric. Engng. Res. 27 45-50

Appendix to the Paper 10 (Baines, Svoboda & Evans)

CALCULATING THE EXTRACTABLE HEAT FROM AEROBIC TREATMENT OF ANIMAL WASTES

For the design of an aerobic treatment system which has the objective of heat recovery, it is necessary to know the volume and the chemical characteristics of the slurry, the type of reactor vessel and the aerator. It is then important to choose a mean treatment time and reaction temperature which not only satisfies pollution control objectives but also provides the optimum quantity of extractable heat.

The quantity of heat released from microbial metabolism can be estimated using the model equations developed earlier (Evans, Svoboda & Baines, 1982). In order to calculate net extractable heat it is necessary to include in the calculation, the heat transferred into the slurry from the aerator and the heat losses in the effluent air, treated slurry and from the reactor surfaces. Thus calculation of net extractable heat for the most appropriate treatment time and reaction temperature is normally very laborious. This report describes a suitable computer programme for such calculations. It has been written for a BBC micro-computer system B using BBC basic.

The Programme

The computer is programmed to calculate the oxygen demand, total metabolic heat released, heat transferred from the aerator into the mixed liquor, heat losses from the surface, losses in the effluent air and in the treated slurry and the net extractable heat for (R) a chosen range of treatment temperatures and a fixed treatment time or for (T) a range of treatment times and a fixed treatment temperature.

To ensure that the mean treatment time and reaction temperature chosen will produce a useful quantity of heat, the programme provides a three dimensional graphical display of the effects of treatment times, between 1 and 15 days, and treatment temperatures between 15 and 50°C. This is normally displayed in the screen only but a screen dump to a printer is also available.

In order to compute the desired data the user is required to answer all the questions listed below.

Type of animal Pig (P) or Cattle (C) ?
Whole (W) or Separated (S) slurry ?

THE SLURRY
Number of animals ?
Volume of the slurry per animal each day (litres) ?
COD of fresh slurry (g/l) ?
Nitrogen concentration of fresh slurry (g/l) ?
Percentage of N oxidised (%) ?

THE AERATOR
Percentage of electrical energy transferred to the slurry (%) ?
Oxygenation efficiency of aerator (kgO_2/kWh) ?
Oxygen transfer percentage (%) ?

THE REACTOR
Insulation factor "U" (W/m^{2}C) ?
Temperature of influent air (C) ?
Ambient temeprature (C) ?
Temperature of fresh slurry (C) ?

The user is then given the option of a graphical display (G) of the extractable heat (kWh/d) for a range of treatment temperatures and mean treatment times; or a table of oxygen demand, heat released, heat losses and extractable heat after treatment at: a constant treatment temperature (T) and a range of treatment times or a constant treatment time (R) and a range of treatment temperatures.

The results are then calculated from the answers supplied, using the model equations described below.

Oxygen Demand

Whole Slurry

The oxygen requirement by the biomass for the oxidation of carbonaceous and nitrogenous compounds in the whole piggery slurry is calculated from equations published earlier,(Evans, et al., 1983). They resulted from the correlation of a model derived from the Monod equation (Monod, 1949) for continuous culture growth, with data obtained from continuous culture treatment of piggery slurry.

Three equations describe the oxygen demand for treatment at temperatures between 15 and 50°C and treatment times from 0.5 to 15 days.

The proportion of biodegradable COD in cattle slurry is lower than that in piggery slurry, therefore the quantity of oxygen consumed is different (Evans, et. al., 1980). The equations for cattle slurry treatment are based on experimental data for a treatment temperature of 15°C only. The oxygen demand at higher treatment temperatures may be higher. Therefore the quantity of oxygen consumed during cattle slurry treatment at higher treatment temperatures may be rather underestimated in the current programme.

Separated Liquor

For the calculation of the oxygen demand for separated liquor the equations have been amended. Aerobic treatment experiments with separated slurries have not covered the whole range of temperatures and treatment times covered with whole slurries. There are insufficient data to develop a further set of equations. The experiments do, however, show, as expected, a higher oxygen demand per unit of COD in the separated liquor than in the whole slurry (unpublished).

Trials with a rotary screen press separator have shown that the separator removed 67% of the coarse solids from both piggery and cattle slurry. The liquid, however, retains 62% and 50% of the COD of pig and cattle slurry respectively (unpublished data). The coarse solids are very slowly degradable at mesophilic temperatures, having a half-life of more than 6 months.

The equations for separated liquor treatment have been amended on the basis of coarse solids concentration. With pig and cattle separated liquors, the programme calculates oxygen demands of 140 to 220% and 154 to 720% of those expected from whole pig and cattle slurry respectively. In each case the span covers the range of treatment times, the higher percentage values being obtained at the shortest treatment times. These amendments then avoid the gross underestimates of oxygen demand that would arise from the use of the original equations.

Heat Evolution

The quantity of heat evolved during treatment of whole piggery slurry is calculated from the oxygen requirement and specific heat evolution per kilogram of oxygen consumed by the biomass (Evans, Svoboda & Baines, 1982). The same value of specific heat evolution as for piggery slurry is used to calculate heat evolved during treatment of cattle slurry and separated liquors from pig and cattle slurries.

Heat Losses

Heat losses from the system are divided into losses from the tank and liquid surfaces, losses into the treated slurry and effluent air. It is assumed that the treatment tank is cylindrical with the diameter equal to its height and the insulation coefficient ("U" value) is the same for all the surfaces. The programme assumes that the heat energy of the discharged treated slurry and effluent air is lost.

Extractable Heat

The quantity of heat that can be extracted from the treatment system is then calculated as the sum of the metabolic heat and heat from the aerator minus the sum of heat lost.

Table 1 and 2 illustrate the actual print-outs obtained from the example given in the paper and illustrated in Figure 4. Table 1 provides the calculated values for oxygen demand, heat evolved, heat losses and net extractable heat from a treatment system using a mean treatment time of 7 days but for a range of temperatures from 15 to 50°C. Table 2 provides similar values from treatment at 35°C but for a range of mean treatment times.

Table 1 Specimen print-out for fixed treatment time

HEAT FROM AEROBIC TREATMENT OF LIQUID ANIMAL WASTES

5000 Pigs $33.75m^3$ Whole slurry/d. Treatment time 7 days

(COD 80g/l; N 4g/l: O_2 transfer 30%; O_2 eff. 2.5kgO_2/kWh; Aerator energy transfer 8¼%; Insulation 0.2W/m^2°C; Influent air 0°C; Ambient temp 0°C; Fresh slurry 10°C)

T	OD	Heat Released			Heat Losses		Extractable
C	kgO_2/d	Metabolic	Aerator	Surface	Liquid	Air	Heat
		kWh/d					
15	931	3751	298	15	196	128	3709
20	931	3751	298	20	393	182	3454
25	1019	4106	326	25	589	271	3547
30	1019	4106	326	31	785	360	3257
35	1019	4106	326	36	981	472	2944
40	1019	4106	326	41	1178	613	2601
45	1019	4106	326	46	1374	793	2219
50	1302	5248	417	51	1570	1309	2734

THE WEST OF SCOTLAND AGRICULTURAL COLLEGE
S. Baines, I.F. Svoboda & M.R. Evans, Oxford 1984

Table 2 Specimen print-out for fixed treatment temperature

HEAT FROM AEROBIC TREATMENT OF LIQUID ANIMAL WASTES

5000 Pigs $33.75m^3$ Whole slurry/d. Treatment temp 35^oC

(COD 80g/l; N 4g/l; O_2 transfer 30%; O_2 eff. $2.5kgO_2/kWh$; Aerator energy transfer $8\frac{1}{4}\%$; Insulation $0.2W/m^2{}^oC$; Influent air 0^oC; Ambient temp 0^oC; Fresh slurry 10^oC)

R	OD	Heat Released			Heat Losses		Extractable
		Metabolic	Aerator	Surface	Liquid	Air	Heat
d	kgO_2/d			kWh/d			
1	613	2472	196	10	981	284	1393
2	756	3047	242	15	981	350	1942
3	847	3413	271	20	981	392	2290
4	910	3666	291	25	981	421	2530
5	956	3852	306	28	981	442	2706
6	991	3994	317	32	981	459	2839
7	1019	$41\frac{1}{4}6$	326	36	981	472	2944
8	1041	4197	333	39	981	482	3028
9	1060	4272	339	42	981	491	3097
10	1076	4335	344	45	981	498	3155
11	1089	4389	348	48	981	504	3204
12	1100	4435	352	51	981	509	3245
13	1110	4475	355	54	981	514	3281
14	1119	4511	358	57	981	518	3313
15	1127	4542	361	59	981	522	3340

THE WEST OF SCOTLAND AGRICULTURAL COLLEGE
S. Baines, I.F. Svoboda & M.R. Evans, Oxford 1984

DISCUSSION

FINSTEIN: Was the operating temperature assumed to have no effect on heat production?

BAINES: There is no effect of temperature in the mesophilic range used as the system is limited by substrates. The temperature never exceeded 50°C. There can be an advantage of higher temperatures in terms of pathogen removal but this is not a priority of these systems. Digesters should not be used as a solution to disease problems. These should be tackled on the farm itself.

BRUCE: Submerged combustion is a particularly efficient technique for using biogas since 95% of the heat is recovered.

HEAT RECOVERY FROM COMPOSTING SOLID MANURE

P. THOSTRUP
Crone & Koch
Faelledvej 1
DK-8800 Viborg
DENMARK

Summary

Composting is a well known method of waste treatment and is done in piles as well as in plants. During composting a considerable amount of heat is released. Some experiments have been done to recover heat from composting processes, mainly on existing plants. However, in Denmark a research project has been done to design a new composting plant, which is more suitable for heat recovery. The plants, in principle, were built for agricultural use, e.g. composting solid manure, but all kinds of solid organic wastes can be used. The first plant was a totally closed and insulated 10 m^3 container, having a mixing unit inside, with forced aeration, and the heat recovery was done in a cooling tower system. The results of the test made on pig manure, showed that six times the electric energy consumed by the plant was recovered as heat. In 6 days retention time, 40% of the organic matter is degraded, giving the production of 1 kW/m^3 manure in the plant. The heat exchanging system delivered a temperature of 60^oC, which is only 2-4^oC below the plant temperature. These results can still be improved and should be shown by a new circular plant, which at the moment (January-February 1984) is being tested.

1. INTRODUCTION

The oil situation in the last 10 years has resulted in a lot of energy projects. At the Institute of Agricultural Engineering, The Royal Veterinary- and Agricultural University of Denmark, the idea evolved to recover energy from aerobically treated manure. The Danish Ministry of Energy agreed to sponsor a project with the aim of designing and building a composting heat plant for solid manure, which should be economicaly profitable in energy terms.

The first pilot plant was ready in summer 1982, and gave the results presented in this paper. The results were positive, and therefore The Danish Ministry of Energy wished also to sponsor a design for the production of plants. This job was given to:

Crone & Koch
Consulting Engineers k/s
Faelledvej 1, DK-8800 Viborg, Denmark

together with manufacturer:

Mr. Thorkild Larsen
Lille Groentvedvej 23, Mygdal, DK-9800 Hjoerring, Denmark

In this project 3 plants were to be built. All were made commercially and the first will be ready for testing in February 1984.

2. IMPORTANT FACTORS IN AN OPTIMAL COMPOSTING PROCESS

A high specific heat generation is the basic condition for achieving the purpose of a composting heat plant, i.e. as much energy as possible should be produced per unit of time and plant volume. In other words the thermophilic microorganismes should be given optimal conditions. In the following, some parameters are discussed which normally describe the composting process rather well.

2.1 Microorganisms.

It follows, that the larger the number, and the more efficient organisms you have, the larger the amount of organic matter that can be converted per unit of time. At the present time no work on composting solid manure has been seen where the species of organisms are artificially regulated. The species and numbers of organisms which occur depend entirely on the environmental conditions created.

However, there has been some success with inoculation, because it can give a quicker start, particularly when running batch processes. In continuous processes, it is doubtful, whether any effect will be noticed. However, if material is mixed in a continuous process, there will be a sort of inoculation of the incoming material.

2.2 Substrate composition.

The composition of solid manure depends on the type of animal and its feed. Analysis normally divides the organic matter into the following groups:

Sugars, lipids, proteins, cellulose and hemicellulose.

The first three can be regarded as easy degradable, while cellulose and hemicellulose are difficult to degrade. Even this type of analysis will not completely describe a substrate. Another important parameter is the C/N ratio. The optimum is considered to be about 30:1.

However, the rate of availability of the substrate is important from the point of view of micro-organisms. Therefore the best test of a substrate is to use it in an actual composting process.

2.3 O_2 - CO_2 requirement.

Aeration of solid manure brings O_2 to the microorganisms and removes the produced CO_2. Many results have given optimal figures for air supply. (41, 48, 54, 55). Our own experience is, that it is difficult to transfer these data because they are heavily dependent on the composting system used. A better way of adjusting the air supply is to follow the O_2 (or CO_2) content in the exhaust air, when using a closed composting plant. If the O_2 is near zero, the system will be O_2 limited. The question is then, how much O_2 should be in the exhaust air. Our tests together with others shows that, even when the O_2 in the exhaust air is down to 7-8%, no limiting effect can be measured. Normally, the O_2 content in the exhaust air must be between 10 and 15% to avoid O_2 limitation.

2.4 Temperature.

Temperature is an important factor for the activity of microorganisms but cannot otherwise be used as a quantitative process parameter. Reaching a certain temperature in a composting plant indicates only that the energy production is equal to consumption (loss through exhaust air, loss of heat by transmission). That is, a high temperature (70° - 80°C) does not indicate a high composting rate.

By changing the temperature the population of microorganisms also changes. The question is, at which temperature do the most efficient organisms occur. References give a range from 50°C to 60°C. Our tests confirm this, but furthermore no significant drop has been found in the composting rate even up to 65°C.

2.5 Water Content.

From a theoretical point of view, optimal water content should be near 100%, because biological decomposition would occur in the absence of any moisture limitation under these conditions. Yet for technical reasons, the practical moisture content must be less than 100%, in order to build up the necessary free air space or structure. Normally solid organic waste with more than 80% moisture cannot be used.

2.6 Structure - Free Air Space (F.A.S.)

This factor is without doubt the most important and the most difficult factor in composting processes. The composting matrix can be considered a network of solid particles. Around these particles there are the substrates and the water films, - the rest is air.

The following can be used to characterise a given matrix of organic waste:

- free air space
- bulk density
- porosity

It is impossible to give exact figures for them. For solid manure our tests shows, that a bulk density of about 300 kg/m^3 corresponding to a free air space of about 70% is optimal.

2.7 Agitation.

Tests have shown that periodic agitation gives a considerably higher and more stable composting rate, particularly when the substrate is deficient in bulking agents. In the case of solid pig manure it is found that, if the manure consists of 50% dry matter as straw, it is possible to build up a rather stable structure by applying agitation once only. Unfortunately most solid manures contain only 25 - 30% dry matter straw and still have a moisture content of nearly 80%. Of course more straw could be recommended, but this would give problems in the animal production system on the farm.

It can also be stated as follows: Production of heat from the composting plant is less dependent on the type and composition of the solid organic waste used when there is thorough mixing.

2.8 Optimal Conditions.

Our experiments show that the following conditions are important in order to achieve maximum heat recovery:

- the temperature in the plant should be kept at about 60°C.
- the O_2 concentration in the free air space should be about 10%
- the water content of the solid manure should be as high as possible, normally 70-80%
- the solid manure should have a bulk density of about 300 kg/m^3 or an free air space of 70%.

3. METHODS OF EXTRACTING HEAT FROM COMPOSTING PROCESSES.

Recovering heat from a composting plant involves heat exchangers in some way. In attempting to find an efficient system, the following principles were tested in pilot plants:

3.1 Water Chamber System

The tests were carried out in a plant, shown in fig. 1

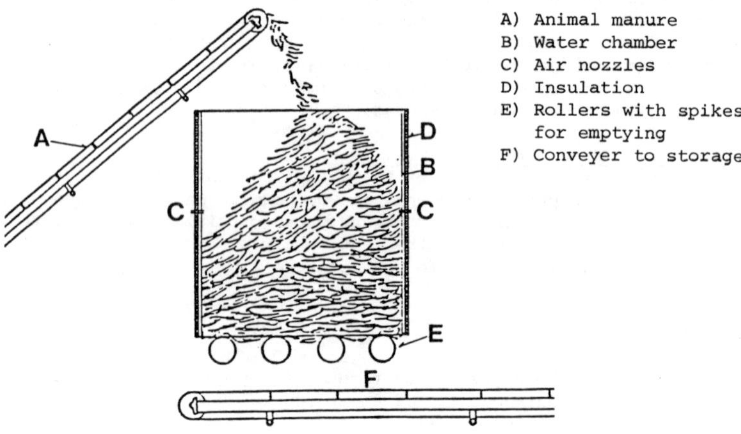

Fig. 1: Pilot plant, water chamber system.

Fig. 2 shows how the water chambers and pipe connections were arranged:

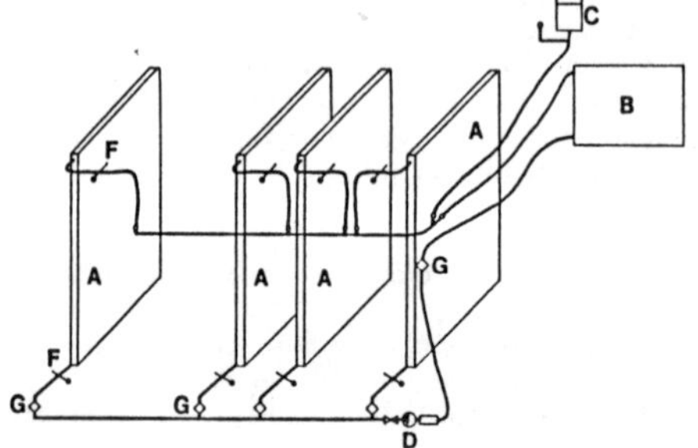

A) Water chambers
B) Central heating system
C) Expansion
D) Pump
E) Calorimeter
F) Thermometers
G) Water meter

Fig. 2: Water chambers and pipe connection for pilot plant in fig. 1.

The results were not encouraging:

- A water temperature of more than 40°C could not be reached. The typical situation was that the temperature of the manure 100 mm from the water chambers was 70°C, and the water temperature was 40°C.
- There was condensation of water along the steel plates, that means that the manure near the plates was wetted, and the manure in the other areas was dried out. The result was an uneven distribution of moisture.
- There will be no room for mixing during the composting process when using heat exchanger plates in the manure.

Even by using different water and air flow patterns, the above results could not be changed.

3.2 The Air to Water System

The idea behind this system is to use the circulating air as an energy carrier. A pilot plant as shown in fig. 3 was set up.

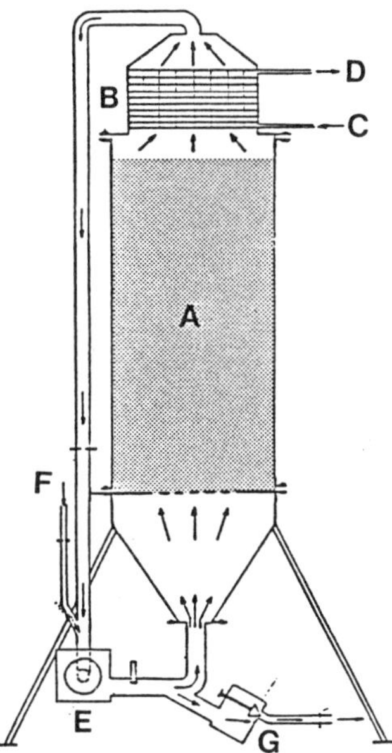

A) Manure
B) Heat exchanger
C) Water inlet
D) Water outlet
E) Blower
F) Air inlet
G) Air outlet

Fig. 3: Pilot plant, air to water system.

The results of this system were positive:

- The temperature of the outlet water could easily reach 55°C (60-65°C in the plant) under normal running conditions.

However there were some disadvantages:

- A lot of water condensed on the heat exchanger, the air was cooled down, which again caused the manure to dry out. It is not possible to moisten the manure by spraying from the top only.

This problem was solved by heat exchanging the exhaust air only. The biological process is not disturbed, but it again means that the heat exchanger should be of the counter flow type, which makes it more difficult to design.

4. COMPOSTING HEAT PLANT, S4

The plant is situated at:
Farmer and Manufacturer
Mr. Thorkild Larsen
Lille Groentvedvej 23, Mygdal, DK-9800 Hjoerring, Denmark.

Mr. Larsen has 200 pigs and 35 sows, giving a manure production of 690 kg per day or 170 kg dry matter per day (including 30% straw, calculated on dry matter basis). If 40% were degraded, he would have a potential for about 16.5 kW from his manure. He uses the energy for heating the farmhouse.

4.1 Description of the Plant

Fig. 4 shows the plant schematically.

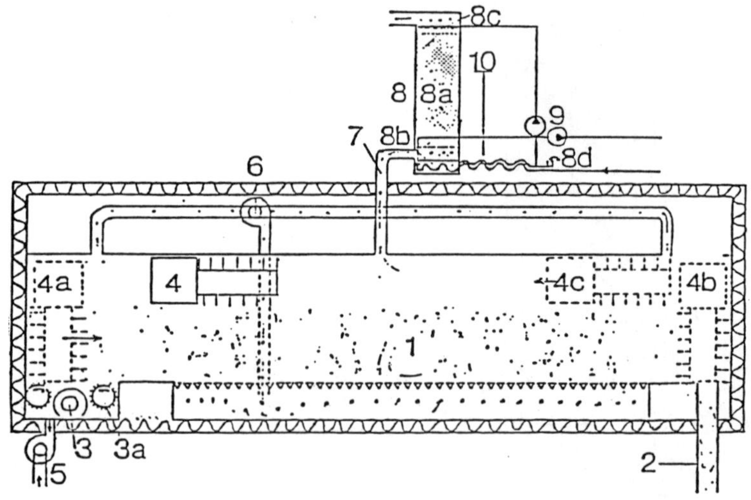

1)	Manure	6)	Fan for air recirculation
2)	Manure inlet	7)	Air outlet
3)	Manure outlet	8)	Heat exchanger/scrubber
4)	Mixing unit	9)	Tandem pumps
5)	Air inlet fan	10)	Water to water heat exchanger.

Fig. 4: Pilot plant type S4.

The following description referring to fig. 4 explains the function:

1. The dry matter of the manure consists of about 30% straw and 70% manure; the water content is about 75% on wet basis and, after mixing, the bulk density is 300 kg/m^3; free air space is about 70%.
2. The manure is fed by a piston type manure compressor. This gives an airtight inlet.
3. The manure outlet is a special auger feed by two rollers. This construction creates an air tight outlet.
4. The mixing unit consits of two vertical contrary rotating spiked cylinders. The timing of 2, 3, and 4 allows the manure to flow through the plant. First, 3 empties its area and 4 tilts down (no. 4a). Then 4 operates through the manure to position 4b, tilts up and returns to its original position. By this operation the maure is moved backwards, freeing an area in the front for new manure, this space is filled by no. 2. Filling is done twice daily and the retention time is 4-6 days.
5. The air inlet fan is adjusted manually to maintain an O_2 percent of between 10 and 15% in the recirculating air.
6. This fan is always running and recirculates 700 m^3 process air per hour.
7. The heat from the composting process is exhausted here. The temperature of the air is 60-70°C with 100% humidity.
8. In this cooling tower, the energy is exchanged, and the air is scrubbed of odour and NH_3. The air will be cooled down from 60°C to about 30°C, which condenses a lot of water. This leaves the sytem at No. 8d. This water is a liquid fertilizer containing about 8% NH_3 (Depending on the type of manure being composted).
9. The pumps give equal flows in the two circuits, so that 10 functions properly.

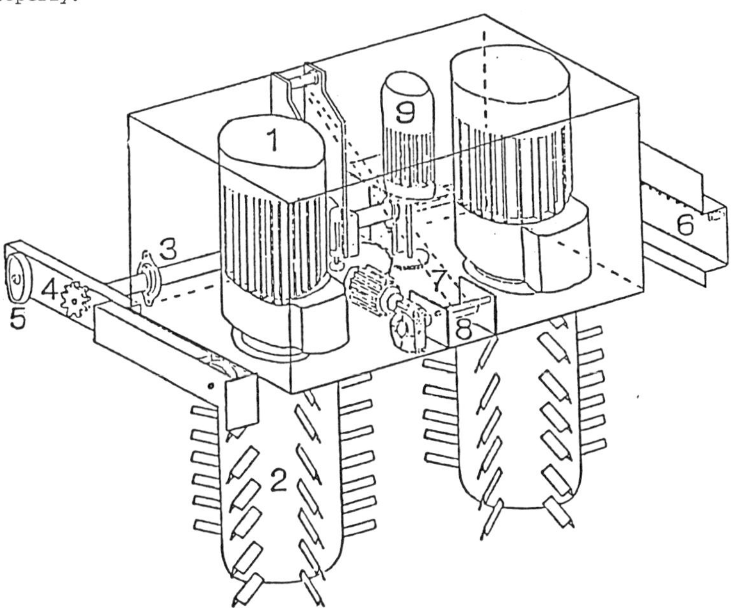

Fig. 5: Mixing unit.

1. Electric motors with direct mounted gears for driving the roller (2)
2. Spiked rollers.
3. Double shaft containing tilt axle and drive-shaft.
4. Chain wheel, which is geared to the chain in the rail (6).
5. Wheels (4 in no.) to carry the entire unit.
6. Rails.
7. Belt to operate the tilting.
8. Winch for the belt 7.
9. Electric motor for propulsion.

The mixing unit is an important part of the composting plant (no. 4 in Fig. 4). It has these two main functions:

- Mixing of the manure
- Transport of manure through the plant

Fig. 5 shows how the mixing unit is designed.

4.2 <u>New design heat exchanger/scrubber</u>

The cooling tower heat exchanger (No. 8, fig. 4) was not quite as good as expected:

- The counter flow water-to-water heat exchanger (No. 10, fig. 4) was not efficient enough.
- The energy loss through 8c, fig. 4 was too high.

A so-called double cooling-tower heat-exchanger/scrubber was made as shown in fig. 6.

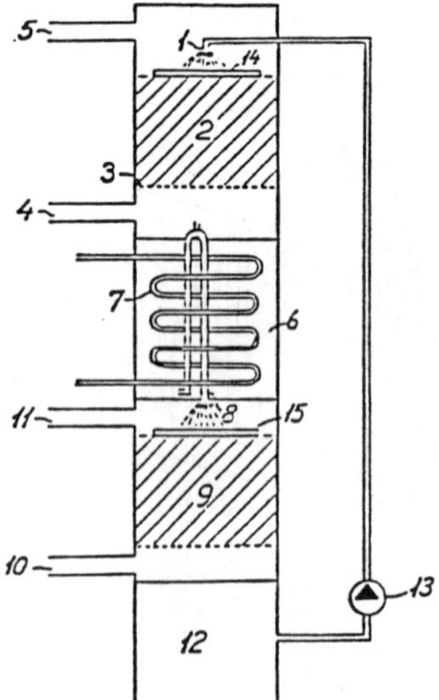

1. Water sprayer
2. Packing material for exhaust air cooling tower.
3. Perforated plate.
4. Air from composting plant.
5. Exhaust air (cooled down).
6. Water.
7. Exchange pipe to central heating system.
8. Overflow pipe.
9. Packing material for incom-ingair cooling-tower.
10. Incoming air.
11. Heated air to composting plant.
12. Water sump (cold water).
13. Pump.
14. Distributing plates.
15. Distributing plates.

Fig. 6: Double cooling-tower

The system shown in fig. 6 has the following advantages:

- The amount of energy which cannot be extracted in the exchanger, (7) heats up the in-coming cold process air.
- The internal water circuits of the cooling towers are cooled (maximum) down at 12 by the in-coming air at 10. This means a minimal loss of energy at 5.

4.3 Results of the S4 Composting plant

A test period of 21 days gave the following results:

- S4 can degrade 40% of the organic matter with 6 days retention time.
- S4 can produce 1 kW/m^3 manure in the plant.
- Temperature of the water of the central heating system is only 2-4 ^oC below plant temperature (normally 60-65 ^oC).
- S4 produces energy equavalent to 6 times the electricity consumed
- Energy produced by the calorimeter:
  8459 MJ = 2350 kWh = burning 300 l of oil = 4.7 kW on average of the 21 day period
- Electricity consumed (numbers refer to fig. 4):

	kWh	% of Total
Mixing Unit No. 4	48	11
Manure Inlet No. 2	22	5
Manure Outlet No. 3	5	1.1
Recirculation Fan No. 6	200	46
Air Inlet Fan No. 5	106	24
Pump No. 9	55	13
Total	436	

436 kWh/21 days = 20 kWh/day = 0.81 kW

- The energy balance is as follows:

Power for heating Farm House	4.7 kW
Loss by Heat Transmission, calculated	1.6 "
Loss in Process Air, calculated	2.8 "
Loss in Hot manure, calculated	0.5 "
Total Biological Energy Produced	9.6 kW

This means:

- that 4.7 x 100/9.6% = 49% is utilized
- that 4.7 x 100/(4.7 + 2.8)% = 63% efficiency for the heat exchanger.
- power factor 4.7/0.81 = 6

5. DISCUSSION

The plant is designed for 10 m^3 of manure but until now only 5 m^3 have been reached because of technical problems with the piston filler .

This is bad for energy economy:

- loss of energy via transmission is too high compared to production.
- consumption of electrical energy is comparatively high because all the motors are designed for 10 m^3 manure. The fact is that the consumption only would rise to about 1 kW with 10 m^3 in the plant.

If the plant contains the planned 10 m^3 manure, it would give the following energy balance:

Total energy production is doubled:	19.2 kW
Loss by heat transmission	1.6 "
Loss in process air	4.0 "
Loss in hot manure	1.0 "
Available for heating	12.6 kW

The results would be:

- Utilized power of total	(12.6 x 100)/19.2%	= 66%
- Heat exchanger efficiency	(12.6 x 100)/(12.6 + 4)%	= 76%
- Power Factor	12.6/1	= 12

As mentioned earlier a plant, type S5, is under construction. In this plant the above results should easily be reached. A schematic diagram is given in fig. 7.

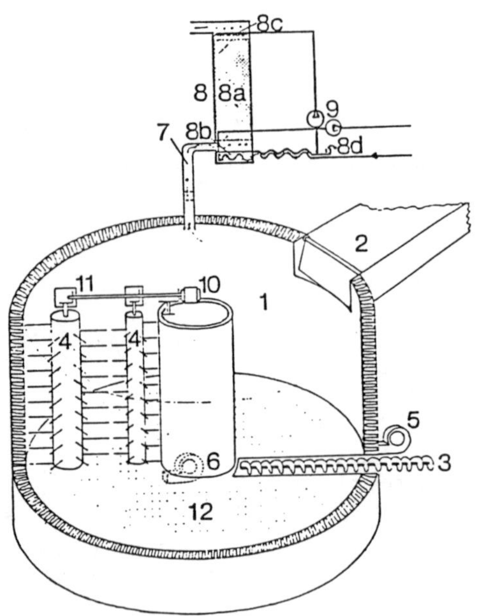

Fig. 7: Composting heat plant type S5. Explanation of numbers corresponds to fig. 4.

The type S5 has several advantages compared to the type S4 plant:

- Smaller surface/volume ratio. This means less loss of heat due to transmission.
- Simpler, cheaper and more reliable power transmission to spiked roller.
- Easier to regulate.
- Cheaper in materials.
- Easier to build.

All this means that S5 is comming closer to a commercial stage.

REFERENCES

1. AHRENS, E., FARKASDI, G., IBRAHIM, I. (1963). Einige Kriterien zur Charakterisierung der Müllrotte. Institut für Landwirtchaftliche Microbiologie der Liebig-Universität Giessen.
2. ANDERSEN, THERPAGER KARL (1971). Varme og fugttransport foredragsmanuskript, SBI.
3. ANDERSEN, PREBEN E., JUST, ARNOLD (1979). Tabeller over fodermidlers sammensætning m.m. Det Kgl. danske landhusholdningsselskab.
4. ASBY, J.D. (1976). An Introduction to Gas Chromatography. Pye Unicam Ltd.
5. ASLYNG, H.C. (1976). Klima, Jord og Planter. DSR Forlag 1976.
6. BAADER, W., SCHUCHARDT F. (1975). Untersuchungen zur Entwicklung eines technischen Verfahrens für de Gewinnung von Feststoffen aus tierrischen Exkrementen. Grundlagen der Landtechnik. Nr. 2 1975.
7. BACKHURST, J.R., HARKER, J.H. (1974). Evaluation of Physical Properties of Pig Manure. Journal of Agricultural Engineering Research 19, page 199 - 207, 1974.
8. BECH, KAJ (1967). Champignonkulturen, dyrkningsvejledning. Særtyk af Gartner Tidende 83: 43-667-674, 47-729-732, 1967.
9. BECH, KAJ (1977). Forsøg med gennemløbet, oprystet kompost, med og uden organisk tilskud, pakket i "små" og "store" afdrivningskasser. Upubliceret.
10. BROCK, THOMAS D. (1978). Thermophilic Microorganisms and Life at High Temperatures. Springer Verlag 1978.
11. CANN, ERIC E., STUMPF, P.K. (1976). Outlines of Biochemistry.
12. CARPENTER, PHILIP L. (1977). Mikrobiologi, Fourth Edition.
13. CHANG, A.C., RIBBLE , J.M. (1976). Particle-size Distribution of Livestock Wastes, Managing Livestock Wastes 1976.
14. CHRISTENSEN, JOHANNES (1976). Håndtering af affald fra husdyrholdet. Landbrugets Driftsøkonomiudvalg.
15. CLIFTON, G.P. (1960). The Self-Heating of wet Wool. New Zealand Journal of Agricultural Reseach, December 1960.
16. COONEY, C.L., WANG, D.I.C., MATELES, R.I. (1968). Measurement of Heat Evolution and Correlation with Oxygen Consumption during Microbial Growth. Biotechnology and Bioengineering Vol XI page 269-281, 1968.
17. COPPAERT, I., VERDONCk, O., DE BOODT, M. (1975). Composting of Bark Pulp Mills and the Use of Bark Compost as a Substrate for Plant Breeding part I. Compost Science Sept-Oct, 1976.
18. COPPAERT, I., VERDONCK, O., DE BOODT, M. (1975). Composting of Bark Pulp Mills and the Use of Bark Compost as a Substrate for Plant Breeding part II. Compost Science Winter 1976.
19. CZERNEY, P., SCHMIDT, B. (1971). Chemisch-Physikalische Veränderungen von Siedlungsabfällen währende Ihrer Biothermischen Entsuchung in Versuch-Rottezellen des IFK Dresden. ISWA Informationsbulletin.
20. DANSK STANDARD (1971). DS hæfte 2. Fysiske størrelser og måleenheder.
21. DIN (1952). Durchflussemessung mit genormten Düsen, Blenden und Venturiedüsen. 1963 udgave.
22. ZÜRICH, EAWAG, ROLLE, G., ORSANIC,,B. Eine neue Methode zur Bestimmung der abbaubaren und der recistenten organischen Substanz.
23. ECKMAN, DONALD P. (1957). Industrial Instrumentation.
24. EILAND, TINN, NISSEN, T. VINCENTS. (1978). Snart bedre udtryk for den mikrobiologiske aktivitet i jorden (ATP måling). Dansk planteavl 13. dec. 1978, page 16.

25. FINSTEIN, M.S., SULER, D.J. (1976). Effect of Temperature, Aeration and Moisture on CO_2 Formation in Bench-Scal, Continuously Thermophilic Composting af Solid Waste. Applied and Evironmental Microbiology. Feb. 1977 Vol 33, No. 2 p. 345-350.
26. FINSTEIN, M.S. (1978). Composting Process Temperature: Conflict between the fastest possible Disinfection and organic Matter Stabilization. Department of Environmental Science, Ruthers University Newo Brunswich, New Jersey.
27. FINSTEIN, M.S., MORRiS, M.L. (1979). Anaerobic Digestion and Composting: Microbiologicul Alternatives for Servage Sludge Treatment. ASM news (American Sociaty Microbiology) 45 (1) page 44-48 1979.
28. GALLER, WILLIAM S., DAVERY, CHARLES B. (1969). High Rate Pultry Manure Composting with Sawdust. Livestock Waste Management.
29. GLATHE, HANS. IBRAHIM, ISMAIL (1972). Stoffwechelvorgänge bei der Lagerung von Frischkompost aus Siedlungsabfällen. Müll und Abfall No. 4, 1972.
30. GLOULD, MICHAEL S., DRNEVICH, RAYMONG F. (1978). Autothermal Thermophilic Aerobic Digestion. Journal of the Environmental Engineering Division, April 1978.
31. GOLUEKE, CLARENCE G. (aprox. 1971). Composting a Study of the Process and its principles.
32. HASSELBALCH, J. (1979). Filtrering af væsker. Smede og Maskinarbejderen (SMEA) nr. 11, 23. maj 1979, page 3-7.
33. HAVE, HENRIK (1979). Regional analysis of Potential Energy Production from Agricultural Wastes. Meddelelse nr. 37 fra Jordbrugsteknisk Institut.
34. HELMER, RICHARD (1973). Thermische Vorgänge und Gasumsetzungen bei der Müllkompostierung - versuche mit Modell - Rottezellen. ISWA Information Bulletin, juni 1973.
35. HENRIKSEN, JØRGEN (1978). In vitro teknik til bestemmelse af kvægfoders energiværdi. Licentiatafhandling fra Husdyrbrugsinstitutet, Den Kgl. Veterinære Landbohøjskole.
36. IPSEN, E.J. (1979) Malkekvægets ernæring. DSR 1979.
37. ISRALSEN, M. (1972). Kemisk og fysisk struktur af tørret grønfoder. Særtryk af Tidsskrift for Landøkonomi nr. 2, 1972.
38. JERIS, JOHN. REGAN, RAYMOND (1973). Controlling Environmental Parameters for Optium Composting, Part I. Compost Science Jan.-Feb. 1973.
39. JERIS, JOHN. REGAN, RAYMOND (1973). Controlling Environmental Parameters for Optium Composting, Part II. Compost Science March-April 1973.
40. KREMER, PER (1976) Komposteringshastighed og -ydelse ved forceret, aerob kompostering, Hovedopgave ved Jordbrugsteknisk Institut, upubliceret.
41. LAI, F.S., MILLER, B.S. (1979). Composting of Grain Dust. ASAE and SCGR paper No. 79-3011.
42. LANDBRUGETS INFORMATIONSKONTOR (1977). Fodermidler og Foderkvalitet.
43. MÜLLER, KARL W. (1979) Elektro Radial ventilatoren. Brochure.
44. NORDISK VENTILATOR CO. A/S (1974). NOVENCO, Ventilation - luft-konditionering. Katalog.
45. PALLIO, FRANK (1977). Energy Conservation and Heat Recovery in Wastewater Treatment Plants. Water and Sewage Works, February 1977.
46. PRICE, W.J. (1978). A Course in the Priciples and Practice of Infared Spectroscopy. Pye Unicam Ltd.
47. POPEL, F. (1964). Das Wesen der Vergärung organischer Stoffe bei der Kompostierung. Müll- und Abfallbeseitigung Band 2.

48. RASMUSSEN, C. RIBER (1962). Komposteringsmetoder, Kuldioxyd-ophobning i champignonkompost, Kan det betale sig at være omhyggelig? Særtryk af Gartner Tidende 79 nr. 6, 11, 21 og 28.
49. RASMUSSEN, R.E. (1964). Elementær måleteori. Gjellerups Forlag 1964.
50. ROTBAUM, H.P. (1960). Heat output of Thermophiles occurring on Wool, June 1960.
51. SCHUCHARDT, FRANK (1977). Einfluss des Ökologischen Faktors "Struktur" auf die kompostierung von Flüssigmist - Feststoff - Gemengen. Dargestellt am beispiel des Feststoffverfahrens. Wissenschaftliche Mitteilungen der Bundesforschungsanstalt für Landwirtschaft Braunschweig - Völkenrode (FAL), Sonderheft 38.
52. SCHUCHARDT, F., BAADER, W., ORTH, H.W., THAER, R. (1978). Untersuchungen über de Möglichkeiten der Steurung der Selbsterhitzung beim mikrobielen mit dem Zeil maximaler Energieansbeute. Institut für Landmaschinenforschung der Bundesforschungsanstalt für Landwirtschaft, Braunschweig-Völkenrode.
53. SCHUCHARDT, FRANK (1978). Einfluss der Haufwerkstruktur auf den Kompostierungsverlauf, dargestellt am beispiel von Flüssigmist - Feststoff - Gemengen. Grundlagen der Landtechnik No. 2, 1978.
54. SCHULZE, K.L (1961). Continuous Thermophilic Composting I. Applied Microbiology. Vol 10, No. 2. 108-122 - Michigan.
55. SOBEL, A.T. (1966). Physical Properties of Animal Manures Associated with Handling. Management of Farm animal Wastes 1966.
56. STEVENS, ROBERT J., CORNFORTH, JAN S. (1974). The Effect of Aeration on the Gases Produced by Slurry during Storage. J. Sci. Fd Agric. 1974, Vol 25, 1249-1261.
57. STOUT, B.A. (1980). Energy for wold Agriculture. FAO publication.
58. THEARET, LINDA, HARTENSTEIN, RAY and MITCHELL, MYRON F. (1978). A study of the Interactions of Enzymes with Manures and Sludges. Compost Science Jan.-Feb.-1978.
59. THOMSEN, K. VESTERGÅRD (1971). Vurdering af analytiske metoder til bestemmelse af fedtfraktionen i forbindelse med fordøjelighedsforsøg. Ugeskrift for Agronomer No. 26-27, 1971.
60. THOMSEN, K. VESTERGÅRD (1971). Vurdering af analytiske metoder til bestemmelse af fedtfraktionen. Ugeskrift for Agronomer No. 29-30, 1971.
61. THOMSEN, K. VESTERGÅRD (1972). Foderstofanalysen. Ugeskrift for Agronomer No. 16, 1972, 316-323.
62. TEICHTER,A., VON MEYENBURG, K. (1968). Automatic Analysis of Gas Exchange in Microbial Systems. Biotechnology and Bioengineering Vol 10, 1968, 535-549.
63. UNKNOWN, Chemische zusammensetzung: Broilerexkremente, Wiederkäerexkremente. Schweineexkremente.
64. VEMMELUND, NIELS, BERTHELSEN, LEIF (1977). Udnyttelse af komposteringsvarme fra staldgødning. Jordbrugsteknisk Institut Meddelelse No. 28.
65. WIBY, JOHN S. (1957). II Progress Report on High Rate Composting Studies. Progress Report from Industrial Waste Conference Purdue University 1957.
66. VILLADSENS, JENS FABRIKKER A/S (1976). ICOPAL Håndbog.
67. WOODS, J.L., CALLAGHAN, J.R. (1974). A Theoretical Description of Aerobic Treatment. Managing Livestock Wastes 1975.

DISCUSSION

TUNNEY: Have you tried to recover the ammonia that is lost to the atmosphere during composting?

THOSTRUP: This has been studied in Holland with sewage. It should be possible to attain 90-95% recovery by controlling pH in the cooling tower. The exit gases have a humidity approaching saturation and ammonia dissolves in the condensate formed on cooling. However, pH soon exceeds 9.2 above which no more ammonia is taken up. This can be countered by the addition of acid.

STENTIFORD: Have any overall costs been produced for the construction and operation of these plants? If so are these costs less than the savings in energy production?

THOSTRUP: We only have experience with the pilot plant. This is not economic but calculations predict that a plant could be nearly economic on energy alone. It is important to consider the other advantages of composting.

SVOBODA: How mechanically reliable has the plant proved to be after about two years of running?

THROSTRUP: The plant was not reliable and was shut down after six months because of high labour costs. There is now a new type of plant but it has not yet been tested.

FINSTEIN: Water is vapourised during composting and the energy produced is used to drive evaporation. What is the view of the Seminar on the concept of composting to produce a dry residue for fuel?

Several members commented that concern was only with composting as a method for waste disposal. If waste is burnt the fertiliser value is lost and hence this approach does not seem applicable to farms.

NAVEAU: A scheme including integrated uses for residues and their by-products should take into account the variable needs on the farm such as heating, which varies not only with the season but from day to day, and carbon dioxide for glasshouses which is only of use in the growing period. Economic viability of a product is very dependent on potential or possible uses.

CHAPMAN: Have any of those involved in heat recovery from compost considered using the type of heat pumps or engines used in producing mechanical or electrical energy from geothermal or solar energy where small temperature differences may be encountered?

THOSTRUP: The possibility has been investigated but suitable machinery is not available at present. A Stirling engine for example requires temperatures of 600-700 ^{o}C.

SESSION IV

COMPOSTS AND CROP PRODUCTION

CONTROLLED MICROBIAL DEGRADATION OF LIGNOCELLULOSE: THE BASIS FOR EXISTING AND NOVEL APPROACHES TO COMPOSTING

J.M. LYNCH and D.A. WOOD
Glasshouse Crops Research Institute, Worthing Road,
Littlehampton, West Sussex, UK, BN16 3PU

Summary

Composting of straw to produce a substrate suitable for the production of edible fungi is a widespread practice but utilises only a small proportion of straw from arable cropping. The composting process is controlled by regulation of stack (windrow) size, turning at suitable intervals and, in the second high temperature stage, by the use of steam and/or restricted ventilation. Composts prepared in this manner are specific for the growth of the otherwise poorly competitive mycelia of edible fungi. Considerable knowledge has now accrued on the microbial ecology of the process and such knowledge is currently being applied to manipulate mushroom production.

Uncontrolled composting to produce organic fertilizers has generally been considered to be uneconomic but recent laboratory observations show that inoculation of straw with cultures of cellulolytic fungi and nitrogen-fixing bacteria can produce a material with fertilizer, soil conditioning and plant protection values. The functioning of the nitrogenase enzyme is expensive in terms of ATP requirement but the process described yields sufficient carbohydrate energy as sugars to give significant nitrogen fixation. The preferred cellulolytic fungi, *Trichoderma* spp., have potential for the control of root diseases.

1. INTRODUCTION

The problems surrounding the disposal of straw in the UK and elsewhere provide a considerable incentive to devise low cost technologies for upgrading the value of straw. A small part of the UK straw collected is already utilised in mushroom cultivation. Although it is unlikely that the UK mushroom industry will dramatically increase its straw consumption in the near future the technologies employed in mushroom cultivation provide a useful model of the biological, chemical, mechanical and economic problems involved in the handling of low or negative value waste plant residues. This article will review some of the recent research on the microbiological processes involved in mushroom cultivation and the applications of those findings and then consider novel approaches to the microbiological treatment of straw and the potential value and uses of such treated material.

2. MICROBIAL PROCESSES INVOLVED IN THE COMPOSTING OF STRAW FOR MUSHROOM CULTIVATION

The current methods used for the production of compost for mushroom cultivation are described in the article by Flegg (9) at this meeting.

From a microbiological perspective composting is used to produce a selective growth medium and in that sense its production can be compared with the production of other selective growth media in traditional microbiological industries such as alcohol fermentation (beer and wine), cheese and yoghurt manufacturing and various other fermented foods and beverages. The success of all these processes normally depends on the manipulation of the materials to be processed by a combination of environmental, biological and chemical procedures. These manipulations render the substrate selective, often in non-sterile conditions, for the subsequent introduction of the desired organism(s). Clearly the ability to produce a selective media in non-sterile conditions represents a considerable economic benefit when treating low value materials. The composting process employed in mushroom cultivation represents the production of a selective growth medium for a slow growing, poorly competitive fungus (22, 23).

The mushroom industry, in the past by trial-and-error methods, and more recently by the use of chemical and visual criteria such as nitrogen content, pH value and other analyses, is normally able satisfactorily to produce a selective growth medium for the mushroom mycelium. No good critera are yet available for the prediction of crop productivity from such compost. The use of other biologically based straw treatments will also require the development of methods for product evaluation.

The basis for the selectivity of mushroom compost is probably due to a number of inter-related factors including chemical composition, nutrient availability, moisture content and bulk density. More work is required to determine the characteristics of each of these factors. The property of selectivity can rapidly be destroyed either by mismanagement of the composting process or by a variety of chemical or environmental treatments of the final product (19). Since the composting process can be reasonably controlled by a variety of environmental and mechanical manipulations this indicates that success in developing other microbiological treatments of straw may also require attention to the details of the environmental and mechanical factor (temperature, moisture content, bulk density of the substrate etc.) which will affect product formation.

The combined effect of the microbial populations involved in the composting process produces several distinct chemical, structural and mechanical changes in the original starting materials. Dry matter loss is extensive, large amounts of ammonia are volatilised off and the carbon: nitrogen ratio at the termination of composting is around 20:1. Microbial growth is at the expense of any initial soluble carbon and nitrogen compounds and afterwards at the expense of the insoluble plant polymers cellulose and hemicellulose. Little, if any, lignin degradation occurs during composting.

There have been several studies of the types of organisms occurring during the composting process (6, 8). The general conclusions are that the microbial flora builds up rapidly after the initiation of the process and that there is an extremely diverse flora formed comprising several genera of mesophilic and thermophilic bacteria, actinomycetes and fungi. Identification of the major dominant organisms present at various stages has been attempted but a complete listing of the species present is not yet complete. These microbiological studies show an initial rise in the population of mesophilic organisms followed by a later rise in numbers of thermophiles as the stack material begins to undergo thermogenesis. The presence of an ecological succession of microbial types has been observed by all the workers examining composting. In the first (stack or windrow) phase of composting these microbial populations will be subject to large gradients of moisture, temperature, oxygen and carbon dioxide. The turning machinery for stack mixing is used to introduce some degree of homogeneity into the process. The second stage of the process, the "peak heat" phase (9) imposes considerably more uniformity on the microbial populations in the substrate. The substrate is now in thinner layers in trays and subject to well controlled temperature and humidity. This stage encourages the growth of a profuse actinomycete flora and also contributes the major part of the selective nature of the final product (20). This ability to manipulate microbial populations by simple environmental and cultural procedures will have implications in other biological processes utilising straw.

It is now known that *Agaricus bisporus*, the cultivated mushroom, can degrade most of the major plant polymers found in the compost including lignin, cellulose and hemicellulose (17, 23, 24). It has also been shown that the microbial biomass fraction of the compost may serve a significant nutritional role in mushroom production. Microbial biomass is defined in this context as both the living and dead tissues of micro-organisms accumulated in the substrate. It has been shown that the mushroom mycelium can degrade and utilise as sole carbon, nitrogen or phosphate source, the dead cells of bacteria, actinomycetes or fungi (5, 7). The carbon part of the microbial biomass has been estimated to be capable of supplying about 10% of the carbon nutrition of the fungus (21). No similar estimates have yet been made for nitrogen, phosphorus or other minerals but it is likely that microbial biomass acts as a concentrated reservoir of these nutrients. The ability to manipulate the quantity of this biomass fraction during composting procedures may be of significance for crop production. In addition, the realisation of this degradative ability of mushroom mycelium has led to work on the use of protein and lipid rich materials including waste microbial cells, which can be used to supplement mushroom compost (18). The production of microbial biomass in other composting procedures may also have nutritional significance for use of such materials as soil additives or animal feeds.

In efforts to further our understanding of the microbial attack on straw we have recently examined the process by scanning and transmission electron microscopy (1) (Figs. 1-8). These studies have complemented the microbiological ones in visualising the ecological succession of micro-organisms on and within the straw tissue. What is of greater significance is that the micrographs reveal that microbial attack on the substrate can occur by a variety of routes (Figs. 1-3) (1). In addition, the studies show that despite the mixing procedures used during composting, attack on the straw is non-uniform. Even at the end of the cropping process (*ca.* 120 days) some straw cells remain completely undegraded. This heterogeneity of attack is likely to be found in other microbiological systems for straw treatment and procedures for producing greater uniformity of degradation will have to be sought.

The retention of structural integrity of the composted straw is also observed visually in the spent compost after cropping. Although considerable dry matter and mechanical strength loss has occurred, the straw is still recognisable. This illustrates that even the accelerated bio-degration occurring during compost preparation for mushroom growth cannot produce complete conversion.

The other major finding of the microbiological studies is that little is yet known about the identity or metabolic relationships of the organisms involved in straw degradation. There is as yet no correspondence between the microbiological and microscopical studies in assigning the roles of individual organisms. It is also likely (see Fig. 4) that part of the degradative microbial flora relies on other micro-organisms for nutrition. The microscope studies show the presence of a group of organisms involved in direct plant cell wall attack with others associated

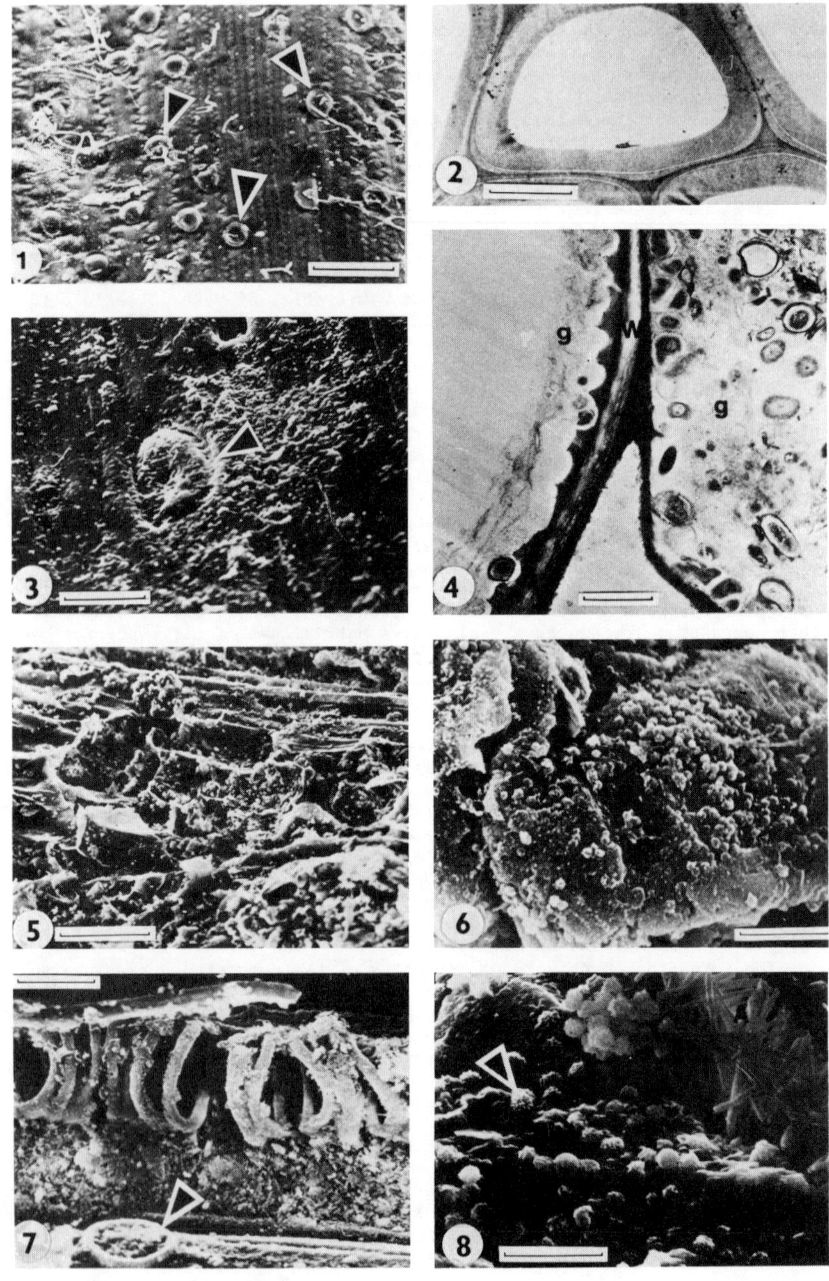
1
2
g
w
g
3
4
5
6
7
8

with them but removed from the primary substrate (Fig. 4). The identity of organisms and their relationships within this community are still poorly defined and worthy of further study. Such studies may lead to improved methods of manipulating such communities for the production of various types of compost.

FIGURE LEGENDS

Fig. 1. Day 0. Raw straw surface. Silica bodies (arrows). Bar mark 50 µm.

Fig. 2. Day 7. Straw showing absence of microbial colonisation. Bar mark 50 µm.

Fig. 3. Day 7. Micro-organisms on straw cuticle. Note erosion around silica body (arrow). Bar mark 25 µm.

Fig. 4. Day 14. Bacteria within straw cells. Note erosion (w, cell wall) and glycocalyx material (g)). Bar mark 1 µm.

Fig. 5. Day 14. Straw surface showing eroded cuticle. Bar mark 100 µm.

Fig. 6. Day 21. Bacteria embedded in walls of straw cell. Bar mark 10 µm.

Fig. 7. Day 14. Straw xylem vessel from which the primary wall has been eroded to reveal the secondary annular thickening. Note separated annulus (arrow). Bar mark 25 µm.

Fig. 8. Day 35. Straw surface with actinomycete spores (arrow) and *Agaricus bisporus* hypha covered with calcium oxalate crystals. Bar mark 5 µm.

3. NOVEL APPROACHES TO AGRICULTURAL/HORTICULTURAL COMPOSTING AT MESOPHILIC TEMPERATURES

In Britain, the majority of the straw produced in arable agriculture is burnt. This is because crop yields are reduced when they have been seeded in the presence of straw, especially in wet autumns and with modern methods of minimal cultivation. The latter methods enable winter crops, which are higher-yielding than spring-sown crops, to be sown where otherwise this would not have been possible. Similarly in the United States, particularly in the Pacific Northwest, minimal tillage is now favoured but the presence of straw, which must be retained for soil conservation purposes, limits crop yields. In these two contrasting farming systems, straw reduces crop yields by about 20% (15). There is thus considerable incentive internationally to find means of handling straw to eliminate this yield penalty. The fundamental problem is that straw left to degrade in the field in an uncontrolled manner becomes a suitable substrate for phytotoxin-producing micro-organisms (15) and plant pathogens, including sub-clinical pathogenic bacteria which invade root cortices (4). Co-operative laboratory studies between the Agricultural and Food Research Council Letcombe Laboratory, the Glasshouse Crops Research Institute and the United States Department of Agriculture at Washington State University have evaluated options to control the degradation of straw.

Notably straw which has been degraded in mineral nutrient solution for 1 month at 20°C has no significant adverse effect on plant growth (Table I). There was a different effect from various types of microbial inocula, although the differences were small. For this purpose alone, these studies showed that provided there is no limit on time or conditions for degradation, there was no particular advantage of inoculation. However, straw being dominantly cellulosic, can be expected to degrade most rapidly when micro-organisms producing copious amounts of cellulase are present.

Table I. Effect of wheat straw on wheat seedling root elongation (13)

Germination substratum	Root elongation (mm)*	
	cv Daws	cv Mardler
Fresh straw	13a	22a
Straw degraded for 1 month	24ab	32b
Soil	38b	32b

In a range of laboratory model systems, it has been shown that there is an advantage of linking the cellulase function to nitrogenase, the enzyme responsible for dinitrogen fixation (10, 16, D.A. Veal and J.M. Lynch, unpublished). The breakdown of cellulose and hemicellulose to simple sugars such as cellobiose and glucose provides energy substrates for nitrogen-fixing bacteria; function of these organisms in soils is normally limited by the availability of such substrates. The nitrogen fixed by the bacteria can become available to the cellulolytic fungi to prevent nitrogen-limitation of the activity. The cycling of C and N thus becomes self-sustaining and hastens decomposition of the straw (Table II).

There is no loss of N from the system and the final product, rich in microbial biomass N is available to plants (11, 12). Trichoderma spp. are now amongst the preferred choices of cellulolytic fungi and Clostridium butyricum, an anaerobe, is the preferred nitrogen-fixing bacterium. The bacteria were isolated from straw where intensive nitrogen fixation, as measured by the reduction of acetylene to ethylene, was occurring, particularly from zones with aerobic/anaerobic interfaces (Table III). The mixture of aerobes and anaerobes appear to co-exist satisfactorily, in part by the aerobes providing respiratory protection to the anaerobes.

Table II. Decomposition of non-sterile straw contained in glass columns at 25°C for 8 weeks (16)

Treatment	Decomposition rate constant, $k(d^{-1})$	N gain (mg)	
		Per g straw lost	Per g original straw
Non-inoculated	0.0096	8.8	2.8
Penicillium corylophilum + Clostridium butyricum	0.0139	11.5	5.0
LSD (P = 0.05)	0.0030	N.S.	0.6

N.S. = not significant

Table III. Acetylene reduction (n moles g^{-1} straw h^{-1}) and numbers of bacteria isolated from straw incubated partially water logged for 4 weeks at 20°C (10)

Position of sample	Mean (10 samples) ± standard deviation	Clostridium butyricum ($x10^{-6}$)
Aerobic zone	20.2 ± 12.6	1.8 (4)
Aerobic/anaerobic interface	34.9 ± 14.4	9.7 (1)
Anaerobic zone	13.2 ± 5.5	5.3 (1)
	LSD (P = 0.05)	2.8

* Figures in parenthesis show percentage of the total bacterial count

A secondary advantage of the community is that polysaccharides (Table IV) are produced which stabilize soil structure. The most useful index of this is when soil aggregates are stable to shaking in water such that little solid is released into suspension (Table V). Volcanic ash from Mount St. Helens in Washington State is a particularly useful model material because it has the same particle size composition as the soil of the region but, because organic matter is absent, it has no inherent soil structure. Initially (14) no advantage could be obtained by microbial inoculation of the straw because the natural microflora associated with the straw were as effective as the inoculants. Subsequent, however, (S.J. Chapman and J.M. Lynch, unpublished) inoculation of straw has improved its soil stabilizing properties.

Table IV. Quantities of alcohol-precipitated polysaccharide extracted from fresh straw and from straw decomposed at various oxygen concentrations (based on the weight of fresh and degraded straw respectively) (2)

Straw sample	Oxygen concentration (% v/v)	Polysaccharide (glucose equivalents) % (w/w)	Carbohydrate in crude polysaccharide % (w/w)
Fresh	-	0.24	35
Decomposed	5	0.48	26
Decomposed	0	0.51	35

Table V. Aggregate stability of Palouse silt loam soil and volcanic ash from Mount St. Helens (14)

Treatment	% solid in suspension*	
	Soil	Ash
Degraded straw	8.5a	25.0c
Undegraded straw	11.0b	38.2d
None (control)	13.6b	40.5d

* Values not followed by the same letter are significantly different (P = 0.05) by Duncan's multiple range test for a variable sample

A tertiary advantage of the inoculants is that the cellulolytic fungi (3) have the potential to control root disease. Thus the inoculum can be multi-functional in minimizing the phytotoxicity of straw, providing plant nutrients and soil stabilizing agents and affording protection against disease.

In testing the feasibility of these studies, horticultural application might take the form of composting in stacks under polyethylene sheeting. This could also be useful in channelling the gas produced to the glasshouse for crops where CO_2 is usually used to enrich the atmosphere. Agricultural application might be by spraying inocula onto straw in the field and composting *in situ*.

4. CONCLUSIONS

The foregoing has illustrated one proven example of composting as part of a successful biotechnological process, that of mushroom production. There is still much scope for further studies in this area, for example supplementation of composts, selection of optimal substrate sources, improvement and selection of *Agaricus bisporus* strains and evaluation of alternative edible fungi for European markets. Basidiomycete fungi might be useful as animal feed and perhaps a process could be developed to produce them by a procedure simpler than that used for conventional mushroom production. This prospect appears realistic as we begin to understand the principles of the manipulation of microbial communities.

Similar basic considerations apply to the development of novel agricultural and horticultural composts. Essentially the inocula would probably be produced in solid-substrate fermentations on lignocellulosic substrates. Ideally these would then be introduced into the larger scale compost, which could be in a polyethylene-sheeted stack or as windrows in the field. An alternative strategy would be to use inocula produced in liquid culture, this might achieve a more even distribution of inoculum but the micro-organisms might be less-well adapted to the substrate. Even if the inoculum were produced on the solid substrate and introduced directly into the field, the remaining substrate that it would carry would give it a better chance of ecological success than conventional microbial inoculants produced in liquid culture where the substrate base is exhausted rapidly. It is our belief that microbial inoculum technology offers exciting prospects in the treament and utilization of lignocellulosic wastes.

REFERENCES

1. ATKEY, P.T. and WOOD, D.A. (1983). An electron microscope study of wheat straw composted as a substrate for the cultivation of the edible mushroom (*Agaricus bisporus*). Journal of Applied Bacteriology 55, 293-304.
2. CHAPMAN, S.J. and LYNCH, J.M. (1984). A note on the fermentation of microbial polysaccharide from wheat straw decomposed in the absence of soil. Journal of Applied Bacteriology, in press.
3. CHET, I., HADAR, Y., ELAD, Y., KATAN, J. and HENIS, Y. (1979). Biological control of soil-borne plant pathogens by *Trichoderma harzianum*. In Soil-borne Plant Pathogens, eds. Schippers, B. and Gams, W., pp. 585-91. Academic Press, London.
4. ELLIOTT, L.F. and LYNCH, J.M. (1984). Pseudomonads as a factor in the growth of winter wheat (*Triticum aestivum* L.). Soil Biology and Biochemistry, in press.

5. FERMOR, T.R. (1983). Fungal enzymes produced during degradation of bacteria. Transactions of the British Mycological Society 80, 357-360.
6. FERMOR, T.R. and WOOD, D.A. (1979). The microbiology and enzymology of wheat straw mushroom compost production. In: "Straw Decay and Its Effect on Disposal and Utilisation", pp. 105-112. Ed. E. Grossbard; Wiley, Chichester.
7. FERMOR, T.R. and WOOD, D.A. (1981). Degradation of bacteria by Agaricus bisporus and other fungi. Journal of General Microbiology 126, 377-387.
8. FERMOR, T.R., SMITH, J.F. and SPENCER, D.M. (1979). The microflora of experimental mushroom composts. Journal of Horticultural Science 54, 137-154.
9. FLEGG, P.B. (1984). Research on mushroom composts in Europe. Seminar on composting agricultural and other wastes. Oxford 1984.
10. HARPER, S.H.T. and LYNCH, J.M. (1984). Nitrogen fixation by cellulolytic communities at aerobic-anaerobic interfaces in straw. Journal of Applied Bacteriology, in press.
11. ITO, O. and WATANABE, I. (1981). Immobilization, mineralization and availability to rice plants of nitrogen derived from heterotrophic nitrogen fixation in flooded soil. Soil Science and Plant Nutrition, 27, 169-176.
12. LETHBRIDGE, G. and DAVIDSON, M.S. (1983). Microbial biomass as a suorce of nitrogen for cereals. Soil Biology and Biochemistry, 15, 375-376.
13. LYNCH, J.M. and ELLIOTT, L.F. (1983). Minimizing the potential phytotoxicity of wheat straw by microbial degradation. Soil Biology and Biochemistry, 15, 221-222.
14. LYNCH, J.M. and ELLIOTT, L.F. (1983). Aggregate stabilization of volcanic ash and soil during microbial degradation of straw. Applied and Environmental Microbiology, 45, 1348-1401.
15. LYNCH, J.M. and ELLIOTT, L.F. (1984). Crop residues. In: Crop Establishment: Biological Requirements and Engineering Solutions. (Ed. M.K.V. Carr), in press. London: Pitmans.
16. LYNCH, J.M. and HARPER, S.H.T. (1983). Straw as a substrate for co-operative nitrogen fixation. Journal of General Microbiology, 129, 251-253.
17. MANNING, K. and WOOD, D.A. (1983). Production and regulation of cellulase of Agaricus bisporus. Journal of General Microbiology 129, 1839-1847.
18. RANDLE, P.E. et. al. (1983). Mushroom compost supplementation. Trials with proprietary and other materials. Mushroom Journal 129, 345-349.
19. ROSS, R.C. and HARRIS, P.J. (1983). An investigation into the selective nature of mushroom compost. Scientia Horticulturae 19, 55-64.
20. SMITH, J.F. (1981). The development of mushroom composting techniques. Report of the Glasshouse Crops Research Institute for 1980, pp. 171-183.
21. SPARLING, G.P., FERMOR, T.R. and WOOD, D.A. (1982). Measurement of the microbial biomass in composted wheat straw, and the possible contribution of the biomass to the nutrition of Agaricus bisporus. Soil Biology and Biochemistry 14, 609-611.

22. WOOD, D.A. (1979). A method for estimating biomass of Agaricus bisporus in a solid substrate, composted wheat straw. Biotechnology Letters 1, 255-260.
23. WOOD, D.A. and FERMOR, T.R. (1981). Nutrition of Agaricus bisporus in compost. Mushroom Sciences 11, part 2, 63-71.
24. WOOD, D.A. and LEATHAM, G.F. (1983). Lignocellulose degradation during the life cycle of Agaricus bisporus. FEMS Microbiology Letters 20, 421-424.

DISCUSSION

FINSTEIN: Is "respiratory protection" a synonym for "anaerobic microsite"?

LYNCH: No. An 'anaerobic microsite' refers to an environment which is generated biologically or abiotically. 'Respiratory protection' is a co-operative action between two or more organisms whereby oxygen is excluded from the immediate environment of the oxygen-sensitive partner.

BOLLEN: In one of your slides you showed a scanning micrograph of mycelium densely set with crystals of Ca-oxalate and you described it as characteristic for Agaricus bisporus. How far is it indicative of Agaricus? Do no other fungi in the same habitat show this phenomenon?

LYNCH: 1) Agaricus will produce oxalate crystals on a variety of media in laboratory cultures; it will produce oxalate crystals in compost cultures (axenic cultures); it is known to produce large quantities of oxalate

2) On the compost surface the bulk of hyphae of Agaricus are covered in oxalate crystals.

3) Other fungi can produce oxalate.

ZUCCONI: You suggested that the biomass is a parameter which may help to assess the success of a process. How would you comment, however, on quantity compared with composition of such biomass, considering that you seem to be confronted by an allelopathic problem (wheat straw affecting wheat culture)? Biomass could represent the success of a single or few organism the metabolic products of which would still remain (with a high degree of probability) incompatible with the growth of wheat.

LYNCH: I agree that we increase total microbial biomass. Qualitatively we increase the population balance of beneficial and harmful micro-organisms. The harmful species include phytotoxin-producers which can be reduced in population size but also beneficial organisms can metabolize the phytotoxins.

DE BERTOLDI: i) Do you have any evidence on the presence of predation between competing microfungi and bacteria?

ii) Why did you choose anaerobic N_2 fixing bacteria in association with cellulose-degrading fungi? Would aerobes such as Azotobacter be suitable for the same purpose?

LYNCH: i) The electron microscopic studies indicate that this is likely in the mushroom compost but we have yet to investigate this in the novel composts.

ii) We have screened a wide range of cellulolytic fungi in association with aerobic and anaerobic N_2-fixing bacteria but only anaerobes appear to be suitable for the N_2-fixation.

RESEARCH ON MUSHROOM COMPOSTS IN EUROPE

P.B. FLEGG
formerly of Glasshouse Crops Research Institute, Littlehampton, Sussex.

Summary

Over the past forty years there has been considerable improvement in methods of preparing composts for mushroom cultivation. Productivity of composts under good conditions has increased during that period from about 100 kg fresh mushrooms per tonne of prepared compost to around 250 kg per tonne.

The main aims of the current research are:-

1. to understand the composting process better in the hopes of making it more reliable and easier to manage
2. to improve the nutritional value of composts to the mushroom and thereby increase crop yields
3. to improve the efficiency of production methods thus increasing the amount of compost prepared from a given quantity of raw materials.

Linked with work on compost production are studies on the nutritional requirements of the mushroom.

Current progress is reviewed in compost microbiology, with special reference to the selectivity of mushroom composts, in the enrichment of mushroom composts and in compost preparation technology. Several areas where future research might usefully be done are discussed.

1. INTRODUCTION

Worldwide production of the mushroom, *Agaricus bisporus*, is currently about 1.0 million tonnes annually. UK production at 65-70,000 tonnes per year is only a small proportion of the total world production yet this country ranks fifth, being exceeded by the USA, France, China and the Netherlands. The UK mushroom industry is valued at about £150 million at retail prices and is one of our most important horticultural crops.

Mushrooms are usually grown on a composted mixture of horse manure and straw bedding. Wheat straw is preferred, but other straws and a variety of animal manures are sometimes used when more traditional materials cannot be obtained. The development over some 300 or more years of methods of preparing mushroom composts has recently been reviewed (7), but the following reviews are commended for a wider understanding of the subject (6,13,15,25 and 29).

The object of composting for mushroom production is to prepare a substrate which will eventually favour the growth of the mushroom in preference to other organisms. During composting micro-organisms are encouraged to utilise the more readily available materials. The resultant substrate, comprising the more resistant components, such as lignin, hemicellulose and cellulose, plus an accumulation of microbial biomass is one on which the mushroom, well-equipped with the appropriate enzymes, can flourish while many of its competitors cannot.

Each year the mushroom industry in the UK consumes about 2-300,000 tonnes of wheat straw in preparing about 4-500,000 tonnes of mushroom compost. Usually composting proceeds in 2 stages. The first, out-of-doors when the wetted straw and manure mixture is made into stacks about 2m wide, 2m high and as long as required. Every 2 or 3 days the compost in the stack is mixed up and re-stacked, a procedure known as 'turning', during which, if required, water and additional materials are incorporated. This stage of composting usually lasts for up to 2 weeks in which time the straw has become much softer and considerably degraded, the water content has reached about 70%, the temperature will have been for much of the time 50°C-80°C and there will probably have been production of ammonia. For the second stage the compost is moved into specially built rooms, either in wooden containers, e.g. trays or shelves 15-20 cm deep, or in bulk when it is 2m deep. During this stage the compost temperature is maintained between 50°C-60°C and large quantities of fresh air are introduced into the building. This stage usually lasts about 1 week and is completed when the evolution of ammonia has ceased, gaseous NH_3 is toxic to the mushroom, and the compost shows little or no tendency to produce heat. The final product, now ready for the introduction of mushroom 'spawn', has, it is hoped, been freed of all mushroom pathogens by the high temperatures experienced, has, to mushroom growers' noses, a pleasant 'sweet' smell, has a water content of about 65-70% and a nitrogen content around 1.8-2.0% on a dry-matter basis. It is dark brown and can be readily broken apart by hand. The physical condition of the compost is very important. For more information on the preparation of mushroom composts see the poster preparation of Gerrits.

Over the past forty years there has been considerable improvement in methods of preparing composts for mushroom cultivation. Productivity of composts under good conditions has increased during that period from about 100 kg of fresh mushrooms per tonne of prepared compost to around 250 kg per tonne. Even today, however, the production of a good mushroom compost requires previous experience and not a little skill and judgement on the part of the composter. To produce mushroom compost ready-to-spawn costs about £12-16 per tonne.

2. RESEARCH AIMS

Because mushroom compost production was well established long before there were serious attempts to understand the biological processes involved, there has been a tendency for new developments to have arisen more from economic and practical pressures than from an improved understanding of the microbiology or chemistry of the process. A large proportion of the research effort on the mushroom crop has been centred on mushroom composts and compost preparation and a serious drawback has been, and to some extent still is, a lack of information on the nutritional requirements of the mushroom.

The main aims of current research are to understand the composting process better in the hopes of making it more reliable and easier to manage, to improve the nutritional value of composts to the mushroom and thereby increase crop yields and to improve the efficiency of the preparation methods thus increasing the amount of prepared compost obtained from a given quantity of raw materials. Linked with work on composting are studies on the nutritional requirements of the mushroom and the enzyme activity exhibited by the mushroom during the various stages of production.

In this review of European work on mushroom composts and composting, while acknowledging the great value of earlier work, I shall be concentrating on work done or in progress in the 1980s.

3. COMPOST MICROBIOLOGY

An aspect often referred to but so far little studied is the 'selectivity' of prepared mushroom composts. Compost selectivity refers to the fact that when properly made, mushroom compost while producing large yields of Agaricus bisporus will support the growth of few other micro-organisms. One of the main reasons for this might be the absence of readily available nutrients (4). Recently Ross and Harris (21) have demonstrated that the selectivity of prepared mushroom compost can be destroyed by heating it to about 60°C or more for several hours, autoclaving at 121°C being the most effective for destroying selectivity. The loss of selectivity was characterised by a rapid increase in the production of gaseous ammonia and the ability of Chaetomium globosum to colonise the compost. Antimicrobial chemicals such as ethanol and chloroform added to the compost and subsequently removed by forced aeration also destroyed the selectivity of mushroom composts. These workers interpreted their results in terms of selectivity being related to the presence of a viable, but dormant, biomass. The effect of high temperatures or antimicrobial chemicals was believed to cause the destruction of many micro-organisms in the compost and so release readily available nutrients on which other micro-organisms could grow and multiply hence removing the selectivity of the compost for the mushroom. Ross and Harris (22) have also shown that inoculating mixtures of moist straw and an organic nitrogen source with cultures of thermophilic fungi greatly aided the development of selectivity. Of four species examined Torula thermophila was particularly effective at lowering the level of gaseous NH_3 level in the substrate and in improving its selectivity. Inoculating conventional mushroom composts during composting with thermophilic fungi was not particularly effective, probably because a large population of suitable composting micro-organisms had already developed.

In addition to the removal of substances toxic to the mushroom, e.g. gaseous NH_3, and to the removal of nutrients, which would be used by the mushrooms competitors, the thermophilic microflora may play yet a further role in the development of mushroom compost selectivity. If I may make a brief reference to some related work from Canada, Tautorus and Townsley (30) have demonstrated that the growth of Chaetomium globosum in mushroom compost can be inhibited by inoculating the compost with a thermophilic Bacillus sp. previously isolated from prepared mushroom composts. Yields of mushrooms were enhanced. The chemical identity of the antibiotic produced by the Bacillus sp. is being investigated.

Studies on the effect of micro-organisms on the breakdown of wheat-straw composts continue, for example, Wood and Leatham (33) and Lynch and Wood (16). A new approach to the subject has been pioneered by Atkey and Wood (1) using scanning electron microscopy (SEM). An ecological succession has been observed with a largely bacterial flora dominating the initial stage and actinomycetes the later stage. Several types of microbial attack on straw were observed, the cuticle and phloem being degraded most rapidly. The distribution of micro-organisms throughout the compost material was very uneven so that, although many of the plant fibres became separated by the non-uniform degradation, the overall structure of the plant materials in the compost was not destroyed. Some areas of straw were found to have suffered little degradation even at the end of mushroom crop production. Clearly the supply of nutrients in the compost for the mushroom cannot be completely exhausted at the end of crop production. Thus the observations made using SEM have confirmed and extended previous work on compost microbiology and promise to add considerably to our understanding of the mushroom composting process in the future.

Micro-organisms are important in mushroom compost not only for the part they play in degradation of the raw material, and in helping to confer selectivity for the mushroom to the compost, but also in serving as a source of nutrients for the mushroom. Wood and Fermor (32) and Fermor and Wood (5) have shown how mushroom mycelium can grow well in cultures in which the sole source of nutrient is acid-killed bacterial cells.

In view of the undoubted importance of micro-organisms in mushroom composts it is extremely useful to have an estimate of the amount of microbial biomass in composts. Sparling, Fermor and Wood (28) comparing several methods of estimating microbial biomass have shown that, based on direct counting of organisms, the total biomass present is 9.23 mg carbon per g of oven-dry compost. Several biochemical methods gave similar values and two methods giving values about 2 and 3 times as large were thought to be overestimates. Atkey and Wood believe that the microbial biomass in composts can provide up to 10% of the nutritional requirement of the mushroom.

4. COMPOST ENRICHMENT

The idea of enriching a prepared mushroom compost with a nutritional supplement has interested mushroom growers and scientists alike for many years. Although the potential for supplementation was amply demonstrated by Sinden and Schisler (23 and 24), practical application has proved difficult. Most materials used as compost supplements e.g. cottonseed meal, soya bean meal, and wheat flour contain readily available nutrients which encourage vigorous microbial activity and a consequent large rise in compost temperature which kills the mushroom mycelium. Work in the USA has led to the development of so-called delayed release supplements which prevent or at least reduce the burst of microbial activity when they are mixed into the compost. Basically these materials are protein-rich meals, e.g. soya bean meal, which have been treated with formaldehyde to denature the protein (2). The production of these substances and, in some countries, their use with mushroom crops is protected by patents.

The American work has led recently to considerable interest in delayed-release supplements for mushrooms in Europe. Randle (17) has recently reviewed the whole field of mushroom compost supplementation and Gerrits (11 and 12) has reported on trials with a range of supplements in the Netherlands as have Randle *et al.* (19) in the UK. The general outlook for these materials is promising, yield increases of up to about 30% have been obtained but the average is perhaps 10-15%. However, results can be variable and the practice of supplementing composts in this way undoubtedly makes additional demands on the farm management. Workers at GCRI in the UK are continuing to evaluate the use of these materials and at the same time are searching for novel and cheaper materials to use as supplements.

5. COMPOST PREPARATION

In view of its practical importance to the mushroom industry it is not surprising that composting methodology has received a major share of the research effort on mushrooms. Among recent developments in this field is the work of Flegg and Randle (8) on the duration of the composting process and on the initial nitrogen content of compost stacks (9) from which has arisen a unifying concept of compost preparation (10). These workers classify all mushroom composting procedures, however diverse, on the basis of the duration of the two main stages of the process and have recently described how this concept should be of use to the practical grower and should allow more logical thinking about mushroom composting. This concept was further discussed recently at the International Conference on Composting of Solid Wastes and Slurries at Leeds University in 1983.

Another recent development, more practical in nature, has been that of bulk treatment of composts. The idea was first put into practice in Italy by Derks (3) and has now been improved and brought to a high level of efficiency in the Netherlands. Briefly in this procedure the compost is moved from the compost yard and loose filled to a depth of about 2 m into a room measuring about 2 m wide 3 m high and 20-30 m long. The floor is slatted and air is blown through the compost via the slatted floor and an air plenum beneath it using a powerful fan. Provision is made for the air to be recycled, with or without a mixing with fresh air, and to be heated and humidified as required. The completion of the second and pasteurising stage of composting under the controlled conditions which can be readily attained in such 'bulk pasteurising tunnels' as they are commonly known, can result in a much more uniform product. The composting process can be more easily controlled than if the compost is stacked in layers of trays or shelves. Although the finished product can be much more uniform in appearance and moisture content, as yet unpublished work at the GCRI suggests that considerable variability in productivity of the compost can occur. Compost from the ends of the tunnel sometimes produces lower yields of mushrooms than compost from the middle. Nonetheless the development has proved successful practically in many European countries and can be expected to be adopted much more widely in future.

Stimulated by the obvious success of the bulk treatment of composts during the second stage of composting there have been attempts to improve the uniformity of conditions during the first stage in stacks on the open yard. Randle and Flegg (20) showed that considerable variation in oxygen content can exist in yard stacks. Trials of forced aeration of yarded compost stacks have been carried out in the UK and continental Europe and probably continue, but so far no published results of such trials have been reported. Attempts by Randle to achieve greater uniformity of compost temperature within the stack by partly covering the stack with insulating 'blankets' have proved only partially successful (18).

Yet another approach to simplify and speed up the composting process is the development of so-called 'rapid' composts pursued both in France and in the UK. One of the great advantages of such composts is the considerable savings which can be achieved in the quantity of raw materials required to prepare a tonne of mushroom compost ready for spawning. Recent research on this topic in France was reviewed by Laborde at the 1979 Mushroom Industry Conference in Yarmouth (14). A low temperature process (L.C.T.) and a high temperature process (H.C.T.) have been the subject of experiment in France. Relatively little time is spent in the conventional first stage of composting on the open yard. The materials are mixed, wetted as required, and moved into the controlled conditions of the specially built pasteurising rooms as soon as practicable. Usually the whole process is completed within a week. While considerable progress has been made in preparing rapid composts, a persistent problem, now being overcome, has been that of obtaining yields of mushrooms comparable to those from conventional composts. Smith (26) reports on the formulation of mixtures suitable for economic short-duration composts and while obtaining commercially acceptable yields of mushrooms points out that the selection of the strain of mushroom could be important in obtaining good yields.

The bulk density of composts is another important factor still awaiting thorough investigation. Generally the greater the weight of compost which can be compressed into a given volume of bed space, the higher will be the yield of mushrooms. There is still uncertainty about the optimum bulk density required to give high yields yet avoiding the harmful effects of poor aeration in excessively dense composts and work on this aspect is now in progress at GCRI. With rapid composts made with wheat straw the problem of obtaining a high bulk density is difficult because the limited duration of composting restricts straw tissue degradation with the result that the composts tend to be 'springy' and difficult to compress.

Obtaining adequate build-up of microbial biomass during the short time required to prepare these composts appears to be another problem. However, the search for consistently high yields from rapid composts continues and success appears to be within grasp.

Even though mushroom composts have traditionally been made from stable bedding for centuries, the raw materials themselves are continually changing. The effects of weather and the storage conditions generally on batches of stable bedding delivered throughout the year are a perennial problem for mushroom composters, but as new wheat varieties are introduced to agriculture the character of their straws may or may not favour composting (31). Changes in horse management can also have this effect. A recent trend towards bedding horses in shredded waste-paper has stimulated studies work on the effect of this material on mushroom composts (27). It seems that a mixture containing a small proportion of paper can be composted quite satisfactorily, but a content of paper-based material much above 10% could be harmful.

6. THE FUTURE

Several exciting new developments for the mushroom farmer are in prospect.

Increased understanding of the nutritional requirements of the mushroom and of the role of micro-organisms in producing selectively could lead to the production of composts more favourable to mushroom growth. The manipulation of the components of the microbial biomass during composting and the enrichment of prepared composts, with microbial products are both possibilities.

The development of criteria for the successful assessment of the cropping potential of composts at the time that mushroom spawn is planted, instead of having to wait until crop production is well under way, would assist compost quality control and facilitate a more rapid correction of errors in compost production. A microbiological criterion to complement the existing chemical and physical criteria may arise from recent and future work on compost selectivity.

The greater precision in control of conditions apparently required in the successful preparation of rapid composts (26) could probably be obtained if bulk pasteurisation techniques were used. A successful amalgamation of the technologies of rapid composting and of bulk pasteurisation combined with the relatively new method of growing mushrooms in deep troughs 1 m deep and 2 m wide could result in a dramatic improvement in mushroom production methods. Following a short period of mixing and wetting, the compost raw materials could be filled into the trough which is in effect a modified version of a bulk pasteurisation tunnel. The composting process should be completed in the trough within a week and the period from assembling the raw materials to planting the spawn could be less than 10 days. Crop production would proceed without moving the compost to a new building. The need for much expensive machinery would be virtually eliminated, crop hygiene would be much improved and production costs drastically reduced.

REFERENCES

1. ATKEY, P.T. and WOOD, D.A. (1983). An electron microscope study of wheat straw composted as a substrate for the cultivation of the edible mushroom (Agaricus bisporus). Journal of Applied Bacteriology 55, 293-304.
2. CARROLL, A.D. and SCHISLER, L.C. (1974). Delayed release nutrients for mushroom culture. Mushroom News, 22, 10-11.
3. DERKS, G. (1973). 3-phase-1. Mushroom Journal, 9, 396-403.
4. EDDY, B.P. and JACOBS, L. (1976). Mushroom compost as a source of food for Agaricus bisporus. Mushroom Journal, 38, 56-67.
5. FERMOR, T.F. and WOOD, D.A. (1981). Degradation of bacteria by Agaricus bisporus and other fungi. Journal of General Microbiology. 126, 377-387.
6. FLEGG, P.B. (1961). Mushroom composts and composting: a review of the literature. Report of the Glasshouse Crops Research Institute, 1960. 125-134.
7. FLEGG, P.B. (1984). Development of the composting process for mushroom cultivation. International Conference on composting of solid wastes and slurries. Leeds University, 1983. In the press.
8. FLEGG, P.B. and RANDLE, P.E. (1980). Effect of the duration of composting on the amount of compost produced and the yield of mushrooms. Scientia Horticulturae, 12, 351-359.
9. FLEGG, P.B. and RANDLE, P.E. (1981a). Relation between the initial nitrogen content of mushroom compost and the duration of composting. Scientia Horticulturae, 15, 9-15.
10. FLEGG, P.B. and RANDLE, P.E. (1981b). A unifying concept of compost preparation for the cultivation of the mushroom Agaricus bisporus. Mushroom Science 11, 341-349.
11. GERRITS, J.P.G. (1982). Enkele niewe produkten voor het bijvoeden. De Champignon cultuur, 26, 463-469.
12. GERRITS, J.P.G. (1983). New products for compost supplementation. Mushroom Journal, 126, 207-213.
13. HAYES, W.A. and NAIR, N.G. (1975). The cultivation of Agaricus bisporus and other edible mushrooms. The filamentous fungi. Vol. I Industrial mycology, (eds. J.E. Smith and D.R. Berry), Arnold, London, pp. 212-248.
14. LABORDE, J. (1980). Rapid substrate making. Mushroom Journal, 94, 349-361.
15. LAMBERT, E.B. (1938). Principles and problems of mushroom culture. Journal of Agricultural Research. 62, 415-422.
16. LYNCH, J.M. and WOOD, D.A. (1984). Microbial degradation of lignocellulose. Seminar on composting agricultural and other wastes. Oxford, 1984.
17. RANDLE, P.E. (1983a). Supplementation of mushroom composts - a review. Crop Research 23, 51-69.
18. RANDLE, P.E. (1983b). The effect of insulated covers on compost-stack temperatures. Scientia Horticulturae, 20, 53-59.

19. RANDLE, P.E. et al. (1983). Mushroom compost supplementation. Trials with proprietary and other materials. Mushroom Journal, 129, 345-349.
20. RANDLE, P.E. and FLEGG, P.B. (1978). Oxygen measurements in a mushroom compost stack. Scientia Horticulturae, 8, 315-323.
21. ROSS, R.C. and HARRIS, P.J. (1983a). An investigation into the selective nature of mushroom compost. Scientia Horticulturae, 19. 55-64.
22. ROSS, R.C. and HARRIS, P.J. (1983b). The significance of thermophilic fungi in mushroom compost preparation. Scientia Horticulturae, 20, 61-70.
23. SCHISLER, L.C. and SINDEN, J.W. (1962a). Nutrient supplementation of mushroom compost at spawning. Mushroom Science, 5, 150-164.
24. SINDEN, J.W. and SCHISLER, L.C. (1962b). Nutrient supplementation of mushroom compost at casing. Mushroom Science, 5, 267-280.
25. SMITH, J.F. (1981). The development of mushroom composting techniques - a review. Report of the Glasshouse Crops Research Institute, 1980, pp. 171-183.
26. SMITH, J.F. (1983a). The formulation of mixtures suitable for economic short-duration mushroom composts. Scientia Horticulturae. 19, 65-78.
27. SMITH, J.F. (1983b). Paper-bedding from horses and poultry. Mushroom Journal, 127. 245-249.
28. SPARLING, G.P., FERMOR, T.F. and WOOD, D.A. (1982). Measurement of the microbial biomass in composted wheat straw, and the possible contribution of the biomass to the nutrition of *Agaricus bisporus*.
29. STOLLER, B.B. (1954). Principles and practice of mushroom culture. Economic Botany, 8, 48-95.
30. TAUTORUS, T.E. and TOWNSLEY, P.M. (1983). Biological control of olive green mould in *Agaricus bisporus* cultivation. Applied and Environmental Microbiology, 45, 511-515.
31. TRUSSLER, M.E. (1982). Wheat straw for composts. Mushroom Journal, 119, 392-394.
32. WOOD, D.A. and FERMOR, T.F. (1981) Nutrition of *Agaricus bisporus* in compost. Mushroom Science, 11, part 2, 63-71.
33. WOOD, D.A. and LEATHAM, G.F. (1983). Lignocellulose degradation during the life cycle of *Agaricus bisporus*. FEMS Microbiology Letters, 20. 421-424.

DISCUSSION

LE ROUX: How much of the significant heat produced during compost production is used and is there further scope for this?

FLEGG: None of the heat produced during composting is recovered for use except in the composting process. The heat required in the indoor, pasteurisation phase of mushroom composting is derived from the heat of the compost and much is vented away during the process.

Heat pumps are used on mushroom farms to recover heat from vented warm air from cropping rooms.

LOPEZ-REAL: Are the yield limiting factors for mushroom production known? If it is dependent on microbial biomass, how important is nitrogen given the considerable losses that occur in composting horse manure and straw with a poor C : N ratio, favouring N losses.

FLEGG: The nutritional requirements of the mushroom during fruit body production in composts are not known with sufficient precision to be able to determine any limiting nutritional factors.

THE USE OF TREE BARK AND TOBACCO WASTE IN AGRICULTURE AND HORTICULTURE

O. VERDONCK, M. DE BOODT, P. STRADIOT and R. PENNINCK
Laboratory of Soil Physics, Soil Conditioning and Horticultural Soil Science
Agricultural Faculty, State University of Ghent, Belgium

Summary.

As a consequence of the increase in demand for organic materials in agriculture and horticulture, it is becoming more and more necessary to recycle the organic wastes of different industries. Pulp and sawmills consume enormous quantities of wood for industrial activities. Recent statistics indicate that about 150-200,000 m^3 of bark is available in Belgium, where the tobacco industry also produces about 20,000 m^3 of organic wastes. As these two organic wastes contain more than 85 % organic matter, their use for agricultural and horticultural purposes is very interesting. The fresh material cannot be used as such because of nitrogen immobilization, phytotoxic compounds and/or high salt content. Therefore a composting process is necessary in order to condition these organic wastes. Composting experiments were carried out with different mixtures of bark and tobacco waste and with the addition of mineral nitrogen. After optimizing the composting process, experiments growing different ornamental plants have been carried out in order to evaluate these mixed composts. These trials indicate that a mixture of 10 % tobacco waste and 90 % bark can be used as a horticultural substrate. 100 % tobacco waste can also be used as an organic soil conditioner.

1. INTRODUCTION.

The high demand for organic soil conditioners in agriculture and growing media in horticulture make it necessary to recycle different organic wastes.

As we have large quantities of tree bark available, this material can be recycled to produce a valuable product. The fresh material cannot be used as such because of nitrogen immobilization and the presence of phytotoxic compounds. Composting is necessary with the presence of a nitrogen source in order to lower the C/N ratio and to eliminate the phytotoxic compounds.

Tobacco waste can be used as a nitrogen source because of its high contents of total and mineral nitrogen. Different mixtures of bark and tobacco waste were composted, to find the optimum mixture for composting. With the results of the laboratory composting simulator, composting experiments on a pilot scale were set up in order to follow the composting parameters as a function of time. These composts were tested as growing media for ornamental plants.

Also trials will be set up using tobacco waste as organic fertilizer.

2. MATERIAL AND METHODS

2.1. Bark.

Ten years ago bark was a waste product of papermills and sawmills. Nowadays it is no longer a waste product but is a good starting material for making substrates and organic soil conditioners, but it cannot be used directly because of nitrogen immobilization and phytotoxic compounds. Therefore composting for 2-4 months is necessary with added nitrogen depending on the kind of bark. After the optimal composting period, this material can be used as a pure growing medium or can be mixed with peat or pine litter. The most important chemical and physical properties are given in tables 1 and 2.

Table 1. Important chemical properties of bark compared with peat.

Material	pH H_2O	Ec 1/25 in µS	% org. matter	% Ntot	C/N ratio	CEC meq/100 g
fresh bark	5.5	250	90-93	0.4-0.6	75-110	40-50
composted hardwood bark	6.5	500	91	1.1-1.3	30-40	70-75
composted softwood bark	6.7	550	93	1.1-1.3	30-40	70-80
peat	4-5	200	94	0.9	± 50	120-140
pine litter	5-5.5	250	75	1.5	± 25	70-80

The results in table 1 show that composted bark has a pH near neutral and that the cation exchange capacity is lower than that of peat. The other chemical properties such as organic matter, C/N ratio and total N are more or less the same. The salt content is a little higher than those of peat and pine litter, but still in the good range.

Table 2. Physical properties of bark compared with peat and pine litter.

Material	Volume weight in g/cm^3	Total pore space	Volume percent water at a tension of			Vol % air	Vol % easily available water
			10 cm	50 cm	100 cm		
composted bark	0.121	94.5	35.7	29.5	28.4	56.8	6.3
peat	0.104	93.1	80.1	43.7	36.2	13.0	36.3
pine litter	0.137	90.6	47.8	38.2	34.4	42.8	9.8
composted bark/ peat 75/25	0.186	87.2	65.5	47.6	43.2	21.7	17.9
composted bark/ peat 50/50	0.146	89.9	64.9	42.7	37.2	25.0	22.2

Table 2 shows that the physical properties of composted bark are completely different from those of peat, but more or less the same as those of pine litter. This means that the volume percent of air of composted bark and pine litter is greater than 40 % and that the easy available water is less than 10 %. In order to ameliorate the physical properties of bark, a mixture with 25 or 50 % by volume is recommended.

2.2. Tobacco waste.

The tobacco industry has 10 % waste as tobacco leaf ribs and 3 % as tobacco dust. In total we have about 2,000 t or 11,000 m^3 leaf ribs and

1,580tor 5,800 m^3 dust. The total tobacco waste in Belgium is 16,800 m^3, which is 10 % of bark. The chemical analysis is given in tables 3 and 4.

Table 3. Important chemical properties of tobacco waste compared with bark.

	Bark	Tobacco leaf ribs		Tobacco dust	
		1	2	1	2
pH	6.37	6.90	7.00	7.27	6.88
Ec (µS 1/25)	240	9,125	7,150	3,675	5,100
% organic material	93.23	84.92	87.94	65.29	77.41
moisture content (%)	60-70	15	16	6	7
acid binding value	-	24	46	13	10
nicotine (%)	-	0.56	0.49	1.06	0.40
N (%)	0.59	3.27	2.51	3.00	3.41
C (%)	51.79	44.46	49.31	35.29	41.84
C/N ratio	88	14	20	12	12

Table 4. Some elements in tobacco waste

		Bark	Tobacco leaf ribs		Tobacco dust		Maximum acceptable in soils
			1	2	1	2	
Ntot	in %	0.591	3.275	2.512	3.003	3.410	-
NH_4^+-N	in ppm	63	4543	4793	3283	3244	-
NO_3^--N	in ppm	42	9136	4626	1585	3512	-
P_2O_5	in %	0.06	1.07	1.23	1.45	1.29	-
K_2O	in %	0.42	20.64	17.32	7.50	10.97	-
MgO	in %	0.09	1.17	1.63	1.43	1.07	-
CaO	in %	0.56	4.05	3.06	3.48	5.13	-
Na_2O	in %	0.03	0.09	0.10	0.31	0.15	-

		Bark	Tobacco leaf ribs		Tobacco dust		Maximum acceptable in soils
			1	2	1	2	
Fe	in ppm	211	669	181	4,734	1,225	-
Mn	in ppm	1,189	166	36	226	219	400
Zn	in ppm	177	105	29	189	145	300
Cu	in ppm	3.6	31.5	17.2	160.1	51.7	50
Pb	in ppm	11.3	6.9	< 2.6	11.3	17.8	300
Ni	in ppm	3.5	4.3	< 2.6	4.5	< 2.7	50-100
Cd	in ppm	1.7	1.6	< 0.5	2.3	1.4	5
Co	in ppm	0.7	< 2.6	< 2.6	< 2.8	< 2.7	-
Cr	in ppm	0.5	< 2.6	< 2.6	< 2.8	< 2.7	25

These results indicate that tobacco wastes:

- contain high concentrations of salts so that they cannot be used as pure material for horticultural substrates
- have a very low moisture content
- have a reasonable acid binding value so that they can be used on acid soils
- still contain a certain amount of nicotine which will act as an insecticide/acaricide
- contain much nitrogen so that this material can be used as a nitrogen source in composts

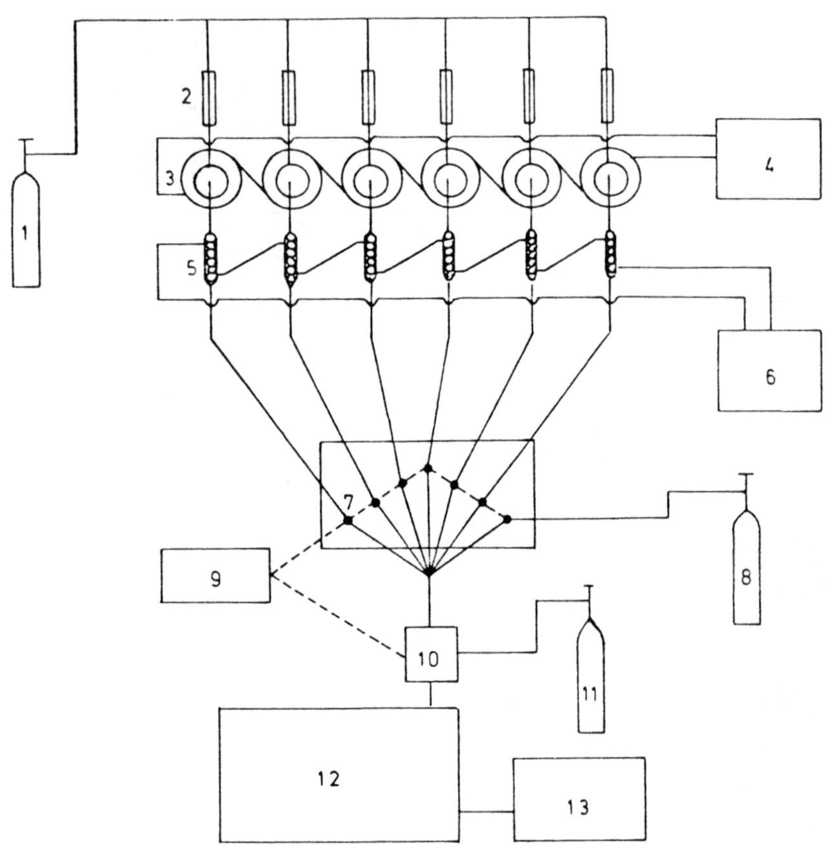

1. Air
2. Gas flowmeter
3. Reactor
4. Heating system
5. Condenser
6. Refrigerator
7. Electromagnetic valves
8. Reference gas
9. Control and steering unit
10. Sample valve
11. High pressure gas
12. Gas chromatograph
13. Recorder

Figure 1. Scheme of the laboratory composting simulator.

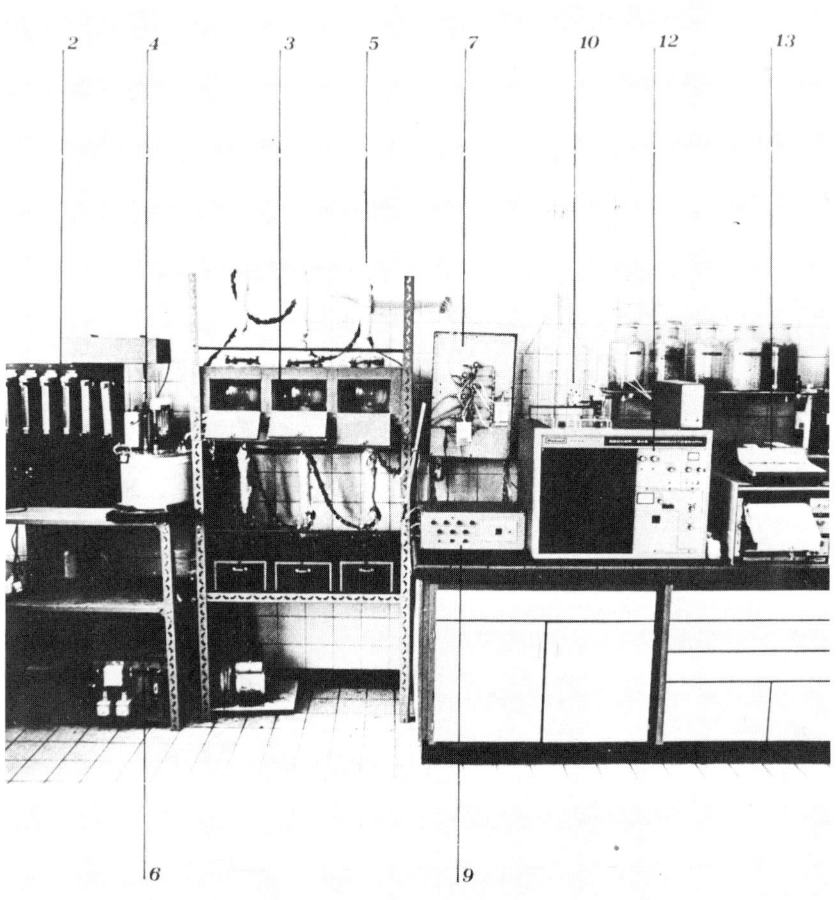

Photo 1. The laboratory composting simulator

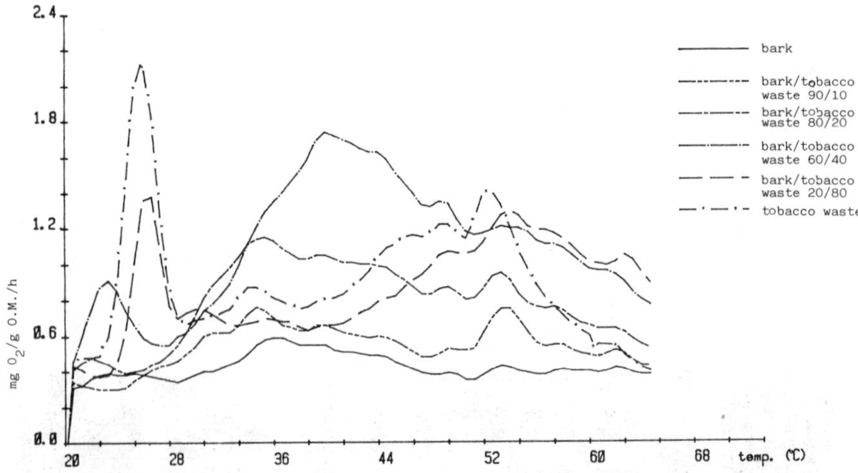

Fig. 2 - Changes in oxygen consumption in the different mixtures of bark and tobacco waste with temperature.

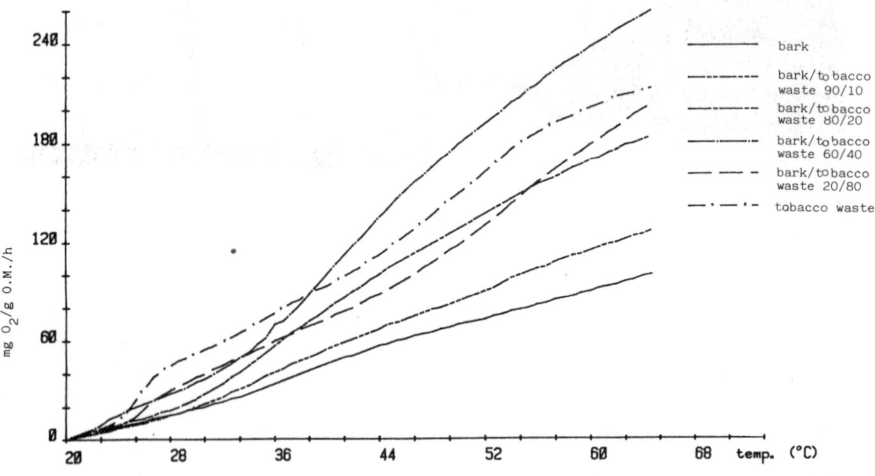

Fig. 3 - Changes of the total oxygen consumption with temperature.

- contain large amounts of potassium which is valuable as fertilizer
- have small concentrations of other elements

2.3. Laboratory composting simulator.

With the help of a laboratory composting simulator we have determined the optimal conditions for different mixtures of bark and tobacco waste. In figure 1 a schematic view of this simulator is shown. Photo 1 shows the apparatus.

2.4. Pilot composting trials.

The pilot composting trials were carried out in containers of 1-2 m^3 volume which are well insulated and protected from sun and rain. During the composting period a large number of measurements were made in order to determine and follow the composting process:

- daily:
  - temperature
  - O_2 and CO_2 content
- after each aeration:
  - chemical composition (pH, Ec, Ntot, NH_4-N, NO_3-N, organic matter)
  - volume and weight
  - biotest with cress
- at the end:
  - total chemical composition
  - physical properties

3. RESULTS.

3.1. Laboratory.

In the laboratory composting simulator a large number of mixtures of bark and tobacco waste were tested:

- 100 % bark + 0.75 % N as urea, being the optimal amount for pure pine bark
- 90 % bark + 10 % tobacco waste
- 80 % bark + 20 % tobacco waste
- 60 % bark + 40 % tobacco waste
- 20 % bark + 80 % tobacco waste
- 100 % tobacco waste

For these mixtures, the oxygen consumption has been measured, which gives a very good idea of the progress of the composting process. The change in oxygen consumption is given in figure 2. The total oxygen consumption is given in figure 3. Summarized results of all these mixtures are given in table 5.

Table 5. Total oxygen consumption of different mixtures of tobacco and bark.

Mixtures	Total oxygen consumption in mg O_2/g organic matter	
	1	2
100 % bark	100	101
90 % bark/10 % tobacco waste	127	122
80 % bark/20 % tobacco waste	184	176
60 % bark/40 % tobacco waste	160	244
20 % bark/80 % tobacco waste	203	-
100 % tobacco waste	213	-

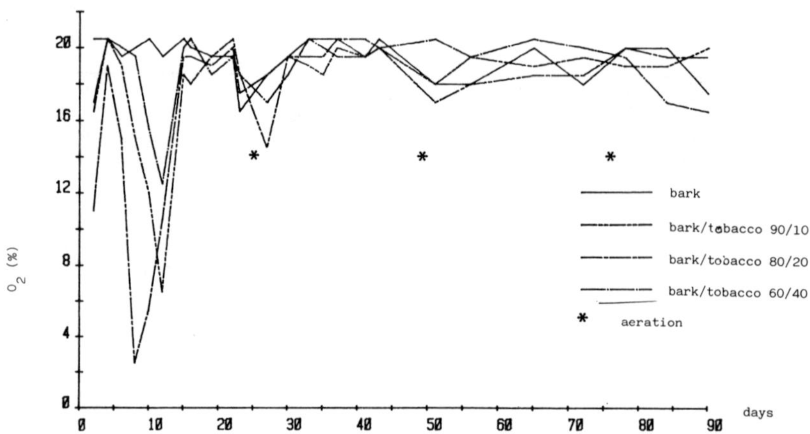

Fig. 4 - Changes in oxygen content as a function of time.

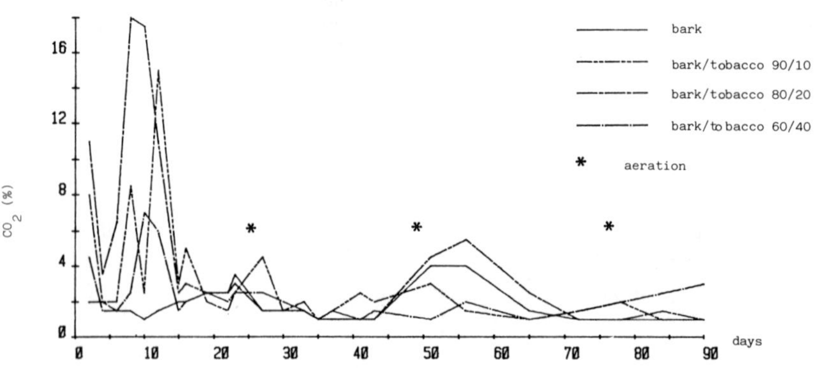

Fig. 5 - Changes in CO_2-content as a function of time.

The results in figures 2 and 3 and table 5 indicate that by mixing tobacco waste and bark activity is increased. The mixture of 60 % bark and 40 % tobacco waste had the highest oxygen consumption, meaning that this mixture is very active microbiologically. This can give the best composting and quickest stabilisation.

The chemical composition of the different mixtures is also important especially for use as horticultural substrate. Therefore the chemical composition is given in table 6.

Table 6. Chemical composition of different mixtures after laboratory composting.

	Bark/ tobacco waste					
	100/0 + 0.75 %N	90/10	80/20	60/40	20/80	0/100
pH	7.4	7.7	8.3	9.6	10.2	10.3
Ec(µS, 1/25)	364	920	1650	3200	6800	7900
Ntot (%)	1.734	1.249	1.448	2.063	2.285	2.296
NH_4^+-N (ppm)	3181	187	98	63	63	84
NO_3^--N (ppm)	64	143	274	421	472	505
% OM on DM	90.9	89.7	88.7	88.1	82.6	78.4
% nicotine	-	0.01	0.01	0.02	0.03	0.05
C/N	27	37	32	22	19	18

These results indicate that:

1. The pH increases with the increasing proportion of tobacco waste, due to the transformation of organic nitrogen to ammonium-nitrogen, which increases the pH of the mixture
2. The conductivity is high when more than 20 % tobacco waste is mixed with bark. Only mixtures of 10 or 20 % of tobacco waste with bark are acceptable as horticultural substrates. The other mixtures must be used as organic soil conditioners.
3. The amount of total nitrogen is sufficient for all the mixtures with tobacco waste so that it is not necessary to add mineral nitrogen. Also the C/N ratio is between 28 and 37, which is nearly optimal.
4. The organic matter content is very high, which is very favorable for use as organic growing medium or organic fertilizer.

3.2. Pilot composting trials.

From the results of the laboratory composting simulator best results were obtained with 60 % bark and 40 % tobacco waste. This will be tested in our trials on a pilot scale. Because of the high salt content of this mixture, causing growth inhibition for ornamental plants, mixtures with 10 and 20 % tobacco waste will also be tested. The changes in temperature as well as the oxygen and carbon dioxide content were measured over three months of composting. These results are shown in figure 4 and 5.

From the results given in figures 4 and 5, the following conclusions can be drawn:

1. The temperature of the mixtures of bark and tobacco waste increases more rapidly than the pure bark, due to the presence of sugars in tobacco waste
2. Also when the temperature is high, a large amount of CO_2 is produced due to the intense microbiological activity
3. Temperature measurements and CO_2 production can be used to automate aeration

As well as these daily measurements, pH and conductivity readings were also made in order to control quality. In table 7 the changes in pH and conductivity are given.

Table 7. pH and conductivity of bark/tobacco mixtures.

Mixtures	pH	Ec 1/25 (μS)
bark	5.6	340
bark + 10 % tobacco waste	7.8	895
bark + 20 % tobacco waste	8.4	1490
bark + 40 % tobacco waste	9.2	3830

The results in table 7 indicate that the larger the proportion of tobacco waste, the higher is the pH and salt content. As 1500 μS is the limit for use as a growing medium, the maximum amount of tobacco waste can be 20 %. With 10 %, the content of tobacco waste will be the optimum for use as horticultural substrates. The other mixtures should be used as organic soil conditioners or should be diluted with other substrates with lower salt content and lower pH.

4. RESULTS OF GROWING EXPERIMENTS.

4.1. Tobacco waste as an organic soil conditioner.

Tobacco waste has been compared as an organic soil conditioner with manure and two kinds of sludges. The following materials have been used:
1. manure 80 t/ha
2. sludge from food industry 40 t/ha
3. sludge from food industry 80 t/ha
4. tobacco waste 25 t/ha
5. tobacco waste 50 t/ha
6. sludge from waterpurification 75 t/ha
7. sludge from waterpurification 150 t/ha

As a testplant *Picea Amorica* is grown for 3 years on a loamy sand with the different organic soil conditioners. The results are given in table 8.

Table 8. Results of growing *Picea Amorica*.

Used soil conditioner	Height of the plant (cm)
manure	89.6 d
food industry sludge 40 t/ha	96.7 b
food industry sludge 80 t/ha	85.1 e
tobacco waste 25 t/ha	101.9 a
tobacco waste 50 t/ha	93.5 c
sludge-water purification 75 t/ha	94.4 bc
sludge-water purification 150 t/ha	89.6 d

The results in table 8 shows that tobacco waste at 25 t/ha gave the best result and was significantly better than the manure treatment. This waste can be used as an organic soil conditioner to replace manure.

4.2. Bark-tobacco waste mixture as substrates.

The different bark-tobacco waste composts were tested as horticultural substrates for ornamental plants. Two plants, *Monstera deliciosa* and *Ficus lyrata* were grown in various media. The results are given in tables 9 and 10.

Table 9. Growth of Monstera deliciosa.

Substrates	Height of plant (cm)	Number of leaves
1. 100 % bark	55.6 b	2.3 a
2. 90 % bark + 10 % tobacco waste	80.2 a	2.6 a
3. 80 % bark + 20 % tobacco waste	58.7 b	2.4 a
4. 60 % bark + 40 % tobacco waste	42.2 c	2.1 b

Table 10. Growth of Ficus lyrata.

Substrates	Height of plant (cm)	Number of leaves
1. 100 % bark	103.7 a	17.75 a
2. 90 % bark + 10 % tobacco waste	100.7 a	15.75 a
3. 80 % bark + 20 % tobacco waste	94.8 a	16.10 a
4. 60 % bark + 40 % tobacco waste	86.8 c	16.70 a

The results in table 9 and 10 indicate that a mixture of 90 % bark and 10 % tobacco waste gave the best results. Sometimes it was better than pure bark compost. That means that this mixture can be used as growing medium for ornamental plants and can replace the normal substrates of peat and pine litter.

5. GENERAL CONCLUSIONS.

Tobacco waste and bark can be composted together without mineral nitrogen. 60 % bark and 40 % tobacco waste gives the greatest microbiological activity which means the quickest stabilization of the compost. This mixtures can be used as an organic soil conditioner, but the salt content is too high for the use as a growing medium. Therefore mixtures with 10 and 20 % tobacco waste were tested as substrates. The results of growth tests indicate that only 10 % tobacco waste gives very good results.

The general conclusion is that a mixture of 90 % bark and 10 % tobacco waste can be used as a horticultural substrate. The other mixtures can be used as organic soil conditioners.

REFERENCES.

1. HOWARD, E.J. (1970). A survey of the utilization of bark as fertilizer and soil conditioner. Pulp and Paper magazine of Canada, 18, 53-56.
2. KOIWA, A., MATSUMOTO, T., NISHIDA, K. and ARIMA, K. (1971). Studies on the fermentation of tobacco. Tobacco Science, 15, 103-105.
3. VERDONCK, O., PENNINCK, R. and DE BOODT, M. (1983). Recycling of organic waste products into horticultural substrates. Symposium Planta et Humus, Prague, 1983.
4. VERDONCK, O., DE VLEESCHAUWER, D. and PENNINCK, R. (1982). Bark compost, a new accepted growing medium for plants. Acta Horticulturae 133, 221-226.

DISCUSSION

NAVEAU: In figures 2 and 3 the amount of oxygen consumed is given as a function of temperature. What is the relationship with time?

In table 7 you comment that the pH increase with increasing % tobacco waste is due to ammonium N, but your NH_4-N figures in table 7 do not increase with increasing proportions from 20 to 100% tobacco waste.

VERBONCK: In figures 2 and 3 we relate oxygen consumption to temperature. In our experiment the temperature is also related linearly with time, because in the laboratory composter we increased the temperature by 5°C per day. The experiment starts at laboratory temperature (20°C) and takes 10 days so that we have a final temperature of about 70°C. Our apparatus can be used also with other temperature patterns, at constant temperatures or with other daily increases.

The pH of the bark increases with higher amounts of tobacco waste for two reasons:

1) the pH of the tobacco waste is higher than that of bark

2) the effect of the large amount of ammonium nitrogen in the pure tobacco waste. The analytical results indicated that there was no increase of the ammonium nitrogen content in mixtures with higher proportion of tobacco waste because of the transformation of the ammonium nitrogen to nitrate nitrogen. The results in the table show an increase of the nitrate nitrogen with higher proportions of tobacco waste.

FINSTEIN: You have developed a useful laboratory-scale experimental apparatus, but it is not a simulator. To simulate, the mechanisms of heat removal must be mimicked. Otherwise the behaviour of moisture differs from that in the field.

VERDONCK: Our laboratory apparatus is specially built to optimise the parameters which influence the composting of organic wastes such as:

- nitrogen economy: amount of nitrogen
  nitrogen source
- moisture content
- temperature
- aeration
- structure of the organic waste

The results indicate that most of these parameters influence the composting process and that the relative importance of these parameters differs from one waste to another.

These laboratory experiments help us to optimize the composting parameters in a limited time.

The optimal parameters are used then for composting on a pilot scale in heaps of 20 to 30 m^3 and we have seen that the laboratory experiments are necessary to help us to optimize the composting process in big heaps.

STENTIFORD: Do the O_2 consumption rates presented in Figures 2 and 3 make allowance for the loss of organic matter with time?

VERDONCK: In our laboratory experiments we calculate the oxygen comsumption on the initial organic matter content of the waste. Our experiments only take about 10 days so that organic matter decreases little in that time. The decrease was found to be about 1 to 2%.

EXPERIMENTAL USE OF ORGANIC BY-PRODUCTS AS CULTURE SUBSTRATES

D. DAUDIN and P. MICHELOT
Chamber of Agriculture of the Vaucluse region
Association for the Development of the Meridionale nursery

Summary

The use of organic wastes and by-products as growing medlia is hindered by the difficulties of preparing mixtures and testing their performances. Some products, mixtures of crushed bark and waste sludge for instance, prove to be worthwhile. However, the wide range of results obtained with all these products, depending on the species used, make them difficult for use by the horticultural industry.

1. INTRODUCTION

Two series of tests have already been carried out on the use of organic wastes on nurseries producing ornamental plant. The results showed the advantages of such materials as grape marc, crushed barks, urban waste compost and sewage sludge, and also the limitations of their use : the importance of the proportions used in mixtures with harmful effects of large quantities. Testing on a large scale proved difficult because of the mixture's lack of homogeneity.

Therefore, this year the situation was remedied by preparing the mixture early in the season and allowing it to compost. The objectives were i) to assess the evolution of the compost through the regular recording of a few simple parameters; ii) to control its agronomic qualities by trying it out on plants; iii) ultimately to achieve a better control in the use of these by-products, which abound in our region, and are often reasonably cheap.

2. DESCRIPTION OF THE EXPERIMENTS : THE DIFFERENT KINDS OF MIXTURES

2.1 Wood shavings and town compost

The compost comes from the household refuse treatment works of Vedene (near Avignon). The compost used was fresh, had been sieved through a 30 mm mesh sieve and put twice through the refining process recently set up (removing heavy materials by gravity, removing plastics by suction).

The sawdust and shavings were a blend in undermined proportions of the various local plant species.

The mixture was prepared using an endless screw by mixing two volumes of sawdust and shavings (15 m^3) with one volume of urban compost (8 m^3). During preparation, water was added and 100 kg urea (46%N) were added, ie about 2 kg N/m^3. This mixture was prepared on 13 December, 1982 and turned each month until the beginning of May, with a possible addition of water to the mixture each time.

2.2 Crushed bark and waste sludge

The barks used come from conifers of the Lozere region. The sludge, provided by the Liebig plant in Le Pontet, is a mixture of primary and biological muds, mechanically dehydrated. The effluents come from the production of pre-cooked dishes.

The mixture was prepared on 3 January, 1983 using a mechanical shovel by mixing 20 m^3 crushed bark with 8 m^3 of sludge. It was then turned each month until the middle of May and moistened when needed.

2.3 Grape marc and vinasse

Marc is the solid residue of fermentation, it is a mixture of grape stalks, seeds and pulp. This product abounds in the whole region. The vinasse is the liquid residue of the distillation.

The mixture was prepared with a mechanical shovel on 20 November, 1982 by mixing : 20 m^3 of fresh marc with 5000 1 of vinasse. After absorption of the vinasse by the marc, the mixture was put into a heap then turned every month until the beginning of May. No addition of water was necessary.

2.4 Other materials

At the request of either the suppliers or the owner of the nursery where the experiment took place, a few other products were also tested.

2.4.1 Lavender straw compost: This compost, 18 months old, comes from the region of Nyons.

2.4.2 Substrate from bag growing culture: Product used after cultivation of melon for two years in the greenhouse. It was mainly composed of crushed barks.

2.4.3 Vermicompost: The material was supplied by the Societe des Eaux de Marseille, was the product from the treatment of sewage sludge by earthworms. This product had the form of a splendid-looking black garden mould, with a fine and regular particle size distribution and a gritty structure.

2.5 Experimental details

In each heap prepared as indicated above, the following measurements were made were recorded during the composting period - temperature, pH, conductivity and the C/N ratio. In this way it is hoped to characterize simple parameters, easy to control and giving a fair idea of the evolution of the compost.

The agronomic qualities were tested as follows : After 5 or 6 months of composting, the mixtures obtained were put into 3 litre containers and some of the most representative plant species of the local production were grown. The repotting was immediately followed by a herbicide application of oxadiazon, at a rate equivalent to 24 kg/ha. Fertilizer was applied in accordance with the normal practice of the nursery where the experiment was carried out.

The quality of the compost was judged by plant growth during the vegetative season and by the general appearance at the end of the season.

The reference mixture was made in equal proportions of sphagnum peat and pozzolana (volcanic gravel).

The different kinds of substrates tested and the species used are summarized in Table I.

TABLE I - SUMMARY OF THE VARIOUS TREATMENTS

Testing place	Substrates tested	Preparation date	Repotting date	Plant species	Final control
Nursery MICHEL Velleron (84)	Shavings-compost 1/2 Pozzolana 1/2	13-12-82	17-05-83	Cotoneaster-Leyland cypress-Grafted and seedinlg blue cypress. Forsythia-Spindle tree-Pomegranate plant-Cherry laurel Oleander-Laurustinus Aleppo pine-Pyracantha-Rosemary Diervilla	30-10-83
	Lavender compost 1/2 Pozzolana 1/2	17-05-83	"		
	Substrate for bag culture	"	"		
	White peat 1/2 Pozzolana 1/2	"	"		
Nursery JOUVE-RACAMOND St Andiol (13)	Bark + sludge	3-01-83	16-05-83	Holm oak-Grafted blue cypress-Euonymus pulchelus-Cherry laurel-Oleander-Lonicera nitida-Pyracantha-Rosemary Santolina	12-10-83
	Vermicomposting 1/4 White peat 1/4 Pozzolana 1/2	16-05-83	"		
	White peat 1/2 Pozzolana 1/2	"	"		
Nursery BELLET Lodeve (34)	Marc + vinasse	20-11-82	6-05-83	Atriplex-Leyland cypress-Escallonia Cherry laurel-Oleander-Laurustinus Lavender-Arborvitae Speedwell	2-11-83
	Fresh marc	6-05-83	"		
	White peat 1/2 Pozzolana 1/2	"	"		

3. MEASUREMENT DURING THE PREPARATION OF THE COMPOSTS

3.1 Wood shavings and town compost

3.1.1 Temperature: Successive readings of the temperature at different heights within the heap showed a rapid rise in the temperature (between a fortnight and a month), then relative stability until repotting. This suggests that the evolution of the compost at that time was far from complete, in spite of the 6 months of composting.

3.1.2 C/N ratio: Because of the difficulties of sampling and the questionable reliability of methods of analysis of this type of organic product, the C/N ratio is not reliable as a parameter for following the evolution of the compost. Only a slight downward trend was noticed in the heap from about 40 to 30.

3.1.3 pH and conductivity: The sawdust and shavings are acid (about pH 5.4): the town compost is alkaline (about pH 7.8). When they are mixed together, the pH first rises to 9.1, which can be explained by the hydrolisis of the added urea to form ammonia. pH then tends to decrease.
The electric conductivity allows the overall salinity of the substrate to be measured. At the start, it is very high, 10 mS, diminishing progressively to 2.9 mS at the time of repotting. This is still a maximum considering the low moisture content of the compost.

TABLE II - Changes in the pH and conductivity in the shavings/town compost mixture.

Date	13 JAN	26 JAN	9 FEB	24 FEB	4 MAY
pH (H_2O)	9.1	9.1	8.85	8.87	/
Conductivity mS	10	9.2	7.3	4.4	2.9

3.2 Compost mainly made from sewage sludge and bark

3.2.1 Temperature: The successive readings of the temperature in the heap showed a rapid rise from about 13° to 36°C, then a stabilization. Towards the end of the composting period, the temperature was still high which means that the compost is not completely mature.

3.2.2 C/N ratio: The same remarks apply as for the previous compost. The results obtained do not allow this parameter to be used for following the evolution of the compost.

3.2.3 pH and conductivity: During composting, the pH of the mixture remains stable, varying only between pH 7.83 and 7.50. The electric conductivity also remains stable, and at very low levels.

TABLE III - Changes in the pH and the conductivity of the crushed bark/sewage sludge mixture.

Date	26 JAN	9 FEB	24 FEB	8 MAR	20 APR
pH (H_2O)	7.68	7.89	7.83	7.52	7.50
Conductivity mS	0.39	0.58	0.51	0.26	0.42

3.3 Compost made from grape marc

3.3.1 Temperature: As previously, a rapid rise in the temperature occured after the formation of the heap, almost to a maximum of 65°C. Towards the end of the composting period, the temperature was still high, indicating that maturity was not complete.

3.3.2 C/N ratio: The same remarks apply as in the previous cases.

3.3.3 pH and conductivity: They both tend to decrease during composting, and, at the end of it, the conductivity finally falls to a value of 1.5 mS.

TABLE IV - Changes in the pH and the conductivity in the grape marc/vinasse mixture.

Date	20 DEC	26 JAN	1 MAR	4 MAY
pH (H_2O)	7.75	7.38	7.28	7.20
Conductivity mS	/	2.42	1.35	1.45

4. MEASUREMENT OF THE AGRONOMIC QUALITY OF THE COMPOSTS

4.1 Wood shavings and town compost

The results shown in Figure I, demonstrate that plant growth on this compost is clearly inferior to that obtained from the reference mixture of sphagnum peat and pozzolana with indexes for all species of 7° and 100 respectively. This difference in growth can be observed for all the plant species tested, although with wide variations between them.

4.2 Bark and sludge

All the results are shown in Figure II. Plant growth is distinctly better than with the reference substrate of sphagnum peat and pozzolana, with the exception of Holm oak and Santolina. Taking the reference mixture as 100, the index of the bark/sludge mixture, on average of all species, was 123 (Holm oak 101, Santolina 52).

4.3 Grape marc and vinasse

The results are shown in Figure III. The reference mixture of sphagnum peat and pozzolana was prepared later than the others and as a result, only a few plant species could be repotted into it. The

FIGURE I.

Growth of the Plants on the composts.

1. - reference mixture sphagnum peat/pozzolana (volcanic gravel)
2. - wood shavings/town compost mixture
3. - substrate for melon bag culture
4. - lavender straw compost.

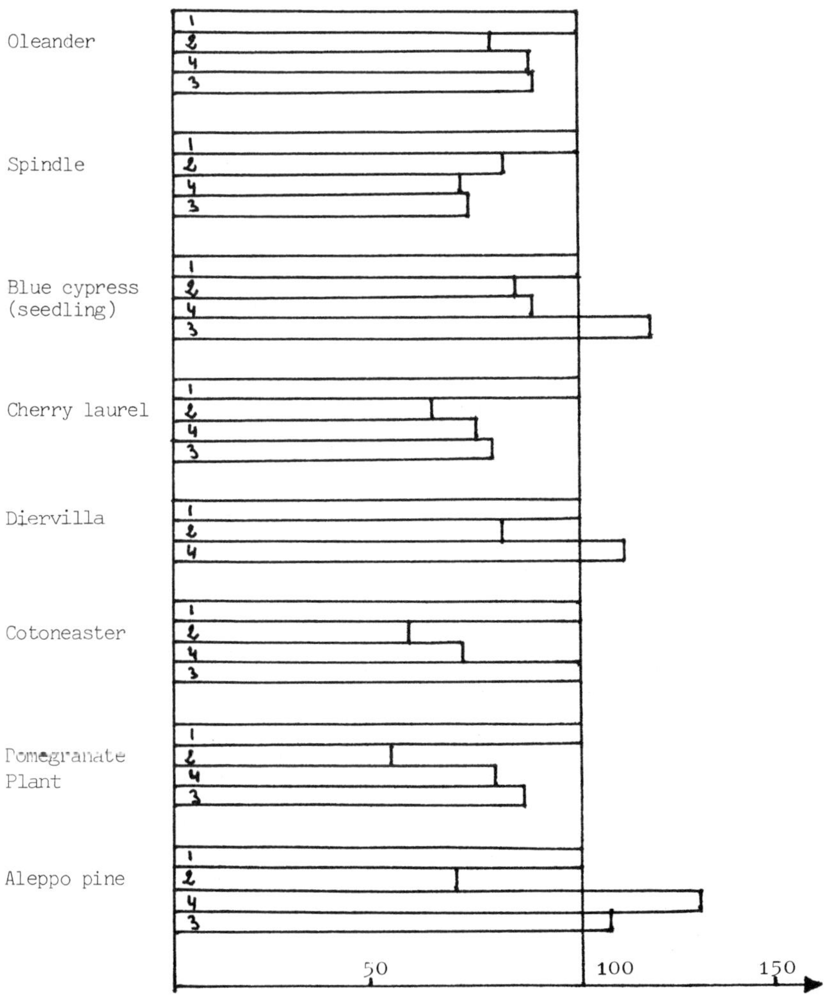

GROWTH RELATIVE TO REFERENCE COMPOST.

FIGURE II.

Growth of the Plants on the composts

1. - reference mixture of sphagnum peat and pozzolana

2. - bark /sewage sludge compost

3. - vermicompost.

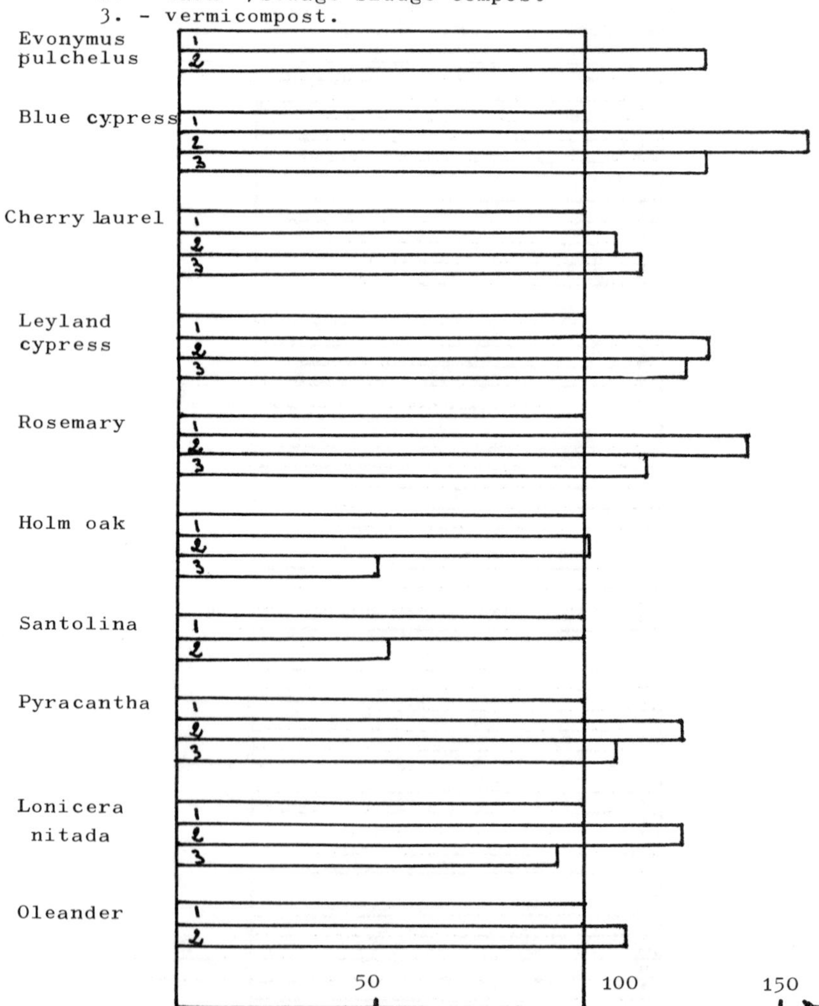

GROWTH RELATIVE TO REFERENCE COMPOST.

FIGURE III.

Growth of the Plants on the composts

1 - reference mixture of peat and pozzolana

2 - composted marc and vinasse

3 - fresh marc.

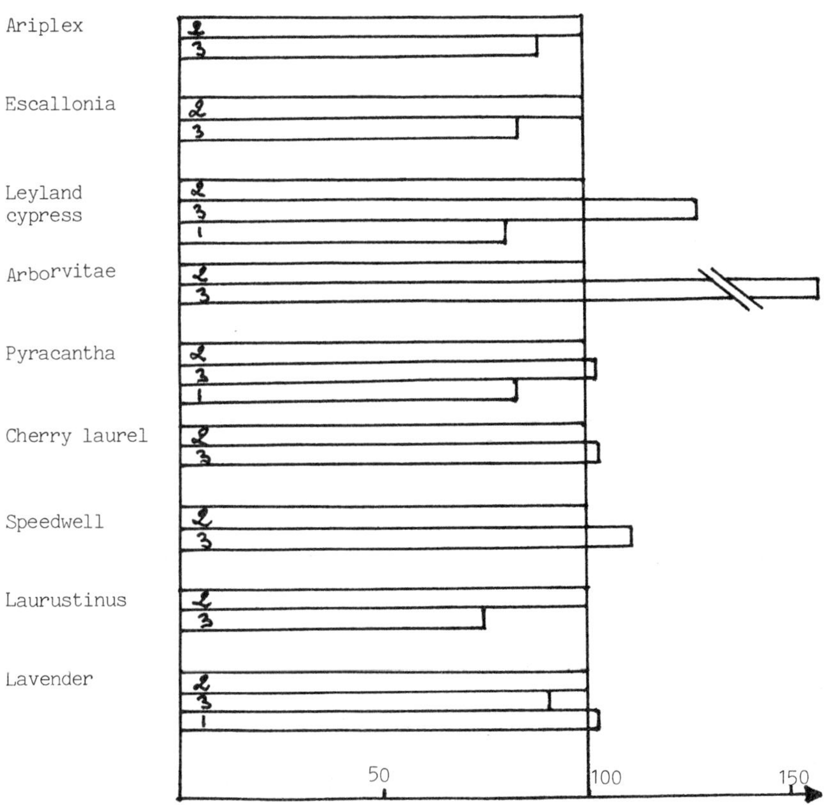

GROWTH RELATIVE TO COMPOSTED MARC.

differences in heights observed for Leyland cypress and Pyracantha were due, at least partly, to the delay between the two repotting dates.

Much chlorosis appeared towards the end of spring especially on Escallonia, Pyracantha and Speedwells. During the growing season, the plants became green again (Sequestrene was applied at the time of the chlorosis) and looked quite normal at the end of the experiment.

4.4 Lavender straw

The results are shown in Figure I. Overall, the plant growth on the compost was inferior to that obtained on the reference mixture. Their respective indices were 81 and 100, on average of all species, however, important varietal differences should be noted.

4.5 Growing-bag culture

All results are shown in Figure I. The performance of this product roughly matched that of the reference mixture (indices were 92.5 and 100, on average of all species) with, here again, some important differences.

4.6 Vermicompost

All results are shown in Figure II. Overall, the performance was equivalent to that of the reference mixture (indices were 99.5 and 100, respectively on average of all species). However, considerable differences between species should be noted : index 130 for the Blue cypress, 49 for the Holm oak and 169 for the Santolina, a number due to a very high mortality rate.

4.7 Fresh marc

All the results are shown in Figure III. Overall, plant growth on substrate is equivalent to that obtained from composted marc and vinasse (indices were 102 and 100 respectively on average of all species), again with important variations between species.

Problems of chlorosis also arose with fresh marc compost. By the end of the growing season, the plants had recovered their normal aspect (Sequestrene was applied after chlorosis appeared).

5. CONCLUSIONS

5.1 Analytical monitoring of the composting

Among the various physical and chemical parameters followed during the composting, conductivity seems the only reliable one.

Mention has already been made that changes in the C/N ratio were not reliable on these composts because of sampling difficulties and methods of analysis.

Temperature does not seem to be a good enough criterion. For example, whereas the bark/sludge mixture matured really well over the composting period at low residual temperatures, it was so at comparable temperatures for the shavings/compost mixture, that is temperatures of about 35°C.

Changes in the conductivity were not always the same in the different heaps but, in case of excessive salinity at the beginning of the composting, it allowed the point from which the mixture may be used to be determined. This parameter, which is easy to calculate, seems therefore worth taking into account when following the preparation of a compost.

5.2 Agronomic quality of the composts tested

5.2.1 Town compost/shavings: There are two possible reasons for the poor performances of this compost : i) the unfavourable physico-chemical properties of the materials composing the mixture, ii) the operations carried out during the preparation of the compost.

The characteristics of sawdust and shavings are rather unfavourable (toxic products, very low retention of water) and they cannot be sufficiently improved by the addition of town compost.

Also, adding nitrogen in the form of urea does not seem to be the best way to speed up the composting process, because hydrolysis occurs very rapidly but the ammonia formed cannot be used as a source of nitrogen by the micro-organisms. Hence the poor maturation of the compost. The repeated turning of the heap only results in draining off the excess ammonia and the high remaining salinity because of the insufficient biological activity. The pH of the compost is also, as far as horticultural crops are concerned, far from being optimum (pH 8.8).

As presently prepared, the shavings/town compost mixture cannot satisfactorily replace the reference substrate containing sphagnum peat.

5.2.2 Bark/sewage sludge: The very satisfactory results obtained with this product again demonstrate that bark is a good material for making composts. The addition of sludge as a binder, with its fine particle size distribution, supplements the mixture and increases its retention capacity. Prepared from sludge rich in mineral salts (conductivity of 1/5 weight extract = 6.09 mS), in proportions, 2/3 bark - 1/3 sludge, the mixture obtained is, after 6 months of composting, homogeneous and has a very low conductivity. This allows the professionals to use the composting that suits them best without having to worry about salinity problems.

5.2.3 Grape marc: The aim of this experiment was to estimate the possibility of using marc, composted or fresh, without adding pozzolana in order to reduce the abrasive character of the compost. The problems of chlorosis which were found with both fresh and composted marc, showed that this is not advisable. Chlorosis was not due to salinity, as was verified by the measurements taken on the spot after the symptoms appeared, but to a form of root asphyxia from the heavy watering made soon after the repotting by the nurseryman who feared a re-starting of the composting with a rise in the temperature sufficient to be harmful to the plants.

Also, composting did not seem to confer any advantages. The performances of the composted marc are very similar to those of the fresh marc, or overall just slightly inferior.

5.2.4 Lavender straw: The questionable reliability of this material (overall growth index 81 as compared with the reference index 100) is considered to be due to its compactness and to its very high K_2O) content. However, this product can be used as basic component of a compost, up to a maximum proportion of 1/3, by mixing it with coarse materials to give the substrate sufficient porosity.

5.2.5 Substrate for growing-bag culture: The satisfactory results obtained with this product (overall growth index 92..5 in comparison with the reference index 100) confirm the value of using bark as a component in a substrate. Pathological problems with the cultivation of melons did not materialize during the test.

5.2.6 Vermicompost: The results show that this can be used provided that it does not exceed one-quarter by volume. In further tests, proportions up to one-half and wholly vermicompost noticeably depressed growth compared with the reference substrate. Also, this product must not be used for plants sensitive to excessive salinity, such as Holm oak and Santolina.

As this product is still at the experimental stage, the Societe des Eaux de Marseille could not give any indication of price.

DISCUSSION

VERDONCK: The reported pH and salt content are very high in many cases, sometimes up to pH 9 and a conductivity up to 10 mS. For ornamental shrubs we need a pH below 7 and conductivity lower than 1.5 mS. This is the reason for selecting organic wastes with these properties. Can you comment on this?

GODIN: The high pH is due to the presence of urea which is hydrolysed to ammonium carbonate and decreases during composting.

EDWARDS: Did you add any nutrients to the reference substance?

GODIN: Only peat and volcanic gravel were used as the test was intended to assay for phytotocixity.

SESSION V

VERMICULTURE

The use of earthworms for composting farm wastes

The economic feasibility of earthworm culture on animal wastes

THE USE OF EARTHWORMS FOR COMPOSTING FARM WASTES

C.A. EDWARDS, I. BURROWS, K.E. FLETCHER & B.A. JONES
Rothamsted Experimental Station, Harpenden,
Hertfordshire, U.K.

Summary

The large amounts of organic wastes produced in intensive agriculture can cause serious disposal problems. A project, begun in 1980 at Rothamsted, has shown that earthworms, especially *Eisenia foetida*, can break down organic wastes into peat-like materials rich in available nutrients and with a good moisture-holding capacity and porosity. These have considerable potential in horticulture as a plant growth medium. The earthworm can produce a valuable food meal suitable for fish, poultry and pigs consisting of 60-70% protein, (rich in lysine) 7-10% fat, 8-20% carbohydrate, 2-3% minerals, and a wide range of vitamins including niacine, and with long chain fatty acids that non-ruminant animals cannot synthesize. The basic biology and ecology of *E. foetida* and five other potentially useful species have been worked out. Conversion ratios of waste to worms of up to 10% were attained and on-farm, pig, cattle, duck, turkey and potato waste systems have been set up. Methods of obtaining maximum waste turnover in 2-4 weeks, under controlled moisture and temperature conditions, are described. Fish and chicken-feeding trials were successful and a wide range of vegetables, shrubs and bedding plants usually grew better in worm-worked compost than in commercial materials. The project has involved collaboration with a wide range of research stations, universities and commercial organizations.

Introduction

Since the Second World War, the increased availability of fertilizers, chemical pesticides and new plant varieties has resulted in mixed farming becoming much less common. Farmers have concentrated either on intensive production of field crops, sometimes in monoculture or close rotations, or on livestock production. Animal husbandry has tended to become increasingly intensive, and concentrated in specific geographical areas. This has led to the localised production of animal wastes in large quantities without nearby crops on which they could be used.

Disposal of these large quantities of animal wastes (Table 1) and vegetable wastes (Table 2) is a serious and increasing problem. Only a relatively small proportion of the fertilizer value of wastes is utilized currently, so there is a considerable loss of potential energy. Moreover, many animal wastes cause serious odour and pollution problems. They are costly to dispose of and, if allowed to drain into water courses illegally cause nutrient enrichment and eutrophication. Holding facilities such as lagoons and tanks are often used to reduce Biochemical Oxygen Demand so that after a considerable period they are suitable for land spreading. Some animal wastes, particularly those from straw bedded poultry, are highly intransigent and take up large areas of land while decomposition takes place. Other forms of organic waste such as spent mushroom compost are disposed of by landfill, again using up useful land areas. At present, the commonest disposal method for farm wastes is still land spreading, usually on grass (Table 3).

Table 1. Excreta from farm livestock in England and Wales (1980)

Livestock	Numbers (millions)	Excreta output (kg per) (animal) (per day)	Excreta from housed animals (million) (tonnes per) (year)	Total excreta all livestock (million) (tonnes per) (year)
Cattle	9.44	18-41	43.06	83.17
Pigs	6.59	1.5-8.0	9.11	9.11
Poultry	109.79	0.115	4.60	4.60
Sheep	22.49	1.5-4.0	-	22.69

Adapted from Nielsen (1982)

Table 2. Plant and vegetable wastes in U.K.

Source of Material	Quantity (millions of) (tonnes of) (dry matter)	Metabolisable Energy (MJ x 10^6)
Straw	9.30	60450
Arable crops	0.85	8460
Sugar beet	0.50	6000
Brewery	0.24	2640
Potato	0.20	1800
Vegetable	0.05	178
Spent mushroom compost	0.20	1300

(from various sources)

Table 3. Percentage of crop area receiving organic wastes (1978)

CROP	% AREA ANY ORGANIC	PERCENT TREATED AREA GETTING			
		FYM	SLURRY	SEWAGE SLUDGE	OTHER ORGANIC
SPRING CEREALS	19	76	19	3	2
WINTER CEREALS	13	81	11	4	4
POTATOES	42	84	8	1	7
SUGAR BEET	29	80	11	5	4
OTHER TILLAGE	21	72	20	3	5
TEMPORARY GRASS	46	65	30	2	3
PERMANENT GRASS	35	73	25	1	1
ALL CROPS & GRASS	29	72	24	2	2

Other methods of utilization of farm wastes include methane production, drying for animal feeds and production of algal, fungal, or microbial protein. None of these methods is used widely as yet and there remains a major problem of utilizing these wastes profitably in order to realise their full potential and avoid pollution. It was in this context that a research programme aimed at using earthworms to break down these wastes was begun in 1980 with the funding of the U.K. Ministry of Agriculture, Fisheries and Food and Agricultural and Food Research Council.

The breakdown of organic wastes by earthworms

Work in the United States has shown that earthworms will break down activated sludges (Hartenstein *et al.*, 1979a) and other organic materials (Neuhauser *et al*, 1979). The worm commonly bred commercially for fish bait is the 'tiger' or 'brandling' worm, *Eisenia foetida* (Savigny) and this is the species that has been studied most as a potential agent in breaking down animal wastes. Its biology is quite well understood (Hartenstein *et al.*, 1979b; Neuhauser *et al.*, 1980b; Kaplan *et al.*, 1980; Hartenstein, 1982) in relation to growth in sewage wastes, but little work has been published in relation to its growth in other organic wastes other than by Neuhauser *et al.*, (1980b). Work at Rothamsted has shown that this species will grow well in a wide range of agricultural wastes including pig and cattle solids and slurries, wastes from laying chickens, broilers, turkeys and ducks, horse manure, rabbit droppings, potato waste (Edwards, 1982), spent mushroom compost and paper pulp Edwards (1983). Not all wastes are as readily acceptable to earthworms or equally productive in terms of worm biomass (Figure 1).

The research at Rothamsted had two main aims: (i) to turn animal and other agricultural wastes into useful materials that could be added to agricultural land to improve soil structure and fertility and also have considerable potential in horticulture as a plant growth medium or component of commercial potting composts. (ii) To harvest worms from the worked waste and process them into a highly nutritious protein feed supplement for fish, poultry and pigs.

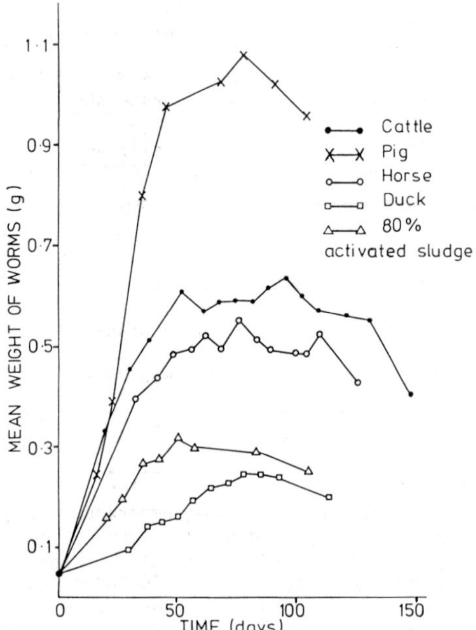

Figure 1. The growth of E. foetida on different animal wastes

To accomplish these aims, research was initiated in the following:

1. A laboratory screening programme into the suitability of different earthworm species and waste products to assess their biological and economic potentials, and studies into the biology and ecology of these worms.
2. Microbiological studies of agricultural wastes and of how earthworms feed on these and produce worm-casts.
3. Scaling up production methods to on-farm systems to maximise the production of worm-worked materials and worm-protein.
4. Characterisation of earthworm protein and its use as an animal feed.
5. Characterisation of worm-worked materials and their use as a plant growth medium.

1. Laboratory studies

The basic biology and ecology of the species *Eisenia foetida* (Savigny) *Dendrobaena veneta* (Rosa), *Dendrobaena subrubicunda*, (Eisen) *Lumbricus rubellus* Hoffmeister, *Eudrilus eugeniae* (Kinberg) and *Perionyx excavatus* (Gates), and investigations into their suitabilities for breaking down agricultural wastes and productivity, has been studied.

These species differed considerably in their growth rates, reproductive rates, size, ability to withstand handling and in their environmental requirements. Some species seem best suited to working organic wastes and producing composts and others more suitable for worm protein production.

For instance, a laboratory comparison between the growth rate of *E. foetida* and *Dendrobaena veneta* (Rosa) in pig waste showed that *D. veneta* grew much faster, and individuals attained a much greater mature weight than *E. foetida*, but each cocoon contained usually an average of only one hatchling compared with about four for the latter species. Another species imported from the Philippines showed considerable promise, growing much faster than *E. foetida* and producing up to 20 cocoons per week. It could also withstand higher temperatures and grew and reproduced well at 30°C, a temperature unsuitable for *E. foetida*.

To date, the most suitable indigenous earthworm in the UK, capable both of processing large amounts of organic material rapidly and which can be readily harvested and transformed into high grade protein, has been *E. foetida*. This species takes 7-8 weeks to reach sexual maturity, after which it can produce two to five cocoons per week either after mating or parthenogenetically. Each cocoon can produce from one to seven hatchlings, giving a potential for each adult worm to produce 15-20 young per week. The preferred pH for the worms is about 5.0 although they will tolerate a pH of 4.0 and grow well up to a pH of 9.0. We have studied the optimum conditions for worm growth and reproduction intensively, particularly in pig and cattle wastes. The temperature and moisture content of the waste are the two most important environmental factors. The upper lethal temperature is about 35°C and they can survive down to freezing. They do not grow very fast below 10°C and growth rates increase up to 25°C. However, although the number of cocoons produced is also greatest at 25°C, the percentage hatch and the number of young hatchlings emerging from each cocoon decreases with increasing temperature, so that the optimum temperature for maximum overall productivity is lower.

Work on optimum moisture conditions has shown that although these species are tolerant of moisture contents between 50% and 90% they grow most rapidly between 80% and 90%. The most rapid increase in total biomass occurs with a ratio of worms to animal waste of 1 : 50. At lower stocking rates down to 1 : 600, cocoon production is higher but biomass was produced more slowly, and at higher stocking rates reproduction becomes limited.

Eisenia foetida responds rapidly to availability of food. With over abundant food large numbers of small worms tend to be produced. In stocking rate studies with different numbers of worms in the same unit quantity of waste, individual worms grew largest at the lowest stocking rate (Figure 2) but total biomass production was greatest at the highest stocking rate (Figure 2). It seems that there is an absolute maximum biomass which can not be exceeded.

2. Microbiological studies

The link between microorganisms is very close although the detailed relationship is still unclear. The microbial populations of several organic wastes have been studied with the aim of identifying the nutrient source of earthworms. These wastes contained large populations of Gram-negative Enterobacteriaceae, up to about 1×10^{10} cells/gram dry weight and these may be important in worm nutrition. Protozoa and fungi were found to be important components of the microbial population, at about 4×10^4 and 1×10^6 cells/gram dry weight respectively. It has been shown recently that certain fungi and protozoa probably constitute a significant proportion of the diet of *E. foetida*, as may the metabolic by-products of some microorganisms.

It has been shown that bacterial numbers in waste increase during passage through the earthworm gut and it is likely that some of these are digested. The mechanical breakdown of the waste material by worms increases its surface area, enhancing microbial activity and further decomposition.

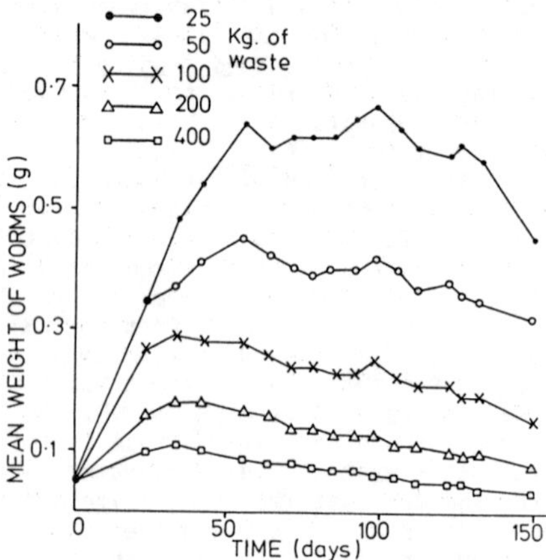

Figure 2. Growth of individual *E. foetida* at different stocking rates.

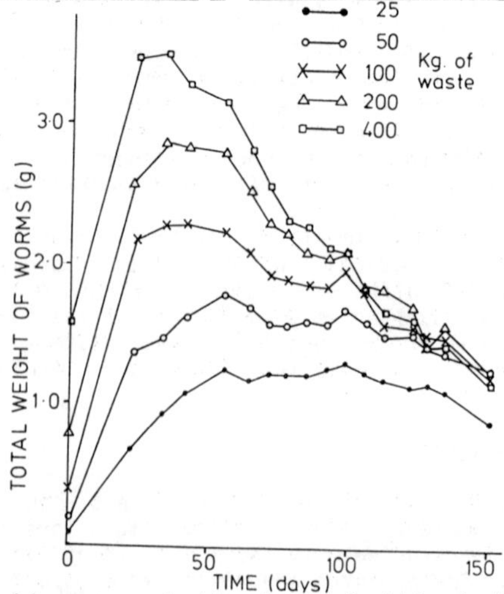

Figure 3. Total biomass production by *E. foetida* at different stocking rates

Work is in progress to assess the nutritional value of different microorganisms in the diet of earthworms and it seems likely that *E. foetida* requires a mixed population of bacteria, fungi and protozoa to sustain normal growth and reproduction.

3. Field studies and large scale vermiculture

Simple systems of breaking down organic wastes in windrows, heaps or boxes have been developed by breeders of worms for fish bait, which is an extensive industry in the United States and Canada (Tomlin, 1983), but these methods have been crude and empirically based. Efficient, scientifically-based, commercially-appropriate methods of production of worm-worked composts and economically viable production of worm protein differ considerably from such methods. The wastes must be worked rapidly under controlled environmental conditions with suitable equipment available for handling wastes, separating worms from wastes and processing both worms and worked composts.

The laboratory studies at Rothamsted on the basic biology of *E. foetida* and other species and optimal environmental conditions and stocking rates for maximum productivity of worms and processed wastes have been extended to larger-scale and on-farm systems. These have included methods of inoculating earthworm beds, optimum stocking rates, bed size and structure and ways of adding wastes. The productivity of earthworm production systems both outdoors and under cover with control of temperature and moisture content have been investigated.

Pig and cattle wastes have been studied on a field scale using both straw-based animal beddings and solids separated from slurries using a commercially available mechanical Farrow slurry separator. Cattle solids become acceptable to *E. foetida* a few days after collection and/or separation, but pig solids may take up to two weeks before the worms will enter them satisfactorily. Straw or wood-shavings-based duck, turkey or chicken wastes pose a greater problem, since they contain considerable amounts of ammonia and until this falls to below 0.5 $mg.g^{-1}$ the worms will not enter or utilise it.

The ammonia content in poultry waste can be decreased by composting or leaching but under natural conditions it can take up to 3 months before it becomes acceptable, making rapid turnover of waste difficult. Poultry wastes, and also spent mushroom compost tend to contain high levels of inorganic salts. When the ionic conductivity falls below 7.0 mmhos usually by leaching, the waste is acceptable to worms. Anaerobic wastes, or wastes with considerable ammonia or inorganic salts become acceptable rapidly if they are fed on to the top of a well-established and productive worm bed.

On-farm vermiculture trials have been set up using pig and cattle waste (Oaklands Agricultural College, St Albans), cattle waste, (Bore Place Farm, Kent), turkey waste (Forge Farm, Diss, Norfolk), and duck waste (Cherry Valley Farms, Caistor, Lincs). The cattle waste was investigated most extensively, and worms were grown successfully on both undigested wastes and on solids that had been separated after digestion for methane production. The liquid fracton of a separated solid could be fed back gradually to the separated solid fraction to utilize all of the waste but application rates required careful monitoring. Worms lived successfully only in aerobic wastes; however, once a worm bed was well-established anaerobic wastes could be added as a surface layer and these become rapidly aerobic and acceptable to the worms.

Batch systems, using containers which could contain up to 2 tonnes of waste, and various kinds of worm beds to a depth of about one metre have been tested. The batch systems tended to be laborious and worm beds seem to be most productive if set up with a small quantity of waste and a large worm inoculum with further layers of waste added in stages subsequently. Careful management of this type of system allowed a bed of separated cattle solids one metre deep to be processed after 30 days.

When layer-feeding was not employed, the maximum bed depth which could be used as a single feed was about 30-40 cm; beds deeper than this, particularly of pig waste, tended to become anaerobic at the bottom and the waste was not processed uniformly. Worms can utilise materials composed of small fragments much more rapidly so separated solids or chopped straw-based wastes are most suitable for a rapid turnover.

On a farm scale, worm production is likely to succeed only if it can be managed with minimal labour and the use of existing machinery and buildings. This requires mechanical loading of beds and, where possible, automated control of temperature and moisture, preferably under cover. Ideally, machinery to separate worms from waste quickly and efficiently with minimal labour is necessary. Machinery using rotating mesh cylinders is available but tends to be slow and laborious. New and much more efficient methods have been developed at the AFRC National Institute for Agricultural Engineering, Silsoe, Beds.

Production of earthworm protein

Earthworms consist of 60-70% protein, 7-10% fat, 8-20% carbohydrate and 2-3% minerals with a gross energy of 4000 $K.cal.Kg^{-1}$. The nutrient spectrum of worm tissue is excellent, at least as good as meat or fish meal and particularly rich in the essential feed amino acid lysine and the sulphyr containing amino acids (Figure 4). They are rich in vitamins particularly niacine, riboflavin and vitamin B_{12}, valuable in animal feeds. In order to use earthworms as an animal feed protein source, they must be separated and cleaned from waste and processed into a powder or paste suitable for mixing into dry feeds. This can be done by blanching or fixing the protein in other ways, then freeze drying, heat drying or ensiling.

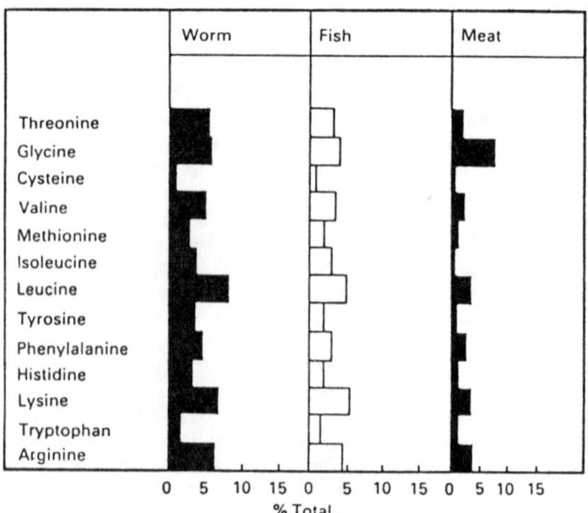

Figure 4. Amino acids in protein sources

Analyses of worm tissues have been made by various workers including Lawrence and Millor (1945), IcInroy (1971), Fosgate and Babb (1972), Schulz & Graff (1977) Yoshida and Hoshii (1978), Sabine (1978) and Taboga (1980). Most have concentrated on the amino acid concentrations, and agree on the beneficial concentrations of essential amino acids present. Work at Rothamsted has shown that they are also rich in long-chain saturated fatty acids such as linoleic acid (18 : 2W6), dihomo-linolenic acid (18 : 3W6) and arachidonic (20 : 4W6). These are essential for feeding non-ruminant animals that cannot synthesize these compounds. The chemical composition of different species of earthworms differs little and is relatively unaffected by the type of waste upon which the worms feed.

The ultimate test of the value of earthworms as animal feed is to include it in animal diets. Guerrero (1983) found that *Tilapia nilotica* fingerlings grew better on worm meal than commercial meal. Hilton (1983) reported that trout did not grow as well on worm meal as on fish meal but Tacon *et al*., (1983) found that trout fed on worm meal from *Allolobophora longa* and *Lumbricus terrestris* grew as well or better than those fed commercial trout pellets. The preparation of *Eisenia foetida* protein affected its palatability to trout. Workers who fed worms to chickens (Harwood, 1976; Sabine, 1978; Mekada *et al*, 1979; Jin-you *et al*., 1982, Taboga, 1980 and Guerrero, 1983) all reported that the birds grew as well or better on earthworm diets than on conventional feed. Sabine (1978) Harwood & Sabine (1978) and Jin-you *et al*., (1982) grew pigs on a worm protein diet and reported that they grew as well or better than on commercial feed protein. Schulz and Graff (1977) fed earthworm meal to rats and mice and found they grew faster than on other sources of protein.

There seems little doubt that if earthworm protein can be produced and processed economically it will provide an excellent animal feed additive.

5. Production and characterisation of worm-worked compost

Most plant growth media are based on peat, to which may be added grit, sand or other bulking agents (about 75 peat to 25 sand). Addition of nitrogen and other inorganic nutrients including phosphorus, potassium and trace elements give a mixture which promotes germination and root growth. The material should have a good water-holding capacity, and porosity together with free drainage and stable structure. Medium grade sphagnum peats are used commercially but these are becoming increasingly scarce; predictions have been made that outside USSR, supplies in Europe are sufficient for only about 10-20 years. Peat-based composts tend to hold nutrients less strongly than soils, thus nutrients are easily leached out by rain or excessive watering. Hence, a cheap, easily-produced material, which possesses many of the properties of peat and is in plentiful supply would be very desirable. Many attempts have been made to produce composts from animal and vegetable wastes, sometimes on a large scale. These products have tended to suffer from variability in structure and nutrient composition, presence of seeds or moulds and lack adequate moisture-holding capacity, resulting in drying and shrinkage.

Earthworms can accelerate the process of breakdown of animal wastes greatly by fragmenting the organic matter, thereby increasing the surface area available for growth of microorganisms and thus promoting further breakdown.

By careful management of earthworm activity, separated animal solids and straw or wood-shavings based animal wastes can be broken down to a peat-like material very rapidly, the time taken varying from 2-4 weeks for separated solids to 2-3 months for straw or shavings-based animal wastes. These materials generally possess a good structure, porosity and moisture-holding capacity coupled with reasonable quantities of plant nutrients (Table 4). The action of worms on the waste material is such that most of the nitrogen is converted to available nitrate by enhanced microbial activity. Similarly, the amount of soluble phosphorus, potassium and magnesium available to plants also appears to be increased. There may however, be some nutrient imbalances resulting from nutrient deficiencies or immobilisation, but this can be corrected easily by addition of suitable inorganic suplements to produce a balanced growth medium.

Table 4. Major Plant Nutrient Elements in Worm Processed Animal Wastes

Waste Material	Element Content (% dry Wt.)					
	N	P	K	Ca	Mg	Mn
Separated cattle solids	2.20	0.40	0.90	1.20	0.25	0.02
Separated pig solids	2.60	1.70	1.40	3.40	0.55	0.03
Cattle solids on straw	2.50	0.50	2.50	1.55	0.30	0.05
Pig solids on straw	3.00	1.60	2.40	4.0	0.60	0.05
Duck solids on straw	2.60	2.90	1.70	9.50	1.00	0.10
Chicken solids on shavings	2.75	2.70	2.10	4.80	0.70	0.08
Levington Compost	1.80	0.21	0.48	0.94	2.20	0.02

Plant growth trials at Rothamsted Experimental Station, the National Vegetable Research Station, Oaklands Agricultural College, A.D.A.S. Wolverhampton and Lea Valley and Efford Experimental Horticulture Stations have shown that worm-processed animal wastes are suitable as plant growing media. A wide range of plants have been grown successfully in both undiluted wastes or a number of mixes including 3 : 1 or 1 : 1 ratios of worm-worked wastes to peat, pine bark or Kettering loam. Results have frequently been better than with recommended growing media such as Levington Compost, and seed germination has been found to be more rapid for most species of plant.

Plants tested, include vegetables such as: aubergine, cabbage, capsicum, cucumber, lettuce, radish and tomato, bedding plants such as Alyssum, Antirrhinum, Aster, Campanula, Calceolaria, Cineraria, Coleus, French Marigold, Plumose Asparagus, Polyanthus, Salvia and Sweet Pea, and ornamental shrubs such as *Eleagnus pungens*, *Cotoneaster conspicua*, *Pyracantha*, *Viburnum bodnantense*, *Chaemaecyparis lawsonia*, *Cupressocyparis leylandii* and *Juniperus communis*.

The waste, when worked by the worms, is at about 75% M.C. Some form of partial sterilization such as by heating to between 60° and 80°C, may be used to kill residual earthworms and their cocoons, and lessen pathogen problems.

There are a number of potential markets for worm-worked composts, including organic growers who currently have no totally organic seedling and potting compost. Worm-worked materials could also be used extensively as tree or shrub planting media as well as all purpose plant growth media.

Table 5. Collaborators in Vermiculture Project

Study	Collaborative organisation	Personnel involved
Vitamins in worm tissue	Roche Products Ltd	M.E. Putnam
Processing of worm protein	National College of Food Technology	G.T. Worgan
Fish feeding trials	Stirling University	A.G.J. Tacon E.A. Stafford
Chicken feeding trials	Poultry Research Centre	C. Fisher
Pig feeding trials	National Institute for Research in Dairying	K.G. Mitchell B.F. Pain
Pathogen transmission	Institute for Research in Animal Diseases	P. Jones
Engineering, Modelling, Economics	National Institute for Agricultural Engineering	V.R. Phillips R.S. Billington J.S. Price A. Wilkins R.S. Fieldson N. Foote I. Jefferies
Plant growth trials	M.A.F.F., Wolverhampton	S.J. Richardson
Mushroom-growing trials glasshouse crops trials	Glasshouse Crops Research Institute	P.B. Flegg A.C. Bunt
Seedling-growth and block-making trials	National Vegetable Research Station	E.F. Cox
Shrub-growth trials	Efford Experimental Horticultural Farm	M.A. Scott
Bedding-plant trials	Lee Valley Experimental Horticultural Farm	J.G. Farthing
Plant growth trials Cattle and pig waste trials,	Oaklands Agricultural College	R.B. Blossom M.J. Hall L.E. Boon A.J. Hawkins
Cattle waste trials	Commonwork Enterprises	N.E. Wates S.A. Crocker
Duck waste trials	Cherry Valley Farm	J.I. Richards B.J. Wilson
Turkey waste trials	Forge Farms	T. Denham-Smith
Potato waste trials	Walker's Crisps	M.A. Kirkman

Table 5 (continued)

Study	Collaborative organization	Personnel involved
Spent mushroom compost trials	Rank, Hovis McDougall	D. Doling J. Delves-Broughton

6. Development of vermiculture, collaborative research and commercialization

The studies described, have been possible only by a broad multi-disciplinary approach with considerable co-operation and collaboration from numerous people, research stations, universities and comercial organisations, as summarized in Table 5.

The compost and protein production system using earthworms has been commercialized by the formation of a company called British Earthworm Technology (B.E.T.) with Government support through the British Technology Group. This company operates profit-sharing contracts with farmers and industrial organizations to grow worms in their organic wastes. B.E.T. will then process and market the wastes and protein.

Acknowledgements

The authors thank the individuals listed in Table 4 for their help in this project. W.F. Raymond, J.K.R. Gasser and D.C.M. Corbett for their administrative and financial encouragement. Also to P. Townley, J.R. Lofty, R. Bardner, J.E. Ashby, M.D. Russell, E. Neale, K. Wilkinson, D. Knight, M.H. Morgan, A.F. Niederer, R.J. Bryson, T. Williams, E. Fisher, A.E. Johnston, V. Cosimini, J.M. Hill, G.R. Cayley and J.E. Beringer for their support in the research and advice.

REFERENCES

1. EDWARDS, C.A. (1982). Production of Earthworm Protein for Animal Feed from Potato Waste. In: Upgrading Waste for Feed and Food. Ed. D.A. Ledmond, A.J. Taylor & R.A. Lawrie, 153-162.
2. EDWARDS, C.A. (1983). Earthworms, Organic Waste and Food. Span, Shell Chemical Co., **26** (3), 106-108.
3. FOSGATE, O.T. and BABB, M.R. (1972). Biodegradation of animal wastes by Lumbricus terrestris. Journal of Dairy Science, **55**, 870-2.
4. GUERRERO, R.D. (1983). The culture and use of Perionyx excavatus as a protein resource in the Philippines. In. Earthworm Ecology. Ed. J. Satchell, Chapman & Hall, London & New York. 309-313.
5. HARTENSTEIN, R., NEUHAUSER, E.F. and KAPLAN, D.L. (1979a). A Progress Report on the Potential Use of Earthworms in Sludge Management. In Proceedings of 8th National Sludge Conference, Florida, 1979. Information transfer Inc., Silver Springs, Maryland.
6. HARTENSTEIN, R., NEUHAUSER, E.F. and KAPLAN, D. (1979b). Reproductive Potential of the Earthworm Eisenia foetida. Oecologia, **43**, 329-340.
7. HARTENSTEIN, R. (1982). Metabolic Parameters of the Earthworm Eisenia foetida in Relation to Temperature. Biotechnology and Bioengineering, **24**, 1803-1811.
8. HARWOOD, M. and SABINE, J.R. (1978). The nutritive value of worm meal. Proceedings of the Second Australasian Poultry Stockfeed Convention, Sydney, 164-171.

9. HILTON, J.W. (1983). Potential of freeze-dried worm meal as a replacement for fish meal in trout diet formulations. Aquaculture, **32**, 277-283.
10. JIN-YOU, X., XIAN-KUAN, Z., ZHI-REN, P., ZHEN-YONG, H., YAN-HUA, G., HONG-BO, T., XUE-YAN, H. and QIAO-PING, X. (1982). Journal of South China Normal College, **1**, 88-94.
11. KAPLAN, D.L., HARTENSTEIN, R., NEUHAUSER, E.F. and MALECKI, M.R. (1980). Physiochemical requirements in the environment of the earthworms Eisenia foetida. Soil Biology and Biochemistry, **12**, 347-352.
12. LAWRENCE, R.D. and MILLOR, H.R. (1945). Protein Content of Earthworms. Nature (London) 3939, April 28 1954, 517.
13. McINROY, D.M. (1971). Evaluation of the earthworm, Eisenia foetida as food for man and domestic animals. Feedstuffs. Feb. 2 1971. 37, 46.
14. MEKADA, H., HAYASHI, N., YOKOTA, H. and OKOMURA, J. (1979) Performance of growing and laying chickens fed diets containing earthworms (Eisenia foetida). Japanese Poultry Science, **16**, 293-297.
15. NEUHAUSER, E.F., HARTENSTEIN, R., and KAPLAN, D.L. (1979). Second Progress Report on Potential use of Earthworms in Sludge Management. Proceedings of 8th National Sludge Conference. Florida, 1979. Information Transfer Inc. Silver-Springs, Maryland.
16. NEUHAUSER, E.F., KAPLAN, D.L., MALECKI, M.R. and HARTENSTEIN, R. (1980a) Materials supporting weight gain by the earthworm, Eisenia foetida, in waste conversion systems. Agricultural Wastes, **2**, 43-60.
17. NEUHAUSER, E.F., HARTENSTEIN, R. and KAPLAN, D.L. (1980b). Growth of the earthworm Eisenia foetida in relation to population density and food rationing. Oikos, **35**, 93-98.
18. SABINE, J.R. (1978). The Nutritive Value of Earthworm Meal. In Utilization of Soil Organisms in Sludge Management. Ed. R. Hartenstein, S.U.N.Y. Syracuse, 122-130.
19. SCHULZ, E. and GRAFF, O. (1977). Zur Bewertung von Regenwurmmehl aus Eisenia foetida (Savigny, 1826) als Eiweissfuttermittel. Landbauforschung Volkenrode, **27**, 216-218.
20. TABOGA, L. (1980). The nutritional value of earthworms for chickens. British Poultry Science, **21**, 405-410.
21. TACON, A.G.J., STAFFORD, E.A. and EDWARDS, C.A. (1983). A preliminary investigation of the nutritive value of three terrestrial lumbricid worms for rainbow trout. Aquaculture, **35**, 187-199.
22. TOMLIN, A.D. (1983). The earthworm bait market in North America. In Earthworm Ecology from Darwin to Vermiculture. Ed. J. Satchell, Chapman and Hall, London & New York, 331-338.
23. YOSHIDA M. and HOSHII, H. (1978). Nutritional value of earthworms for poultry feed. Japanese Poultry Science, **15**, 308-311.

DISCUSSION

TUNNEY: You speculate that the very good plant growth results obtained with worm-worked are due to phytohormones. Could you tell us what plant hormones or related compounds are involved?

EDWARDS: I may have overemphasized this aspect. We have obtained very good growth of a wide range of species of plants in worm-worked wastes. The results often exceed that which might be expected due to nutrients. It might be due to physical factors such as water-holding capacity. Some of our results which have shown acceleration or delay of flowering, stem elongation or change of form of plant could be accounted for by the action of phytohormones. It is well known that microorganisms can produce such materials but most of them act on plants at a critical concentration. More research, which we are just beginning, is needed to resolve this question.

VERDONCK: Have you analysed the physical properties of the earthworm compost because this can be the parameter influencing the growth of plants?

EDWARDS: We have only looked at the more obvious physical characteristics such as moisture-holding capacity and pore space. We plan to do a full analysis of the physical and chemical characteristics which could influence the usefulness of a compost.

WOOD: What are the major microbial groups present in the worm worked wastes?

In other composted materials the Pseudomonads often dominate at this stage. This may have implications for plant growth.

EDWARDS: I am not a microbiologist. You will meet my research assistants this afternoon and I am sure they can answer this question. They have done thorough microbial analysis of all organic wastes upon which we have worked.

LYNCH: We have been disappointed in our own studies of the effect of exogenous phytohormones. Pseudomonads can penetrate roots and any products may therefore have a potentially greater effect. It is possible that you have a population of this type giving a plant growth response.

STENTIFORD: Is the moisture content of the waste in the range 60-90%?

EDWARDS: The optimal moisture content is 80-90%. Initially it was thought that conditions that favoured the worms were also best for waste treatment. It now seems that the worms can be made to work harder in a less nutritious waste.

THE ECONOMIC FEASIBILITY OF EARTHWORM CULTURE ON ANIMAL WASTES

R.S. FIELDSON
Agricultural Economist, NIAE

Summary

Problems and practices of animal waste disposal are reviewed, and it appears that slurry is quantitatively and qualitatively a greater problem than solid wastes, especially on intensive livestock units. Large pig units experience greater difficulty in disposing of waste, and cause more water pollution and odour problems, than dairy or poultry units. The processes necessary for a vermiculture system are outlined, from initial preparation of the animal waste to processing of the end products. Required operations and alternative methods of performing them are briefly described. The potential benefits of vermiculture are the worms, the worm-worked waste, and reductions in waste disposal costs. Four models of farm-based vermiculture systems were developed which showed that worm-worked waste is likely to be the largest contributor to project benefits. The assumed system was profitable only on large-scale units, and smaller farms require a system which is not based on separated slurry. The existence of economies of scale suggest that centralised operations could be more attractive than farm-based systems, especially in areas with high livestock densities. However, their profitability is sensitive to variations in transport costs.

1. INTRODUCTION

The paper presented by Dr. C.A. Edwards outlined the trials which have been performed into earthworm culture on animal wastes at Rothamsted Experimental Station, and went some way towards specifying the requirements of a practical on-farm system. The potential benefits of earthworms as a source of protein, of worm-worked waste as a source of good quality compost, and of savings in waste disposal costs resulting from the system were mentioned. However, few farmers are going to invest their capital in an innovative waste processing system such as this unless they consider that there is a good probability of economic returns to their investment. In addition, although much primary research is inevitably speculative, it is appropriate that agricultural research funds for applied work be directed towards projects which offer the prospect of substantial benefits to the industry in the reasonably near future.

A study was undertaken in 1982 by the Operational Research Department of the National Institute of Agricultural Engineering (NIAE) with the objectives of assessing the economic feasibility of earthworm production on British farms, specifying which components of the system are particularly crucial to its success, and identifying assumptions or operations which appeared to require further research before a more precise evaluation of the process could be undertaken (1).

Even when an economic evaluation is made of a clearly specified project using established technology there are a number of uncertainties, for example concerning future values of inputs and outputs. But the uncertainties are much greater in an evaluation of a system which is still at the research stage. In 1982, when the study of the economics of vermiculture was undertaken, many of the input-output relationships were not precisely quantified, the effects of environmental factors on worm growth and reproduction were still being investigated, and work had scarcely begun on methods of farm-scale operation of a worm-production system - techniques used at the research station are frequently not reproducible for large-scale production within the constraints of existing farm enterprises. Therefore rather arbitrary assumptions had to be made in the modelling,some of which can now be refined as efforts to expand commercial worm production have helped to identify optimum techniques and processes.

The results of the study are summarised in sections 2 to 4 under the following headings:

Section 2 The Problem

" 3 The Processes

" 4 The Profitability

The first of these examines the problems connected with animal waste disposal in Britain, including the costs of disposal and pollution problems. It analyses briefly the structures of the main livestock industries and considers the methods of housing and waste handling which determine the forms of waste which are available - the raw material of the vermiculture system.

The next section looks at the processes which have to be performed in a vermiculture system, from preparation of the animal waste through to packaging of the end products, describing briefly what operations are necessary and what alternative methods of achieving them are available.

Section 4 outlines the main potential benefits of the system. Conclusions are drawn from the case studies of the 1982 report on the principal determinants of the profitability of farm-based vermiculture systems.

2. THE PROBLEM

There are two main reasons for looking at the problems of farm waste disposal in the context of an economic study of vermiculture. First, it is necessary to obtain information on types of livestock housing and feeding and systems of slurry handling, so as to establish the form the waste is in and how a vermiculture system could be integrated into the unit. Secondly it is important to identify the types of unit and locations where costs of waste disposal or other problems with animal wastes are such that a method of waste treatment such as vermiculture is particularly attractive.

Vermiculture has been shown to be possible on cattle, pig and poultry manure and the distribution, and handling and disposal practices of each of these types of waste are examined in this section. Sources of information include MAFF Agricultural Statistics, the Meat and Livestock Commission, ADAS Farm Wastes Unit, and the Department of Agriculture and Fisheries for Scotland. However, there appears to be a lack of information on present practices for the disposal of wastes from intensive units.

2.1 Dairy cattle

Dairy cattle are the largest single source of animal wastes in

Britain. In 1976 an estimated 22 million tonnes of excreta were produced during the winter (2). On most of Britain's dairy farms the main problem with animal waste is not how to dispose of it, but the costs of storing it prior to disposal, and the related management problems including the avoidance of pollution. Only a small proportion of dairy herds are kept on farms which do not have sufficient land on which to spread the manure produced by the herd: if there is enough grassland to feed the herd there is almost certain to be enough to permit the disposal of the waste. However, much of the waste produced while the cattle are housed over winter cannot be spread until the spring and the costs of storing it are likely to exceed its fertiliser value when spread on the land.

The typical dairy unit consists of cubicle housing with excreta handled in liquid or semi-liquid form. It has been estimated that 70% of dairy herds produce the majority of their excreta plus bedding in the form of slurries, and the remainder as a solid consisting of excreta plus straw (3). The majority of British dairy farms produce most of all their bulk foods, for summer and winter feeding, on the farm. Typical stocking rates range from 1.5 to 2.5 cows per ha (4) at which level it is quite feasible to apply all the waste produced while the cattle are housed in winter to the grassland. The average period of winter housing is around 180 days, so at about 40 litres of excreta per cow per day some 7200 litres or 7.2 tonnes of excreta will be produced by each animal over winter. On a farm with a very high stocking ratio of 2.5 cows per ha and half its grass area producing conserved grass (the area to which most manure will be applied), this represents 36 tonnes of waste per ha of grass for conservation.

According to ADAS recommendations (5), 20 cu m of slurry per hectare can be applied after the first cut of grass and a further 14 cu m after the second cut. Recommendations by the Scottish Agricultural Colleges (6) indicate a maximum of 55 cu m per hectare in one application, and work in Northern Ireland (7) has shown that applications of up to 56 cu m per hectare can have beneficial effects on herbage yields, though weather and soil conditions may limit the spreading operation. So the wastes accumulated on even a highly stocked farm can readily be applied to conserved grassland, as well as in more modest quantities to grazed grassland, and to other crops.

The potential fertilizer values of cattle manure and slurry were calculated by ADAS (5) at £1.72 and £1.52 per tonne respectively (1982 prices) i.e. £12.38 and £10.94 per cow per winter. This total value is only realised if the manure is spring applied and if all the available nutrients are fully utilised by the crop. In fact many farmers do not balance their fertilizer use according to crop needs and manure application so in effect undervalue the waste.

Even where slurry can be spread daily so that little storage is needed, the costs of disposal, at about £22 per cow per winter, exceed the manurial value. More usually waste has to be stored for 4 to 6 months before it can be applied to the land, in which case a lagoon or above-ground storage system is required. Costs are then likely to be at least £70 per cow per year if all handling, storage and spreading facilities are included. So the costs of slurry disposal are considerably higher than the benefits gained by its application to the land.

Water pollution and odours can cause problems for dairy farmers, and in some cases have led to the installation of separators and other equipment to treat slurries (8). While such problems are rarely severe in the dairy industry, they could provide an added incentive for the adoption of vermiculture on those farms affected.

2.2 Pigs

Waste production from pigs totals about 9 million tonnes per annum (2), and most of this is produced indoors. Some parts of Britain such as Suffolk and Humberside have particularly high concentrations of pigs. About 80% of pig farms handle their waste as liquid slurry compared to 20% dealing with scraped slurry (9).

The pig industry is characterised by large-scale intensive units, many of which do not have sufficient land on which to dispose of their waste. In 1980 there were 1962 herds with over 1000 pigs, containing 58% of the pigs in England and Wales. Of these herds 27% have less than 10 ha of land and 205 have less than 2 ha.

The amount of excreta produced per pig varies widely according to the size of the animal and its diet, averaging about 5 kg per day. Thus for each pig place the annual output of slurry is approximately 1.75 tonnes.

The maximum level of slurry which can be applied per ha depends on soil conditions, crops grown, etc. but it should usually be feasible to apply up to about 50 tonnes per ha without serious effects on soil structure or build-up of toxic elements. This is equivalent to the production of excreta from almost 30 pigs. It is calculated from 1979 Agricultural Census figures that 18.6% of holdings with 45% of total pig numbers exceeded this level of stocking. A 1973 survey of pig farms in Scotland (10) covering over half of the pig population showed that 49% of the sample farms had stocking rates which were considered to represent a pollution risk (fixed at 25 "pig units" per ha). It also showed that 35% of the effluent produced on sample farms was spread on the land of other farms. Informal contracts between pig producers and neighbouring farmers are common in Britain, the receiving farmer usually meeting the cost of transport and spreading. Dung for straw exchange agreements are also found.

If slurry can be spread by tanker direct to the land regularly throughout the year then there is no need for long term storage, odour risks are reduced, and costs of disposal are kept down to about £3 per pig place per year (ADAS estimate (3)). More usually additional storage is required such as an above ground retention tank with associated pumps, etc. raising the cost to £6.50 per pig place.

One of the main problems of pig slurry storage and disposal is the control of pollution, especially odours released during agitation or emptying of the store. There were 2095 complaints to environmental health officers about odour from pig farms in 1980/81, making pig slurry by far the largest subject for such complaints against agriculture. There were also 253 warnings about water pollution resulting from pig farms and 25 prosecutions in England and Wales during 1981. Pollution problems have been so severe in some cases that farmers have installed expensive treatment plants such as slurry separators, aerators, and anaerobic digestors. Odour control was one of the main reasons reported by pig farmers for installing slurry separators in an ADAS survey (8). Vermiculture may appeal to these units as a means of solving their pollution problems.

2.3 Poultry

Production of poultry manure was about 1.8 million tonnes in 1976, over 90% from chickens (2). There has since been a slight decline in poultry numbers. Most broilers are housed in large buildings with some form of litter system. The chicken droppings are added to this throughout the growing period, and the waste is then removed when the house is emptied. It is either applied directly to the land or stocked to await later land application. Laying hens are generally housed in battery cages, either tiered with some form of mechanical cleaning or stepped with the waste from the cages simply falling into a deep pit from which it is removed once a year or less.

Chicken manure has a high dry matter content of around 25% making it relatively easy to handle and stack. Further drying during storage may raise this to 40% or more. Because of its high dry matter content and high fertilizer value, poultry waste can usually be applied to land without difficulty. Expensive storage facilities are not necessary for chicken waste which can be stacked in the open before spreading, though sometimes it is washed out with water and thereafter treated as a slurry.

In common with the pig industry, the poultry industry is characterised by large scale highly intensive units, many of which have insufficient land for the disposal of their wastes at safe levels of application. For example, 53% of broilers are on units with over 100,000 birds, and two-thirds of these have less than 10 ha of land. With an average output of excreta per 1000 birds of 125 kg per day or over 40 tonnes per annum, it is clear that many poultry farms have large surpluses of waste. These can usually be disposed of to neighbouring farmers, some of whom pay for the waste. Brownlie (10) found that 89% of the poultry farmers in his sample said that it was easy to arrange contracts for the disposal of manure. However, some intensive units cannot find sufficient local farmers willing to take their surplus, or find that organisational, storage and transport problems become acute, and in such cases the vermiculture system may be appealing.

The potential of vermiculture as an odour controlling measure is also relevant to poultry units, as complaints are received about odours from manure storage and spreading.

2.4 The animal waste problem - Conclusions

As most pig excreta and about two-thirds of the waste from dairy farms is stored as slurry, liquid or semi-liquid waste is quantitatively a much greater problem than solid waste. Slurry is also a more serious problem qualitatively as it presents a greater risk of both water pollution, and of offensive odours which result from microbial activity in anaerobic slurries.

Dairy farms usually have sufficient land on which to spread the excreta produced by the herd when housed, and dairy herds have not produced large numbers of odour complaints. However, costs of storing and spreading cattle wastes are typically much higher than the fertiliser benefits so vermiculture may be attractive on such farms to reduce the costs of waste handling.

Large herds on small areas of land are more common in the pig than dairy industry, necessitating arrangements for off-farm disposal of slurry and in some cases costly anti-pollution systems. Slurry handling and treatment equipment is again expensive. If vermiculture can reduce the volume of waste and hence the cost of disposal, and also ease odour problems, then the pig industry could offer a large potential demand for the system.

Poultry waste is readily stackable and so less costly to store. Most units appear to find ready outlets for their manure, though there are cases where this is not true, e.g. on very large units. Odour problems do occur, and especially if a poultry unit is near a residential area this may provide some justification for vermiculture as a means of odour control.

3. THE PROCESSES

The main processes in a system of worm production are shown in Fig 1,

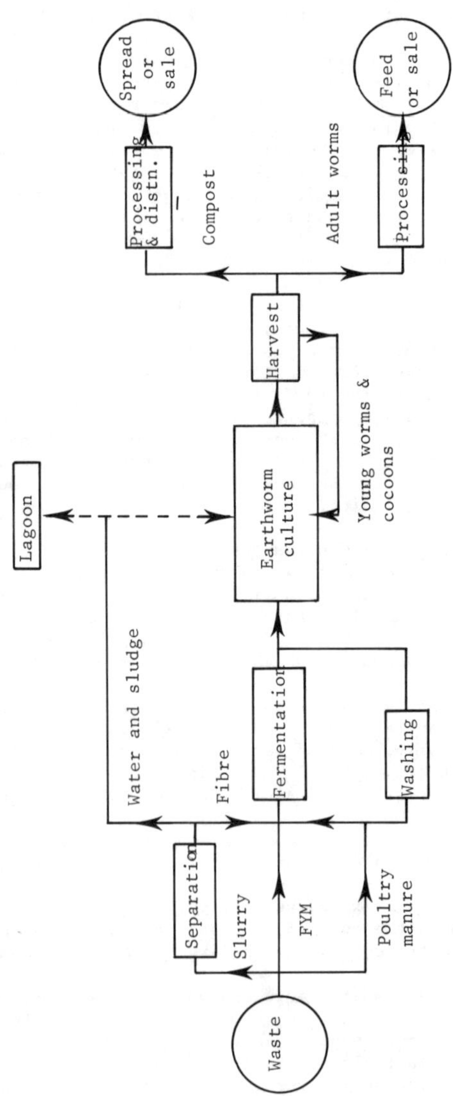

Fig 1 Vermiculture system processes

and each operation will now be described and the methods of achieving it outlined.

3.1 Preparation of waste

Studies have been carried out on the optimum and critical levels of various important factors which influence worm survival, growth and reproduction and these are discussed in Dr. Edward's paper. It is possible to stipulate, for example, nitrogen concentrations, moisture content and pH, which must be met by the waste on which worms are to be bred. The processes which are necessary to prepare the waste depend on its initial state.

For example, the dry matter (d.m.) of whole slurry is around 13% for dairy cattle and 10% for pig slurry, and is commonly reduced further by additional water entering the slurry. This is well below the level required for earthworm culture. To increase the d.m. of slurries they must either be mixed with a bulk material such as straw or wood shavings, or some form of separation of the liquid is required. The feasibility of worm culture on whole slurry plus chopped straw has now been demonstrated at Rothamsted. Such a method increases both the length of time the waste mix needs to compost before the addition of earthworms and the time earthworms take to work the compost. In addition costs of acquiring and transporting the bulking material must be borne in mind. Separation of the fibrous part of the slurry from the liquid fraction can be achieved by filtration, evaporation, or by machine. Filtration methods, including seeping wall compounds and hessian strainers, give slow results and are unsuitable for vermiculture systems involving daily separation. Evaporation requires a high energy input so is expensive. Mechanical separators are therefore the most suitable type for vermiculture as they provide stackable solids quickly. However, they are still uncommon (less than 500 in use in England) and there have been some reliability problems, particularly relating to the ancillary equipment (8). Capital costs of installing a separator and associated equipment are considerable.

Farmyard manure has a high enough d.m. to allow the breeding of worms without further treatment. However, poultry manure, though it has a high d.m., also has a high ammonia content which would be toxic to earthworms, and so requires treatment to lower this level before worms are introduced.

3.2 Breeding environment

Some form of bed is required in which the separated solids or FYM can be deposited and the worms introduced. Most trials have been conducted with on-floor beds with a concrete base and walls of bricks, sleepers, or similar materials. Maximum depth is around 0.5 m for pig waste and 1.0 m for cattle waste, otherwise the deeper waste becomes anaerobic. The bed may be filled initially and then emptied (perhaps by tractor if access is possible) when the waste is fully worked. Alternatively waste can be added gradually in layers so that the worms work their way up through it. Experiments at Rothamsted Experimental Station have shown that earthworms will move to new waste once the old waste has been worked. A system commonly used in small-scale earthworm production in developing countries is to deposit the waste in boxes which can be emptied easily when the waste is worked. There is also the possibility of a more sophisticated automated continuous system, using a compartmentalised conveyor, but the engineering problems of such a method have not yet been tackled.

Whatever type of bed is used, some control of the breeding environment is required. For example, research has demonstrated that *Eisenia Foetida* grows and reproduces best at temperatures between 15°C and 20°C and outside the range 10°C to 30°C worms lose weight and eventually die. Therefore if worm beds are to be in use all year round some heating and insulation is necessary between the months of about October to March in Britain. Worm culture could be carried out only over the summer months, but it is in winter that slurry storage from housed animals is a particular problem, so additional storage would be needed. Moreover, the worm beds, when active, would have to cope with double the daily production of waste.

Various methods of temperature control have been used, including polythene tunnels, fibreglass insulated tunnels, well insulated outdoor beds and underbed heating.

Additional environmental controls include moistening of beds to prevent drying out (this can be done manually or by sprinklers), and possibly the provision of lighting to avoid worms migrating from the beds.

3.3 Harvesting

The length of time needed to convert the waste and the final worm population depends on the type of waste, the inoculation level and environmental factors. But when the waste has been fully worked it will be necessary to harvest the worms i.e. to separate them from the worm worked waste, in order to obtain young worms for inoculation into new waste, and to process the adults for feed.

Methods of harvesting earthworms are still at the development stage. Interest has so far been focussed on mechanical methods of separation. It is known, however, that earthworms will respond to various stimuli, such as heat, light, electric current and moisture, and that they also react to certain chemicals such as formalin, ammonia and potassium permanganate. It may therefore be possible to use some stimulus to remove earthworms from compost. Alternatively the tendency of earthworms to migrate to new sources of food might be exploited. Additional processes may be necessary in order to bring the waste into a form which allows easy separation of the worms. For example, at high moisture contents "balling" of the waste may occur, necessitating drying of the waste and/or breaking up of clods before worms can be removed efficiently.

Research currently underway at the NIAE should soon result in conclusions on the most effective and economic method of worm separation.

3.4 Worm processing

Earthworms require processing for four reasons:

1. To wash, sterilise, and remove pathogens and parasites
2. To preserve and maintain a uniform product
3. To ensure palatability
4. To simplify distribution and marketing

External washing is necessary to remove any residue of waste from the worms, and waste should also be removed from the worms' intestines. This can be achieved by soaking them in water. The risk of pathogens in the worms can be reduced by sterilisation and this has the additional advantage of stabilising the worms against decomposition.

Without processing, earthworms deteriorate quickly after removal from the compost, so unless feeding is to follow soon after separation some method of preservation must be adopted. Possible methods include freezing, freeze drying, air and heat drying and pickling or ensiling. The choice of method affects the value of the end product.

3.5 Processing of worm worked waste

The processing of the worm worked waste depends on its condition after the removal of the worms and the end use to which it is to be put. If it is applied directly to the land as a soil conditioner no further processing is needed, but if it is to be marketed it is likely to require drying, blending and packaging. Drying is necessary because the worked waste will be at a high moisture content of around 75% at which it will not store without deterioration, and blending because sale to the horticultural market requires a standardised and consistent product which cannot be guaranteed by the vermiculture system, especially if the quality and types of waste used vary.

4. THE PROFITABILITY

The three potential economic benefits of vermiculture are the worms, the worm-worked waste and reductions in waste disposal costs.

Comparisons of earthworms with other animal feedstuffs have shown them to be high in protein and essential amino acids, and that when suitably processed they appear to be palatable to various animals (11, 12, 13). Feeding trials have shown that animals will thrive on earthworm meal, growth rates being similar to those from conventional feeds. Assuming that dried earthworms can substitute for existing protein sources in animal feed rations, their value per tonne is likely to be from £200 to £400, though if they prove suitable for certain specialised markets (such as starter rations) their value could be higher.

The worm-worked waste is an odourless compost, similar to peat, which is potentially valuable in horticulture. Though varying slightly according to the type of animal waste from which it was produced, the worm-worked waste has higher nutrient content than peat and good slow-release characteristics. Growth trials have been carried out comparing worm-worked waste with other growth media (14). Results have been encouraging, suggesting that it may form a good basis for horticultural compost though it will require blending to ensure consistent quality. There is scope for processing and marketing of worm-worked waste as a higher value product, though this is beyond the scope of the individual farmer. For example, in Italy several companies have succeeded in marketing their product as an organic-based pot plant additive at a very high price. However, such openings could not absorb a large volume of material as would be produced if the system were widely adopted. The value of worm-worked waste on the farm is therefore likely to be between £20 and £40 per tonne, though in the shorter term individuals may achieve higher prices for their production. Even if it became uneconomic to market compost to the horticulture industry it could be land spread as a soil conditioner or fertiliser, but its value would then be very low.

The third area of benefit which could result from vermiculture is reduction of waste management costs. As costs of handling, storage and disposal of animal wastes are substantially greater than their manurial value, the reduction in volume that occurs when worms are grown on it brings about a reduction in waste disposal costs. There may also be qualitative improvements in animal waste resulting from vermiculture and its associated processes. For example, worm-worked waste is odour free and readily stackable, and separated liquid is more easily pumped and less prone to crusting and stratification than raw slurry. Such qualitative improvements may also reduce storage and handling costs.

The profitability of a vermiculture system depends on a number of factors which have not yet been precisely quantified, so it cannot be estimated with precision. However, using the best information available four case studies were constructed of possible farm vermiculture systems, based on medium and large dairy herds (100 head and 200 head) and medium and large pig units (500 and 5000 pigs). Details of cost and output assumptions are given in (1). A number of conclusions emerged from these case studies.

The worm-worked waste was clearly the most important component of project benefits, and that on which the viability of the vermiculture system largely rests. On average the compost made up 63% of total benefits, the worms 29% and waste disposal savings 8%. While the figures vary according to the assumed values of worms and compost, they are more-or-less the same for all sizes of unit. The savings which could result in costs of waste disposal, including pollution control, were small in all cases.

There were substantial economies of scale in the assumed system, particularly relating to slurry separation and separation of earthworms from the compost, and as a result the models demonstrated satisfactory profits only for the large-scale units. For smaller units slurry separators would clearly be under-utilised and the introduction of a smaller scale cheaper slurry separator than those currently available on the UK market would make the economics of vermiculture more favourable on such holdings. Alternatively vermiculture systems for smaller farms will have to be based on farmyard manure, or on mixing raw slurry with chopped straw or similar material. The latter requires a larger bed area, and the rate of conversion of waste would be lower, but on the other hand the utilisation of surplus straw may in the future become a major advantage.

The existence of economies of scale suggest that some form of centralised unit would have advantages over scattered farm-based operations. Centralised operations would be particularly attractive in areas with high densities of livestock farms, where waste disposal problems are most prounounced and transport costs minimised. A simple model of one possible centralised system showed it to be economically attractive if the separated liquid fraction could be utilised within the system, but it was highly sensitive to variations in transport costs. Additional work will shortly be undertaken at the NIAE to investigate the economics of alternative methods of centralised worm production.

Labour costs had a major impact on profitability, and only on the 5000 pig unit could a vermiculture system remain profitable if it required an additional full-time employee. Where operations are to be performed by existing staff, much depends on the opportunity cost of their time and this is likely to vary seasonally.

Finally, the case studies were useful in drawing attention to a number of key relationships within the vermiculture system which are crucial to its economic viability but which had not been quantified in earlier research. More detailed economic analysis will be undertaken when the results of subsequent experimental work provide data on such issues.

REFERENCES

1. FOOTE, N. and FIELDSON, R. (1982). A study of the economic feasibility of earthworm culture for feed production and animal waste conversion on livestock units in England and Wales. Div. Note DN/1141, Natn. Inst. Agric. Engng, Silsoe, UK, (unpubl.)
2. LARKIN, S.B.C., MORRIS, R.M., NOBLE, D.M. and RADLEY, R.W., (1981). Resource mapping of agricultural wastes and residues. NCAE, Silsoe, UK.
3. NEILSEN, V.C., ADAS Farm Wastes Unit, Reading, UK. Personal communication.
4. NIX, J. (1980). Farm Management Pocketbook,Wye College, Ashford, Kent, UK.
5. ANON (1982) Farm Waste Management: Profitable utilisation of livestock manures. ADAS Booklet 2081, MAFF, UK.
6. ANON (1980). Handling and utilisation of animal wastes. The Scottish Agricultural Colleges, Publication No. 16.
7. STEWART, T.A. Applying slurry to grassland. Agric. in Northern Ireland, 44, 1, 12.
8. REDMAN, P.L., ADAS Liaison Unit, Silsoe, UK. Personal communication.
9. MORRIS, R.M., and LARKIN, S.B.C. (1981). The energetics of livestock manure management; In Vogt, F. (Ed) Energy conservation and use of renewable resources in the bio-industries, Pergamon.
10. BROWNLIE, T.G., Department of Agriculture for Scotland, Personal communication.
11. McINROY, D.M. (1971). Evaluation of earthworm "E. foetida" as food for man and domestic animals. Feedstuffs, Feb 1971, pp 37-46
12. MEKADA, H.; HAYASKI, N.; YOKOTA, H.; OKUMURA, J. (1979) Performance of growing and laying chickens fed diets containing earthworms (E. foetida). Japanese Poultry Sci., 16, 293-297.
13. TABOGA, L. (1980). The nutritional value of earthworm for chickens. Brit. Poultry Sci., 21 (5), 405-410.
14. FOSGATE, O.T.; BABB, M.R. (1972). Biodegradation of animal waste by Lumbricus terrestris. Aus. J. Dairy Sci., 55 (6), 870-872.

DISCUSSION

GARRETT: At several points in your paper you comment that handling and spreading costs for animal wastes exceed their nutrient value. This is very sensitive to the figure which is used for nitrogen use efficiency - a small increase in this figure can result in fertilizer replacement value exceeding disposal costs.

FIELDSON: That is correct, except to add that storage costs should be included. However, I based my figures on the best studies that are available at present so I hope that they are as reliable as existing information allows.

BIDDLESTONE: In view of the need for heat during the winter is a two-stage system possible? Stage 1 of true composting would provide a heat source for a second stage of vermiculture.

FIELDSON: Yes. It has been striking how several speakers have stressed the problems of overheating during composting, yet in the vermiculture system there is a problem of temperatures falling too low in the winter. I am sure there is scope for some sort of heat exchange between composting material and worm-beds.

NAVEAU: If liquid/solid separation is done, there is an opportunity to digest anaerobically the liquid phase to produce energy for the vermiculture process or any other use.

FIELDSON: Yes, that is correct. On one farm where vermiculture has been introduced (Bore Place in Kent) just such a system is used. Worms are bred on the solid fraction of separated cattle slurry, and the liquid fraction enters the digester, energy from which has been used for generating electricity.

DE BERTOLDI: I wonder if there is any correlation between the fertilizing value of compost prepared from worm-worked waste and its price on the market.

FIELDSON: There appears to be very little relation between the fertilizer value of any composts and their market price. This makes it very difficult to assess the value of worm-worked waste by comparing it with other composts. Market value seems to depend on subjective criteria such as appearance and smell.

FINSTEIN: Is it fair to say that earthworm culture is a management intensive enterprise, compared to "conventional composting", whatever that might be?

FIELDSON: Yes. Environmental control is more important in earthworm production, and an additional produce, the worms, have to be processed, and marketed. The current move to commercialise worm production in Britain attempts to "centralise" some management roles. The farmer will be provided with a straight-forward manual on how to operate and regulate the system, and certain operations such as worm separation and worm and compost processing will be carried out centrally.

SESSION VI

EFFECTS OF TREATMENT ON ODOUR AND ON DISEASE SURVIVAL

A REVIEW OF THE EFFECTS OF ANAEROBIC DIGESTION ON ODOUR AND ON DISEASE SURVIVAL

M. DEMUYNCK, E.-J. NYNS and H.P. NAVEAU

Unit of Bioengineering, University of Louvain, B-1348 Louvain-la-Neuve, Belgium

Summary

At the present time, anaerobic digestion or biomethanation is currently applied for the stabilization of sewage sludges and farm wastes and for the production of energy. It is interesting to know to what extent this process allows the reduction of odour and the control of disease. Based on a review of the literature published in this field, this paper attempts to answer the question.
Concerning odour control, anaerobic digestion appears to solve significantly the odour problems. Compared to aerobic digestion, its effect on odour reduction is to stabilise the waste more.
Concerning the effect of anaerobic digestion on disease survival, it appears that it is essentially a function of 2 factors : first, the type of pathogen considered and second, the scale -laboratory or full-scale equipment- for which this effect is analysed. In theory, the effect of pathogenic reduction by anaerobic digestion is significant for bacterial pathogens, viruses, parasitic cysts and plant pathogens whereas it is poor for parasitic eggs. In practice, especially in completely mixed digesters, this effect is reduced by short-circuiting and by simultaneous drawing off and feeding the digester. Nevertheless, the beneficial effects can be improved by digesting in batch or plug-flow systems or by working at thermophilic temperatures.

1. INTRODUCTION

Anaerobic digestion is by far the most common method of sludge stabilization and is certain to remain so for the foreseeable future (Bruce et al., 1983). The heated process operating at around 35°C is normal, especially for large works. Mesophilic anaerobic digestion of agricultural wastes appears also to becoming much more popular in Europe (Demuynck and Nyns, 1984). However, there is until now little known about the effects of digestion on disease survival. Based on a review of the literature and on known experimental results, this paper attempts to stress the extent to which the anaerobic digestion process allows the reduction of odour and the control of disease.

2. REDUCTION OF ODOUR

Both for sewage sludge and for manure, smells during utilization (spreading) probably cause more complaints than any other aspect of the operation. Odour problems can spoil good public relations and in the case of sludges, make them less acceptable to farmers. These problems can be minimized and controlled by spreading stabilized effluents. As well as aerobic digestion or lagooning for 2 years, anaerobic digestion is able to produce an effluent of relative inoffensive odour (Pike and Davis, 1983). Although quantitative determination of odours is very difficult, attempts at odour measurements have already been tried. Van Harreveld reported by Klarenbeek (1982) measures odour emission during and after land spreading. Air samples are taken after spreading of untreated and anaerobically digested slurry. It is shown that digested slurry reaches threshold levels of detection within 24 hours while for untreated slurry it takes 72 hours (see Fig. 1.).

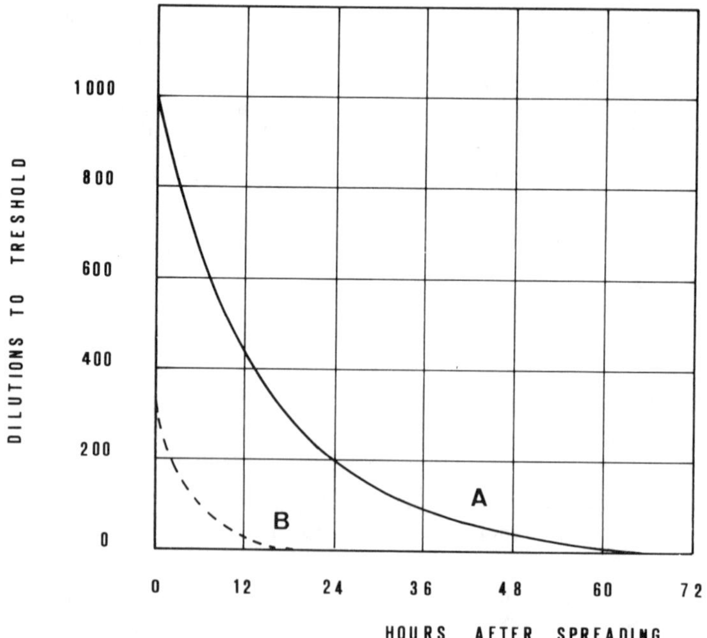

Fig. 1. Decay of odour concentration of two types of manure (A = untreated pig slurry; B = anaerobically digested pig slurry)

It is also calculated that the emission from 1 ha spread with digested slurry is 10% of the emission from 1 ha spread with untreated slurry. According to Offermans <u>et al</u>. (1982), the reduction of smell could justify the fact that cattle prefer grazing on pastures spread with digested manure.

Wenglaff reported by Wellinger (1983) finds that, during anaerobic digestion of pig manure in a full-scale continuous digester, odour reductions of 50 per cent are achieved.

Comparatively, aerobic digestion may be superior : indeed, odour reductions may reach approximately 50 per cent with intermittent aeration and 98 per cent with continuous aeration in oxydation ditches. Measurement of odorous compounds is another way to estimate odour reduction. The main substrates responsible for malodours are compounds such as volatile fatty acids, phenol, p-cresol, 4-ethylphenol,... In continuously aerated manure, p-cresol is completely degraded and the volatile fatty acids are oxidized to a level of approximately 200 ppm (Wellinger, 1983). During the anaerobic digestion process, the malodorous compounds are also drastically reduced (Rygole et al., 1983).

It appears, however that if aerobic digestion may be superior in reducing malodours, its effect is of lower value long term. Indeed, well aerated manure has been reported to smell again after two to three weeks. The oxidized intermediates are relatively quickly reduced again to malodorous compounds as soon as the redox potential allows the growth of anaerobic bacteria. On the other hand, well anaerobically digested manure remains fairly stable over a long period of time. Odour remains unchanged even after 120 days of storage and according to Van Velsen (1981), the concentration of malodorous compounds may even be reduced during that time.

3. CONTROL OF DISEASES

3.1. Pathogens in sludges and animal wastes

3.1.1. Types of pathogens

Sewage sludges and animal wastes may contain a variety of pathogens which present disease hazards to man and food animals. These pathogens may be bacteria, viruses, helminths, fungi or protozoa.

Among the bacterial pathogens, *Salmonella* species are those which most frequently infect man and animals. The most well-known serotypes are *Salmonella typhi*, specific to man, *Salmonella dublin*, specific to cattle and *Salmonella typhimurium* which may infect man and a wide range of animals. The main source of infection in man is by eating infected food, particularly meat and dairy products, contaminated either during production or handling or storage. It is recognized that the main routes of transmission in food animals are directly from animal to animal, or indirectly through feedstuffs and slurry. Consequently vegetables and grass contaminated by non-hygienic sludges or manures can cause salmonellosis.

Viruses are also widespread pathogens which can induce diseases in man, animal, crop and bacteria. According to Carrington (1980), and although the demonstration of the infective agent from patients and environment is more difficult than for bacteria, there seem to be only a few viruses which can be found in sewage or manure. One may cite the poliomyelitis virus which can be found in sewage most of the time, although this is usually linked with vaccination of a population and is not indicative of actual infection. The other enteroviruses, *Coxsackievirus* and *Echovirus* as well as *Adenovirus* and *Reovirus* can be demonstrated in polluted waters and consequently in sewage sludges as well as in faeces of man and animals. The virus of infectious hepatitis can also be found in water and so in sludges.

Among the third category of pathogens, we find the helminths, parasites which are of veterinary importance. Taenia saginata (beef tapeworm) and Taenia solium (pork tapeworm) both infect animals, the latter species being almost unknown in man, although with severe consequences. The mature segments each containing about 5000 eggs are passed out by infected humans and can therefore be present in sewage sludge. The eggs, if eaten by their specific secondary hosts, cattle and pigs respectively, develop to a cyst stage in the muscles of the animal. The cycle back to man is completed, if infected meat is eaten uncooked. Then the cysts attach themselves to the small intestine of man and develop to the adult worm. Ascaris suum, a roundworm, is also a pathogenic parasite of pigs which can infect humans and consequently can be present in sewage sludge since it survives in water for very long periods. Because they are extremely resistant, Ascaris ova are widely recognized as indicator organisms of parasitic contamination and of the survival of parasites in sludge or in manure (Golueke, 1983).

The fungi which can be encountered in sewage sludges can be divided in two groups, namely yeasts (Candida) and filamentous fungi (Aspergillus). According to Golueke (1983), fungi are not potentially as pathogenic to man and animals as bacteria, viruses, and parasites. Nevertheless, certain species have been implicated as secondary agents of disease, as agents of hypersensitivity reactions, and as producers of mycotoxins.

Finally, the protozoa Entamoeba histolytica can be conveyed by water and may cause amoebic dysentery. In tropical and sub-tropical areas, amaebiasis is the most common fatal parasitic infection (Golueke, 1983).

Besides human and animal pathogens, plant pathogens may be present in plant residues (Turner et al., 1983) as well as in wastewaters of processing industries (Carrington, 1980) and in sludges from sewage plants treating these waters. Potato cyst eelworms i.e. Heterodera rostochiensis and Globodera pallida, viruses, fungi and bacteria may among others destroy the root tissues of their host plants or cause them to wilt.

3.1.2. Density levels of pathogens

Until now, it is difficult to give values of density levels of the whole range of pathogens which have been found in sludges or in faeces. In fact, the number of pathogens present will depend on the degree of infection already existing in the area concerned and will also be function of environmental and seasonal factors, being greatest in the warmest months. On the other hand, pathogens' concentrations vary widely according to source and depend also on the effect of the different sewage treatment processes. For example the activated sludge process and percolating filters can generally remove 90 per cent or more coli-aerogenes bacteria whereas the primary treatment has little effect on bacterial or even viral pathogens (Carrington, 1980). Moreover, variability in counts of pathogens are often reported, either for bacteria (Pike, 1979) or for viruses (Lund, 1979).

Besides the inhomogeneity in sampling, this variability in counts is caused by differencies in accuracy of the counting methods and in their reproducibilities.

Nevertheless, some density levels of pathogens are reported by various authors. In raw sludges, values of concentrations of Salmonellae reported by Pike (1979 and 1980) vary from 22 to about 10^4 Salmonella counts per 100 ml raw sludges. Pike (1979) shows the influence of the sewage plant size on the Salmonella concentration. Indeed he observes that Salmonellae are more prevalent in raw sludges from medium-sized works, than from the very small (population served less than 1000) or very large (100 000 or greater). Concentrations of Salmonellae in slurries also vary widely. From a Danish survey of 156 herds, Larsen and Munch (1983) report that Salmonella concentration is generally less than 10 per ml of slurry and sometimes 100 to 200 per ml. Only in one herd, they found about 3.5 x 10^6 Salm. per ml of slurry. Jones (1983) reports values of 10^4 to 10^6 Salmonellae per ml of slurry.

3.2. Beneficial effects of anaerobic digestion

Many papers or research works deal already with the beneficial effect of anaerobic digestion but they contain often conflicting results. Nevertheless, some conclusions may be already drawn from them.

In a first step, we will examine the effect of the anaerobic digestion process, under normal operation conditions, i.e., at mesophilic temperatures (around 35 °C) and in completely mixed methane digesters (one-step, once-through digestion system which homogenizes the digester content, the loading can be continuous or intermittent). These running conditions are frequently encountered both in the digestion of farm wastes and of sewage sludges.

3.2.1. Effect on the bacterial pathogens

Although some authors claim that anaerobically digested effluents are free from pathogens, it can be agreed that digestion reduces their numbers, but does not eliminate them. From experimental results, it appears that mesophilic anaerobic digestion may remove 98.5 % of Salmonellae (duesseldorf) with a 15 day retention time (Pike, 1980). From laboratory experiments on slurries, Willinger and Thiemann (1983) report that, if all bacterial strains are still alive in the digested slurries after 14 days, most of the strains (Staphylococcus aureus, Salmonella typhimurium, Proteus vulgaris) could not be re-isolated from the digested slurries after 21 days with the exception of S. typhimurium which was still viable on day 21 in slurry from pigs and from pigs plus poultry. Only one strain, Pseudomonas aeruginosa remains detectable until day 28 except in slurries from cattle and from cattle plus pigs. Digestion of cattle manure appears consequently to have a higher killing effect on bacteria than for other manures.

Nevertheless, results of the beneficial effect of the anaerobic process, when obtained in full-scale digesters are not so significant. Indeed, from an English survey on 7 sewage plants, it appears that the remaining fractions of Salmonellae after treatment are respectively 0.38, 0.13, 0.46, 0.30, 0.084 and in two other plants, they are even higher after than before anaerobic digestion (Pike, 1979). Even if the imprecision of the counting methods and of the sludge sampling can be allowed for, these results show that the effectiveness of the process in real scale is less obvious than in laboratory experiments. Different reasons can be advanced. First of all, "short-circuiting" can occur in full-scale completely mixed plant. That means, that in such digester, there is always a risk that part of the mixed liquor does not remain inside it during the whole retention time. With daily loading for example, a small part of the mixed liquor may remain not more than one day and consequently a significant number of infectious organisms may be discharged and recontaminate the treated effluent. Secondly, in full-scale biogas plants, the drawing off of the digested sludge does not always precede the entry of feedstock (Munch and Schlundt, 1983), whereas it is often the case in the laboratory experiments. Thirdly, the laboratory experiments are based on inoculation with pure cultures of bacteria which differ from the indigenous bacteria present in sludges or animal wastes. Since the rate of death depends upon the physiological state of cells, the first may not show inactivation kinetics characteristic of the second. Finally the wastes digested can be rapidly recontaminated if precautions are not taken during further storage.

So it can be concluded that if in theory, mesophilic anaerobic digestion can destroy most of the bacterial pathogens, in practice, its effect is less significant. This is valid for all types of bacterial pathogens.

3.2.2. Effect on viruses

Decimal reduction times T_{90}, that is the time it takes to reach a 90 % decrease in numbers, have only been established for certain types and strains of viruses. The T_{90} values reported by Lund et al., (1982), Eisenhardt et al., (1977) and Bertucci et al., (1977) go from 24 to 48 h, which are nearly identical to those calculated for bacterial pathogens. These viruses decimal reduction times can however vary considerably from type to type and even from strain to strain. Lund (1982) notes also that some of the endogenous viruses may be solid-associated instead of being free as in aqueous suspension. These latter are reported to be more sensitive compared to the solid-associated ones. According to Lund, this would mean that the endogenous viruses protected by solid-association could be about 10 times more resistant than the bacteria *Escherichia coli* which has a T_{90} of 1-2 days. Taking into account this observation together with those restricting the treatment effect in full-scale plants (see 3.2.1), it can be concluded that the viruses are only partially destroyed by anaerobic digestion, at least in practice.

3.2.3. Effect on helminths

Most of the parasitic helminths seem to be resistant to digestion. They may either be partly inactivated or survive the digestion process. Concerning the roundworm *Ascaris*, Carrington and Harman (1983) find that, on average, 62 % of the ova added to the 35 °C digesters are not recoverable or are destroyed by digestion. The effect of the digestion process by itself, as compared to control experiments, consists in a reduction of the effective viability of the ova to about half that of the control. In the experiments of Black et al., (1982) *Ascaris* ova viability is found to be reduced by only 23 %. The difference in the results can be explained by the fact that, depending upon their state (immature, mature, embryonated), the eggs used in the experiments are more or less resistant. With a higher proportion of mature eggs, known to be the most resistant stage, the number of eggs destroyed will obviously be lower than it would be with immature or embryonated eggs.

The roundworm *Trichinella spiralis* (Golueke, 1983) and the tapeworm *Taenia saginata* (Alderslade (1980), Pike and Davis (1983)) seem to be completely destroyed, although there is at present little confirmation of this. Ova of two other roundworms, *Trichuris* and *Toxocara* are found to survive the mesophilic anaerobic digestion treatment (Black et al., 1982).

Many authors reported by Carrington and Harman (1983) point out that helminths become more sensitive to heat at temperatures above 39 °C. Indeed these pathogens have evolved and adapted as parasites to mammalian hosts but, it would appear, cannot tolerate more elevated temperatures.

3.2.4. Effect on fungi and protozoa

Little is known of the effect of anaerobic digestion on fungi and parasitic protozoa. Apparently, the number of cysts of protozoa can be reduced by mesophilic anaerobic digestion but not the number of ova. This is suggested for Entamoeba histolytica (Carrington, 1980).

3.2.5. Effect on plant pathogens

From the few experiments dealing with plant pathogens, it appears that mesophilic anaerobic digestion reduces them signigicantly. Williams (1979) suggests that potato cyst eelworm (Globodera rostochiensis and G. pallida) are killed. Turner et al., (1983) show that none of the plant pathogens studied i.e. Fusarium oxysporum, Corynebacterium michiganense and Globodera pallida are detectable 10 days after introduction into mesophilic anaerobic digester. But since they work in batch conditions, it can be expected that the reduction rate is less effective in a completely mixed digester. This will be examined further (§ 3.3.3.).

3.3. Influence of the running conditions of anaerobic digestion on its treatment effect

Variations in the running conditions of anaerobic digestion have an influence on the treatment effect. Three factors are considered : temperature higher or lower than 35 °C, the retention time and the type of digestion system (batch, plug-flow, continuous systems with accumulation of active biomass). The effect of pre-or post-treatments of the digested effluent is also analysed.

3.3.1. Influence of the temperature

Whereas most of the methane digesters are running at around 35 °C, some can also be operated at around 55 °C. This thermophilic temperature improves significantly the treatment effect. Since most of the pathogens have their fatal threshold around 55 °C, they will be rapidly killed at thermophilic temperatures. Considering the bacterial pathogens, the decimal reduction time is of the magnitude of minutes to a few hours at thermophilic temperature whereas it is of the magnitude of days under mesophilic temperatures (Munch and Schlundt, 1983). Plym-Forshell (1983) concludes in his paper that the Salmonella bacteria are killed at 55 °C after 24 h since they are not very heat tolerant. The same is true for viruses. For example, Poliovirus, one of the most resistant strains, has at 55 °C a decimal reduction time (T_{90}) of 60 minutes. But even at thermophilic temperature, the T_{90} could attain 6 hours for Coxsackievirus. Lund (1982) attributes this phenomenon to the solid-association of the virus which apparently protects it in sludge. By treating eggs of helminths (Ascaris, Monieza and Fasciola) at thermophilic temperatures (50°, 55° and 60 °C), Kiff and Lewis-Jones (1983) demonstrate that there is complete inhibition of normal egg development. Carrington and Harman (1983), studying the evolution of Ascaris eggs at 49 °C show that very few ova develop into larvae; they note that initially the action of digestion, increasing with increased temperature, inhibits the mechanism controlling the development of organs or larvae within the egg without inhibiting initial cell division.

These observations allow us to conclude that thermophilic anaerobic digestion produces an effluent almost free of pathogens. But also in that case, precautions have to be taken against regrowth. On the other hand, thermophilic digestion is not often used for practical reasons : difficulty in starting up of the digester, resistance of the materials utilized at high temperature, higher energy requirement for heating the waste to be digested...

On the contrary, there is a tendency, in these last years to operate the digestion at ambient temperature which is called cold or psychrophilic anaerobic digestion. Under temperatures equal to or lower than 20 °C, the inactivation rates are reduced, independently of the type of pathogen. In a review reported by Pike (1982), it is concluded that at temperatures of 20°-25 °C, 90 % reduction of respectively bacteria, parasite ova and viruses require one month, six months and more than 2 months. At temperatures below 20 °C, these times are increased to more than 6 months, at least 6 years and greater than 8 months. The observations are confirmed by other authors. For example, Plym-Forshell (1983) finds that Salmonella survives 35 but not 42 days in a manure pit maintained between 22 °C and 27 °C. Lund (1983) shows in her study on viruses that at 5 °C, it would take approximately 2 years to make stored manure harmless, whereas it would take 3-4 months at 20 °C.

Nevertheless, it may be added that in cold or psychrophilic digesters, the liquor remains in general several months before emptying. The beneficial effect of this type of digestion will thus be valuable for bacterial pathogens. This effect will be increased under higher ambient temperatures, as in summer.

3.3.2. Influence of the retention time (duration of stay in digester)

Apparently the retention time does not have a significant influence on the treatment effect. Carrington and Harman (1983) show that it does not appear to influence the numbers of ova of Ascaris recovered. Concerning Salmonella inactivation, they find that the decay rate is lower when the retention period is 10 days than when it is 20 days but the rates for retention periods of 16 and 20 days at 35 °C are similar.
Lengthening the retention time to more than 15-20 days (retention time which is normally applied in mesophilic anaerobic digesters) could allow the kill of most of the more resistant pathogens.

3.3.3. Influence of the digestion system

The preceding observations are valid for completely mixed digesters, and it has been seen that the inactivation effect could be reduced by "short-circuiting" and by simultaneous drawing off the effluent and feeding (see 2.2.1.). This problem is even worse when the digesters are fed continuously : in this condition part of the liquor may remain less than one day in the digester, which is certainly too short for even partial pathogen inactivation. Only real batch (system characterized by a single loading at the beginning of the process and a single unloading at the end of the process) and plug-flow (one-step, once-through biomethanation system with horizontal flow, which may be mechanically assisted) systems allow obtaining the greatest degree of inactivation. Indeed in these conditions, the digester contents will remain in the reactor during the whole retention period.

Considering the digestion systems with accumulation of active biomass, there is at present no evaluation of the degree of pathogen deactivation. Retention times in these systems are much shorter : from 2-3 days to less than one day, depending on the system. In these conditions, the treatment effect will be reduced. Nevertheless, it can be expected that the pathogens could be trapped inside the digester together with the active biomass and that they consequently remain subjected for long times to the effect of the anaerobic digestion. Carrington (1980) reports that sand filtration and carbon adsorption used as tertiary treatment for sewage effluents can remove from 50 to 100 per cent of the viruses and helminths ova but he concludes that these forms of treatment are less effective in removing bacteria; viruses may be adsorbed on the support but not inactivated, and later may be eluted.

3.3.4. Influence of the post-treatment

Among the possible pre-or post-treatments of digested effluents, some can be quite effective in allowing the number of pathogens to decline, although not completely; some other treatments produce pathogen free effluents. Pasteurization before anaerobic digestion is now currently practised in Switzerland for producing sewage sludges free of pathogens (Clements, 1982). Lime treatment at pH > 11, or irradiation both produce good effluents (Pike, 1980), although these 2 latter treatments are less effective for _Ascaris_ ova (Carrington, 1980). Air and bed drying or lagooning may have beneficial effects. Golueke (1983) suggests that if the sludge is dewatered and dried, some pathogens will be inactivated. For example, _Salmonella_ is reduced by 90 per cent when the sludge is dried to 95 per cent solids at room temperature. Pike (1982) reports that the number of _Salmonella_ in 100 ml domestic sludge, mesophically digested for 10-15 days falls from 325 to 2.6 after 6-8 weeks lagooning and he concludes that lagooning of sludge can be quite effective in allowing numbers of _Salmonella_ to decline, particularly for further treatment of mesophically digested sludge. Nevertheless, according to Carrington (1980), neither drying nor lagooning seem to eliminate helminth ova and viruses. On the other hand, these two post-treatments allow the volatilization of ammonia and hence produce an effluent with a lower fertilizing value.

3.4. Aerobic digestion versus anaerobic digestion

Mesophilic anaerobic digestion may be related to conventional aerobic digestion at ambient temperature. However, the effect of aerobic digestion seems not to be increased by increasing temperature of aeration (Pike, 1982).

The removal of _Salmonella_ during aerobic digestion is very similar to that during anaerobic digestion. From a survey in English sewage treatment plants, Pike (1979) notices that the fraction of _Salmonella_ remaining after aerobic digestion varies from 0.18 to 0.84 and that the rate of reduction of the numbers of bacterial pathogens decreases with increasing digester retention time. Lund (1983) notes that under aerobic conditions, bovine enteroviruses are killed within 2,5 weeks instead of 8 weeks for anaerobic digestion at the same temperature. She concludes that at temperatures below 25 °C, aerobic processes are better than anaerobic. But from preliminary results, it appears however that at 33 °C, anaerobic stabilization may be superior to aerobic stabilization for destruction of viruses. In their experiments comparing the effects of aerobic and anaerobic digestion on the survival of parasite eggs, Black _et al_. (1982) found that aerobic digestion is sensibly superior to anaerobic stabilization although still not completely efficient.

It may be concluded that aerobic digestion has in general nearly the same treatment effect on pathogens as anaerobic digestion.

4. CONCLUSIONS

Anaerobic digestion allows to solve the odour problems encountered during spreading of sludges or manures. Compared to aerobic digestion, its effect on odour reduction is stable while that of aerobic treatment is of short duration.

The effect of mesophilic anaerobic conditions on pathogens reduction is summarized in Table 1.

Table 1. Treatment effect of mesophilic anaerobic digestion (in theory)

Type of pathogen	Relative reduction
Bacteria	moderate to good
Viruses	moderate
Helminths ova	poor
Parasitic cysts	good
Plant pathogens	good

In theory its beneficial effect is significant for bacteria, viruses, parasitic cysts and plant pathogens. In practice, especially in completely mixed digesters, this effect is reduced by short-circuiting and by simultaneous drawing off and feeding the digester. The disinfection effect can nevertheless be improved by digesting in batch or plug-flow systems and by working under thermophilic conditions. Although thermophilic digestion is seldom employed at present, in the future, the most infected wastes, as for example abattoir waste should preferably be digested at 55 °C.

On the other hand, good handling rules during the utilization of the digested effluent can allow further disinfection. Two restrictions are for example proposed in the directive of the European Commission (1982). A no-grazing period between applying the sludge or the manure to grazing land and returning cattle to them, allows pathogens to decay and become non-infective or to disappear in this interval; the risk of contamination is practically eliminated. Another type of rule is banning the application of effluent to growing crops or on land used to grow them, when these crops are eaten uncooked by man or animals or are brought raw into the kitchen (thereby possibly contaminating other foods).

This study is being supported by grant ECI-1012-B7210-83-B from the Commission of the European Communities.

REFERENCES

ALDERSLADE, R., 1980. The problems of assessing possible hazards to the public health associated with the disposal of sewage sludge to land : recent experience in the United Kingdom. In Characterization, Treatment and Use of Sewage Sludge, Proceedings of the 2nd European Symposium, Vienna, 21-23 October 1980; P. L'Hermite and H. Ott eds, D. Reidel Publishing Company, Dordrecht, Holland, 372-388.

BERTUCCI, J.J., LUE-HING, C., ZENZ, D., SEDITA, S.J., 1977. Inactivation of viruses during anaerobic sludge digestion. Jour. Water Poll. Control Fed., 1642-1651.

BLACK, M.I., SCARPINO, P.V., O'DONNELL, G.J., MEYER, K.B., JONES, J.V., KANESHIRO, E.S., 1982. Survival rates of parasite eggs in sludge during aerobic and anaerobic digestion. Applied and Environmental Microbiology, 44 (5), 1138-1143.

BRUCE, A.M., CAMPBELL, H.W., BALMER, P., 1983. Developments and trends in sludge processing techniques. In Treatment and Use of Sewage Sludge, Proceedings of the Third European Symposium, Brighton, 26-30 September, P. L'Hermite and H. Ott eds, D. Reidel Publishing Company, Dordrecht, Holland, in the press.

CARRINGTON, E.G., 1980. The fate of pathogenic micro-organisms during wastewater treatment and disposal. Technical report TR 128, Stevenage Laboratory, Water Research Centre, 58 pp.

CARRINGTON, E.G., HARMAN, S.A., 1983. The effect of anaerobic digestion temperature and retention period on the survival of Salmonella and Ascaris ova, presented at WRC Conference on stabilisation and disinfection of sewage sludge, Paper 19, Librarian, WRC Processes, Stevenage, 13 pp.

CLEMENTS, R.P.L., 1982. Sludge hygienization by means of pasteurization prior to digestion. In Disinfection of Sewage Sludge : Technical, Economic and microbiological Aspects, Proceedings of a workshop, Zürich, May 11-13, Bruce A.M., Havelaar A.H. and L'Hermite P. eds, D. Reidel Publishing Company, Dordrecht, Holland, 37-52.

COMMISSION OF THE EUROPEAN COMMUNITIES, 1982. Proposal for a council Directive on the use of sewage sludge in agriculture. Off. J. Europ. Comm. 8 October 1982, No. C/264, 3-7.

DEMUYNCK, M., NYNS, E.-J., 1984. Biogas plants in Europe - a practical handbook. Palz, W., ed., D. Reidel Publishing Co., Dordrecht, Holland, in the press.

EISENHARDT, A., LUND, E., and NISSEN, B., 1977. The effect of sludge digestion on virus infectivity. Water Research, 11, 579-581.

GOLUEKE, C.G., 1983. Epidemiological aspects of sludge handling and management. Biocycle, 24, (3), 52-58.

JONES, P.W., 1983. The survival and infectivity for cattle of Salmonellas on grassland. In Characterization, Treatment and Use of Sewage Sludge, Proceedings of the third European Symposium, Brighton, 26-30 September 1983. Commission of the European Communities, L'Hermite P. and Ott H. eds, Reidel D., Publishing Company, Dordrecht, Holland, in the press.

KIFF, R.J., LEWIS-JONES, R., 1983. Factors that govern the survival of selected parasites in sewage sludges. Pollution Research Unit Umist, Manchester.

KLARENBEEK, J.V., 1982. Odour measurements in Dutch agriculture : current results and techniques. Research Report 82-2, Institute of Agricultural Engineering, Wageningen, The Netherlands.

LARSEN, H.E., and MUNCH, B., 1983. Practical application of knowledge on the survival of pathogenic and indicator bacteria in aerated and non-aerated slurry. In Proceedings of a point workshop of expert groups of the Commission of the European Communities, German Veterinary Medical Society (DVG) and Food and Agricultural Organisation; D. Strauch ed, Institute for Animal Medecine and Animal Hygiene, University of Hohenheim, Stuttgart, Federal Republic of Germany, 20-35.

LUND, E., 1979. Detection of viruses in sludges. In Treatment and Use of Sewage Sludge, Proceedings of the First European Symposium, Cadarache, 13-15 February 1979, D. Alexandre & H. Ott eds, Commission of the European Communities, 204-208.

LUND, E., LYDHOLM, B. and NIELSEN, A.L., 1982. The fate of viruses during sludge stabilization, especially during thermophilic digestion. In Disinfection of Sewage Sludge : technical, economic and microbiological aspects, Proceedings of a Workshop, Zurich, May 11-13, 1982. A.M. Bruce, A.H. Havelaar, P. L'Hermite, eds., D. Reidel Publishing Company, Dordrecht, Holland, 115-124.

LUND, E., 1983. Inactivation of viruses under anaerobic or aerobic stabilization in liquid manure and in sludge from sewage treatment plants. In Proceedins of a joint workshop of expert groups of the Commission of the European Communities, German Veterinary Medical Society (DVG) and Food and Agricultural Organisation; D. Strauch (ed.), Institute for Animal Medicine and Animal Hygiene, University of Hohenheim, Stuttgart, Federal Republic of Germany, 199-209.

MUNCH, B., SCHLUNDT, J., 1983. The reduction of pathogenic and indicator bacteria in animal slurry and sewage sludge subjected to anaerobic digestion or chemical disinfection. In Proceedings of a joint workshop of expert groups of the Commission of the European Communities, German Veterinary Medical Society (DVG) and Food and Agricultural Organization; D. Strauch (ed.), Institute for Animal Medicine and Animal Hygiene, University of Hohenheim, Stuttgart, Federal Republic of Germany, 130-149.

OFFERMANS, H., TUIN, B., DE VRIES, H., DE VRIES, U., 1982. De mogelijkheden van biogas in de landbouw. Serie studentenverslagen n° 26, of the Milieukundig Studiecentrum Groningen (The Netherlands).

PIKE, E.B., 1979, Salmonellae in sewage sludges. A survey of sewage works in England and Wales. In Treatment and Use of Sewage Sludge, Proceedings of the 1st European Symposium. Cadarache, 13-15 February 1979, D. Alexandre & H. Ott eds, Commission of the European Communities, 189-200

PIKE, E.B., 1980. The control of Salmonellosis in the use of sewage sludge on agricultural land. In Characterization, Treatment and Use of Sewage Sludge, Proceedings of the 2nd European Symposium, Vienne, October 21-23, 1980; P. L'Hermite and H. Ott eds, D. Reidel Publishing Company, Dordrecht, Holland, 315-327.

PIKE, E.B., 1982. Long term storage of sewage sludge. In Disinfection of sewage sludge : technical, economic and microbiological aspects, Proceedings of workshop, Zurich, May 11-13, 1982, A.M. Bruce, A.H. Havelaar, P. L'Hermite, eds., D. Reidel Publishing Company, Dordrecht, Holland, 212-224.

PIKE, E.B., DAVIS, R.D., 1983. Stabilisation and disinfection - their relevance to agricultural utilisation of sludge, presented at WRC Conference on stabilisation and disinfection of sewage sludge, Paper 3, Session 1, Librarian, WRC Processes, Stevenage, 30 pp.

PLYM-FORSHELL, L., 1983. Survival of Salmonella bacteria and Ascaris suum eggs in a thermophilic biogas plant. In Proceedings of a joint workshop of expert groups of the Commission of the European Communities, German Veterinary Medical Society (DVG) and Food and Agricultural Organization; D. Strauch ed, Institute for Animal, Medicine and Animal Hygiene University of Hohenheim, Stuttgart, Federal Republic of Germany, 35-45.

RYGOLE, R., POELS, J., VERSTRAETE, W., 1983. Biogaswinning op de boerderij. Eindverslag 1 oktober 1981 - 30 september 1983 van het programma IWONL nr 3699A COVAL sectie 3 "Energetische Valorisatie", RUG Fakulteit Landbouwetenschappen, Laboratorium voor Algemene en Toegepaste Microbiële Ecologie, Coupure Links, 653, 9000 Gent, België.

TURNER, J., STAFFORD, D.A., HUGHES, D.E., CLARKSON, J., 1983. The reduction of three plant pathogens (Fusarium, Corynebacterium and Globodera) in anaerobic digesters. Agricultural Wastes, 6, 1-11.

VAN VELSEN, A.F.M., 1981. Anaerobic digestion of piggery waste. Ph. D. Thesis of the Agric. University, Wageningen, 103 p.

WELLINGER, A., 1983. Anaerobic digestion - the number one in manure treatment with respect to energy cost ? In Proceedings of a joint workshop of expert groups of the Commission of the European Communities, German Veterinary Medical Society (DVG) and Food and Agricultural Organization D. Strauch, ed, Institute for Animal Medicine and Animal Hygiene, University of Hohenheim, Stuttgart, Federal Republic of Germany, 163-183.

WILLIAMS, J.H., 1979. Utilisation of sewage sludge and other organic manures on agricultural land. In Treatment and Use of Sewage Sludge, Proceedings of the 1st European Symposium, Cadarache, 13-15 February 1979, D. Alexandre & H. Ott eds, Commission of the European Communities, 227-242.

WILLINGER, H., THIEMANN, G., 1983. Survival of resident and articifially added bacteria in slurries to be digested anaerobically. In Proceedings of a joint workshop of expert groups of the Commission of the European Communities, German Veterinary Medical Society (DVG) and Food and Agricultural Organisation; D. Strauch ed, Institute for Animal Medicine and Animal Hygiene, University of Hohenheim, Stuttgart, Federal Republic of Germany, 210-217.

DISCUSSION

LE ROUX: Do you have data relating the quantity and quality of residual odours with the retention time of the sludge in the digester, and the amount of bio-gas produced?

DEMUYNCK: No, we do not have this information.

BRUCE: I cannot agree with your general conclusion that the effect of mesophilic anaerobic digestion on reduction of helminth ova is "poor".

Recent studies carried out in the UK by the Water Research Centre and the Institute for Research on Animal Diseases have shown that anaerobic digestion at 35° for 10 days and 15 days cold storage reduced the infectivity of eggs of *Taenia saginata* by 100%. In the experiments, *Taenia* eggs were subjected to the digestion conditions and then fed to calves. After a period of 3 months, the calves were examined for cysts and none were found. In the control calves receiving untreated eggs, heavy infection with cysts was observed.

I would agree that it is important to ensure that no short-circuiting occurs in digesters but this can be done by appropriate mixing and feeding regimes. If this is done, then mesophilic anaerobic digestion should be effective in destroying pathogens and parasites.

DEMUYNCK: I think that the effect on helminth ova depends on the maturity of the eggs and consequently on their resistance to the deactivation effect of anaerobic digestion. This might explain the difference in results observed.

DE BERTOLDI: As far as the conclusions of project COST 68 Working Party 3 are concerned, anaerobic digestion of sewage slude does not seem to give an adequate guarantee of pathogen reduction particularly for bacteria such as *Salmonella*.

DEMUYNCK: I agree. Both with untreated and mesophilic digested waste a period of time must be left between spreading and grazing.

BRUCE: I would agree that odour problems can arise from storage of sludges and slurries which have been only *partially* stabilised by aerobic digestion. But if aerobic digestion is carried to *completion*, a very stable and non-odorous product is obtained and further storage does not usually give rise to problems.

COMPOSTING : ODOUR EMISSION AND ODOUR CONTROL BY BIOFILTRATION

K.W. VAN DER HOEK
Rijks Agrarische Afvalwaterdienst
(Government Agricultural Waste Water Service)
Kempenbergerweg 67, NL-6816 RM ARNHEM

J. OOSTHOEK
N.V. Vuilafvoer Maatschappij VAM
(Waste Disposal Company VAM)
Marijkeweg 11, NL-6709 PE WAGENINGEN

Summary

Composting is a technique for converting organic material into a product suitable for land application or disposal without adverse environmental effects. Composting is based on aerobic breakdown, mainly by bacteria and fungi.

This paper first describes the appearance of odour with windrow composting and with accelerated or aerated composting systems. The paper continues with odour reduction by biofiltration. Finally the results of an aerated static pile pilot-plant for greenhouse wastes are presented. The pilot-plant experiment was carried out to check the odour emission of this composting technique. The positive results led to the development of a full-scale plant with an ultimate capacity of 160,000 m^3 yearly.

1. WINDROW COMPOSTING AND ODOUR

One of the main requirements for composting is the availability of oxygen inside the composting mass. Haug (1980) examined the literature for reported oxygen consumption rates for various composting mixtures like municipal solid waste (MSW), sludge and mixtures. He concluded that a supply of 7-12 mg O_2/g of volatile solids/hour would appear to be sufficient in all but the most extreme cases. The oxygen supply by windrow composting is achieved by natural ventilation, so there can be anaerobic spots inside the windrow.

Jäger and Jager (1980) described the odour emissions from MSW composting. During the first stage of the composting process volatile fatty acids are formed as intermediates followed by alcohols, aldehydes and ketones. The natural ventilation induces an ascending airflow, containing *inter alia* ethanol, diacetyl and acetoin. The two last compounds have a very low odour threshold. During the thermophilic stage also thermal-chemical reactions take place, resulting in volatile compounds like pyridine and pyrazine. The main volatile sulphur compounds are dimethylsulphide, dimethyldisulphide and dimethyltrisulphide; H_2S is only found under completely anaerobic conditions.

The authors also mention that the outer layer of the windrow acts as a biofilter. In this way the ascending airflow from the (anaerobic) interior of the windrow can be partly deodorized. By turning the windrow the interior is directly exposed to the environment, resulting in an increasing odour emission; this is confirmed by odour measurements.

Haug (1980) cites odour measurements during windrow composting of digested sludge blended with recycled compost. The odour emission rate drops significantly during the first stage and then fluctuates somewhat during the remainder of the composting period. The odour emission rate increases immediately after turning, but within about one hour returns to the rate before turning.

2. DYNAMIC ACCELERATED OR AERATED COMPOSTING AND ODOUR

These composting systems are characterized by an adjustable mechanical air supply to the composting mass. Often the exhausted gas can be collected at one spot before it enters the environment. As a result of this forced aeration the active composting stage is shorter than with windrow composting. An adequate oxygen supply is not always possible and this implies that the exhaust gas can create odour nuisance.

The compost plant Duisburg-Huckingen (West Germany) is equipped with two DANO rotary drums for the composting of MSW and dewatered digested sludge. To overcome odour complaints a biofilter has been successfully in operation since 1966. This biofilter consists of a layer 1 m deep of compost and is operated at an air loading of 60 m^3/m^2/hour (Helmer, 1974a). The total carbon content of the exhaust gas is reduced from 230 to 8 mg/m^3 (Gust, 1979).

Composting of MSW and dewatered digested sludge in "Etagenrottetürmen" in the compost plant Heidelberg-Wieblingen (West Germany) also led to odour nuisance. "Etagenrottetürmen" are vertical towers, consisting of a number of floors. A biofilter, filled with compost, has been installed and operates successfully. The exhaust gas volatiles consist mainly of ethanol, diacetyl, limonene and acetoin. Measurements revealed that the organic carbon content of the exhaust gas is reduced from 557 to 40 mg/m^3 (Jäger and Jager, 1978).

The aerated static pile technique has been developed for composting of sewage sludge (Haug, 1980). The direction of ventilation can be vacuum-induced or forced-pressure. With the vacuum-induced ventilation, the exit air has to be cleaned in a biofilter. Using the forced-pressure ventilation the pile has to be covered with a layer of compost, acting as a biofilter (Miller, 1982).

3. ODOUR REDUCTION BY BIOFILTRATION

Biofiltration means that an odorous airflow is passing through a layer of filter material (compost, filamentous peat and so on), followed by biodegradation of the captured odour components. The odour components are transferred from the gas phase to the liquid and solid phase of the particles in the filter material. On these particles a microbial degradation of the odour components takes place. This results in a continuous driving force for mass transfer of the odour components from the gas phase to the particles of the filter material.

In most biofilters a layer about 1 m deep of filter material is used. The odorous airflow is divided evenly on the bottom of the biofilter. The air load (expressed in m^3 air/m^2 filter/hour) is important for the size of the biofilter. A low air load gives an adequate residence time for odour removal and a low pressure-drop, but needs a large filter area. Increasing the air load can result in inadequate residence time, so that not all odour components are removed. However, the airflow leaving the biofilter will never be odour-free, it will always carry the odour of the filter material.

In Table I, some data are collected on plant scale biofilters. The pressure-drop across the biofilters not only depends on the applied air load but also on the nature and compostion of the filter material (different types of compost, filamentous peat).

Odour measurements can be performed by an odour panel. The panel-members are exposed to odour samples that have been diluted with odour-free air. The number of dilutions required to achieve a 50% positive response by panel members is termed the threshold odour concentration. For example, 1 m^3 air of 100 odour units requires dilution with 99 m^3 of odour-free air to reach the threshold concentration (Haug, 1980, Jäger and Jager, 1980).

4. AERATED STATIC PILE COMPOSTING OF GREENHOUSE WASTES

In Holland a lot of horticulture activities are concentrated in a small area south of the The Hague, the so-called Westland. In this area vegetables and flowers are produced in greenhouses. As a first estimate approximately 150,000 m^3 of greenhouse wastes are produced yearly. These organic wastes consist mainly of plants which have finished cropping like tomatoes and carnation stems. Normal windrow composting of all the Westland greenhouse wastes at one location is not acceptable because odour complaints are expected when the windrows are turned. To overcome this odour problem the use of the aerated static pile composting process was investigated in a pilot-plant experiment (van der Hoek, 1983). This pilot-plant experiment had two aims:

a) to check the odour emission,
b) to gain experience with this new composting technique.

4.1 Design of the pilot-plant experiment

Two piles were formed, pile I consisted of 1,300 m^3 and pile II of 850 m^3 of finished tomato plants. The piles had different slopes, pile I 60° and pile II 45°; the base of both piles was equal, viz. each 12 x 22 = 264 m^2 (see figures 1 and 4). Each pile was surrounded by a small earthen wall, and a plastic lining covering the bottom was dug into the crown of these walls. About 40 cm of tree bark was placed on the plastic lining. The function of this bark is to facilitate the removal of the leachate and direct airflow towards the perforated pipes. These perforated pipes are connected with a fan (pile I : 4 kW and pile II : 3 kW). The exhausted air is not odour-free and has to be cleaned before it enters the environment. In the pilot-plant experiment we used two biofilters, filled with a layer about 1 m deep of specially prepared MSW compost.

4.2 Composting aspects

During the making of the pile from tomato plants the temperature increased and the piles steamed until the fans were started. At that time, temperatures at some points inside the piles were 80°C. The specific ventilation rate is expressed as m^3 air/hour/m^3 fresh waste and was kept equal for both piles. The specific ventilation rate during the first 10 days of the experiment was about 2 and during the next 60 days reduced to 1.2 m^3 air/hour/m^3 fresh waste. It was calculated that this specific ventilation rate corresponds to 5.9 and 3.6 m^3 air/hour/1,000 kg organic matter.

The decline of the composting activity was observed as follows:

- the slow decline of the temperature of the exhaust air (see figure 2 and 3, the black bars represent periods of standstill of the fan),
- the decreasing temperature in the piles themselves,
- the chemical composition of the leachate from the piles: the decreasing content of ammonia.

Table I. Some data about biofilters on plant scale

	Filter material	Filter area	air load	Pressure-drop	Chemical or sensory evaluation		gas in	gas out	Lit.
Compost plant Heidelberg-Wieblingen	MSW-compost from own plant	250 m^2	10 $m^3/m^2/h$	4.5 mbar	ethanol	mgC/m^3	391	-	7
					diacetyl	"	16	-	
					limonene	"	16	5	
					acetoin	"	64	-	
					rest	"	70	35	
					total	"	557	40	
Compost plant Duisburg-Huckingen	MSW-compost from own plant	264 m^2	60 $m^3/m^2/h$	16 mbar	total C	mgC/m^3	230	8	2
Aerated static pile composting of greenhouse wastes	MSW-compost, specially prepared	25 m^2	45 $m^3/m^2/h$	2 mbar	odour*	ou/m^3	1,000	75	13
		35 m^2	45 $m^3/m^2/h$	2 mbar	odour*	"	2,400	70	
Rendering plant	compost from sewage sludge	42 m^2	21 $m^3/m^2/h$	37 mbar	H_2S	ppm	reduction more than 99.8%		12
Rendering plant	MSW-compost from Duisburg	288 m^2	88 $m^3/m^2/h$	16 mbar	total C	mgC/m^3	45	4	2
					odour*	ou/m^3	reduction 94%		
Rendering plant	filamentous peat	800 m^2	125 $m^3/m^2/h$	1.5 mbar	odour*	ou/m^3	reduction up to 98%		10
		160 m^2	188 $m^3/m^2/h$	2.5 mbar					
Pig house	filamentous peat	34 m^2	no data	± 1 mbar	odour*	ou/m^3	750	30	1
		30 m^2	no data	idem	odour*	"	430	19	

odour* : ou means odour units, see Chapter 3.

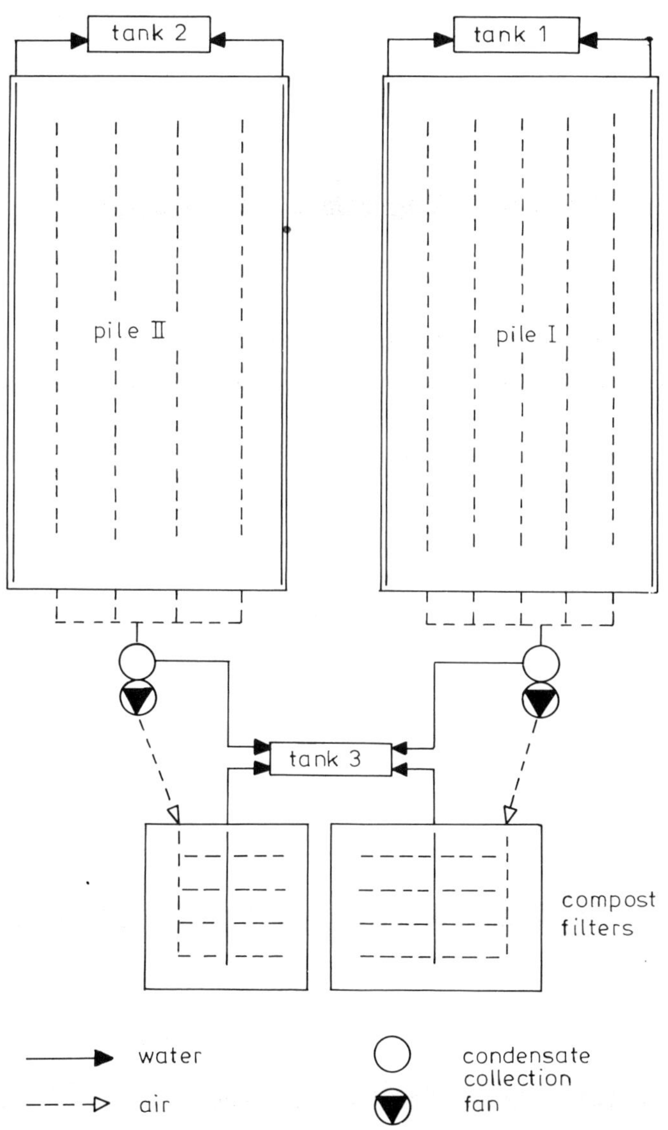

Fig 1. Lay-out of the aerated composting pilot-plant

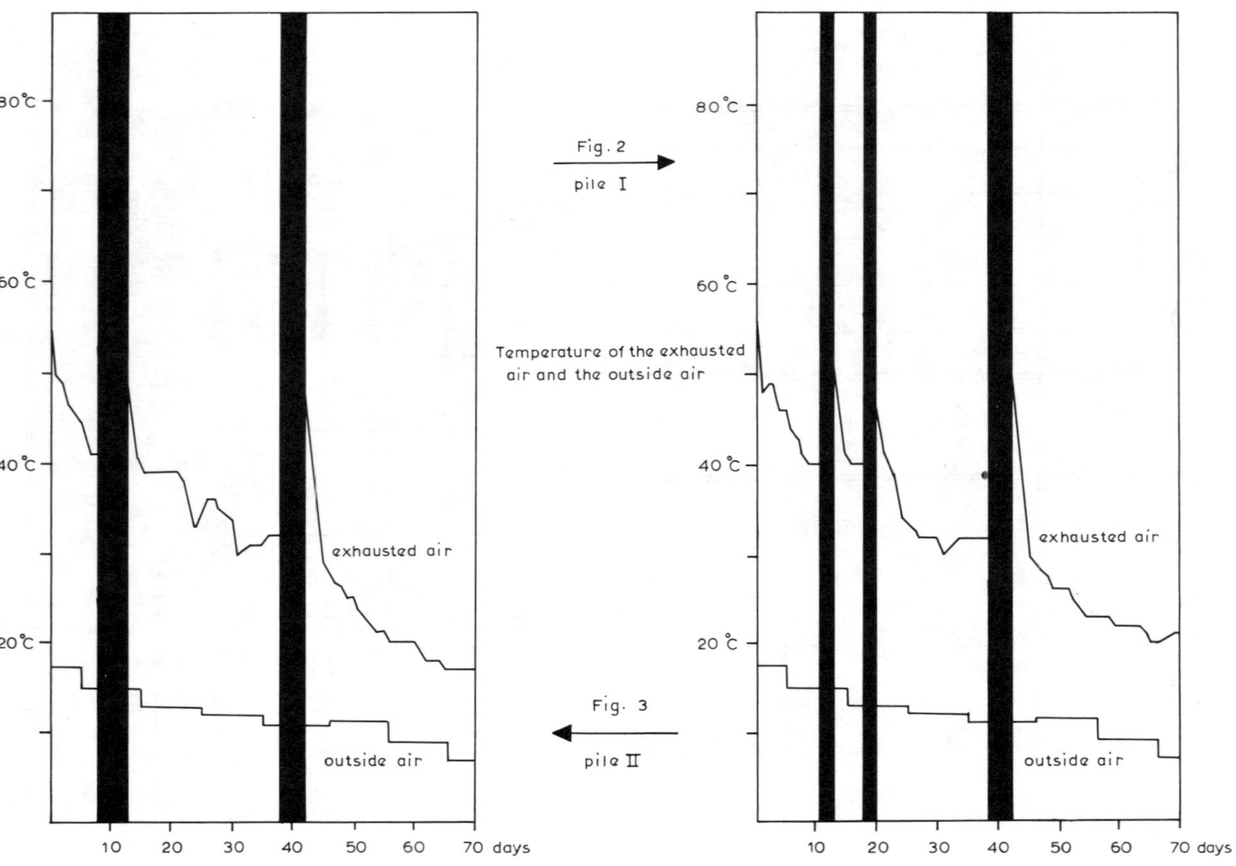

80°C
60°C
40°C
20°C
10
20
30
40
50
60
70 days
exhausted air
outside air
Fig. 2
pile I
Temperature of the exhausted air and the outside air
Fig. 3
pile II

Fig. 4 - View of the pilot-plant for aerated static pile composting.

To have a visible check on this decline of the composting activity, on day 48 we dug in one of the piles a hole down to the bark. During the digging we observed odour-free composted tomato plants at temperatures of 20-30°C, without wet or dry spots.

The chemical composition of the tomato plants before and after composting is given in Table II; the increase in the ash content is substantial. The specific weight of the fresh tomato plants and the composted tomato plants was about 450 kg/m^3 and 600 kg/m^3. The pilot-plant experiment continued for 70 days and resulted in approximate reductions of 60% in volume-, 50% in weight- and 40% in dry-matter.

Table II. Chemical composition of tomato plants before and after composting

	before composting*	after composting**
dry matter (d.m.)	28% (27- 29)	33% (27- 46)
ash in g/kg d.m.	255 (209-283)	487 (357-587)
N in g/kg d.m.	19 (19- 20)	16 (15- 18)
P in g/kg d.m.	6 (5- 6)	7 (5- 9)
K in g/kg d.m.	39 (30- 49)	39 (29- 46)

* 3 samples, the range is shown in brackets
** 6 samples, idem

4.3 Odour emission

The aerated static pile composting process can give rise to two sources of odour:

a) the exhaust air from the piles,
b) the transfer of the composted material from the aerated pile to a non-aerated maturation pile.

The exhaust air was cleaned in a biofilter, filled with a 1 m deep layer of MSW-filter compost. In the pilot-plant experiment each pile had its own biofilter. Each biofilter was operated at an air loading of 45 m^3/m^2/hour, resulting in a pressure-drop of about 2 mbar. The odour measurements were carried out by TNO, Netherlands Organisation for Applied Scientific Research, Division of Technology for Society. Both biofilters showed 88-97% odour reduction. After passing the biofilter, the cleaned exhaust air contained about 75 odour units/m^3, including the odour of the filter compost (Roos, 1982).

The odour emission arising from the transfer operation mentioned at b) has not been measured directly. A comparison has been made between the odour emission of 1 kg of composted tomato plants and 1 kg of stabilized MSW from a DANO rotary drum. This comparison turned out very clearly to the advantage of the composted tomato plants (Roos, 1982).

Afterwards the odour emission was estimated for a full-scale plant with a yearly capacity of 100,000 m^3 fresh tomato plants. Assuming regular deliveries of tomato plants from August to February, about 33,000 m^3 fresh waste is present at any time in the aerated stage; this results in about 50,000 m^3 exhaust air/hour. The odour emission from the cleaned exhaust air is about (50,000 : 3,600) x 75 odour units/m^3 = 1,050 odour units/sec. It has been calculated that a reasonable odour emission from the transfer operations is of the same magnitude.

Dutch research in odour emission from animal fattening units showed that the average odour emission in summer amounts to 2 odour units/sec per fattening pig. This concerns pig houses with slatted floors and effluent handling as liquid manure (Klarenbeek, 1982). This means that the estimated odour emission from the 100,000 m^3 compost plant equals a house with (1,050 + 1,050) : 2 = circa 1,000 fattening pigs.

4.4 A full-scale aerated composting plant

Based upon the positive results of the pilot-plant experiment a full-scale plant has been designed. This plant consists of 4 identical units, each having a yearly capacity of 40,000 m^3 fresh tomato plants. In this way one can start with one or two units and have an extension later on.

Based upon the pilot-plant experiment the composting period has been fixed at 45 days, with a specific ventilation rate of 1-2 m^3 air/hour/m^3 fresh waste. For the present it is not planned to use tree bark in the full-scale plant. Instead, prefabricated ventilation ducts covered with grating are used for simultaneous removal of leachate and air. The leachate from the aerated piles, the biofilters and the maturation piles has to be purified in a water purification plant.

REFERENCES

1. BRÖKER, G. and GLIWA, H. (1983). Olfaktometrische Messungen bei Intensivtiehaltungen - Abschätzung des Verbesserungseffektes nach Einbau von Filteranlagen. Stab - Reinhaltung der Luft, 43: 478-481.
2. GUST, M. and GROCHOWSKI, H. and SCHIRZ, St. (1979). Grundlagen der biologischen Abluftreinigung. Teil V. Abgasreinigung durch Mikroorganismen mit Hilfe von Biofiltern. Staub - Reinhaltung der Luft, 39: 397-402.
3. HAUG, R.T. (1980). Compost Engineering, principles and practice. Ann Arbor Science, Michigan. 655 pages.
4. HELMER, R. (1974a). Abluftreinigung in Müllkompostwerken mit Hilfe der Bodenfiltration. Müll und Abfall, 140-146.
5. HELMER, R. (1974)b. Desodorisierung von geruchsbeladener Abluft in Bodenfiltern. Gesundheits-Ingenieur, 95: 21-26.
6. HOEK, K.W. VAN DER (1983). Compostering van tuinbouwafvallen met behulp van geforceerde beluchting. Raad - VAM publicatie, Arnhem. 52 pages.
7. JÄGER, B. and JAGER, J. (1978). Geruchsbekämpfung in Kompostwerken am Beispiel Heidelberg. Müll und Abfall, 48-54.
8. JÄGER, B. and JAGER, J. (1980). Ermittlung und Bewertung von Geruchsemissionen bei der Kompostierung von Siedlungsabfällen. Müll und Abfall, 22-28.
9. KLARENBEEK, J.V. (1982). Odour measurements in Dutch agriculture: current results and techniques. IMAG Research Report 82-2, Wageningen. 14 pages.
10. KOCH, W., LIEBE, H.G. and STRIEFLER, B. (1982). Betriebserfahrungen mit Bio-Filtern zur Reduzierung geruchsintensiver Emissionen. Staub - Reinhaltung der Luft, 42: 488-493.
11. MILLER, F.C., MACGREGOR, S.T., PSARIANOS, K.M., CIRELLO, J. and FINSTEIN, M.S. (1982). Direction of ventilation in composting waste water sludge. J. Water Poll. Control Fed., 54: 111-113.
12. RANDS, M.B., et al, (1981). Compost filters for H_2S removal from anaerobic digestion and rendering exhausts. J. Water Poll. Control Fed., 53: 185-189.
13. ROOS, C. (1982). Onderzoek naar de geuremissie bij de compostering van tuinbouwafvallen met behulp van geforceerde beluchting. Rapport 82 - 016572, MT - TNO, Apeldoorn. 19 pages (is included in ref. 6).
14. SAMMET, D. (1980). 27. Informationsgespräch des ANS : Entstehung und Vermeidung von Gerüchen bei der Kompostierung. Müll und Abfall, 165-166.

DISCUSSION

LE ROUX: You used special compost in the biofilters. Why did you not use the tomato waste compost?

VAN DER HOEK: We did not have tomato compost available at the start of the experiment. Also, the special compost had a low-back pressure. The tomato compost appeared to be made from very fine material and therefore, we expected a high back pressure.

STENTIFORD: i) Why did you not aerate from the first day rather than waiting for a high temperature to develop?

ii) Did you try shredding the tomato material?

iii) Did you try temperature feedback control of aeration?

VAN DER HOEK: i) There were some technical difficulties in the pilot-plant experiment, for example the delivery of the waste took 14 days.

ii) We did not try to shred the tomato plants because shredding involves energy costs and in addition, to aerate a pile of shredded material requires more energy than unshredded plants. We are not sure that it is possible to aerate such a pile of shredded plant material.

iii) No, that was not the first priority. The main aim was to demonstrate that without turning the windrows, an odourless compost could be produced.

FINSTEIN: I would like to comment on the terminology. 'Static pile' refers only to the configuration of the pile. It is essential to describe the process control strategy which can drastically alter the performance. For example fixed aeration rate and temperature feedback control can greatly alter process performance.

VAN DER HOEK: This was the first time that we tried this type of compost and we had no idea of the amount of air needed or the time for which aeration was necessary. In future, control of aeration is planned.

HOVSENIUS: I am interested in the odour of leachate collected before ventilation. Was the carbon content of the leachate investigated and how did the chemical characteristics compare with the volatile compounds?

OOSTHOEK: We analysed the leachate flowing to tank 3 on day 21 and day 23. The NH_4-N content ranged from 156-1810 mg/l, the volatile fatty acids from 0-120 mg/l and pH was between 8.7 and 9.0.

NAVEAU: The odours which are the result of anaerobic conditions within your pile, are reduced by an aeration process. Would it not be easier to aerate correctly the pile from the beginning so that no odour is formed and there would be no need for the biofilter?

OOSTHOEK: The finished tomato plants are sometimes stacked beside the greenhouse for a couple of days before they are removed, so that when delivered, tomato plants sometimes undergo anaerobic fermentation.

Also it is not possible to aerate correctly all the composting material in the pile, so there are always anaerobic spots inside the pile.

GASSER: Was the C:N ratio sufficiently low that additional odorous compounds were formed?

VAN DER HOEK: From table II it can be calculated that the waste had C:N ratio of 19:1 before composting. The ratio of microbial available C to microbial available N is not known. The exhaust gas contains NH_3, the content decreases with time: so that on day 15 it contains about 100 ppm NH_3 and on day 65 about 10 ppm NH_3.

DE BERTOLDI: I think that an air-flow of 2-3 m^3/min/tonne during the initial composting phases is not enough to satisfy the oxygen demand of the composting organic matter.

VAN DER HOEK: The specific ventilation rate during the first 10 days of the experiment was 5.9 m^3/hour/tonne organic matter. (see chapter 4.2.) During this period the CO_2-c ontent of the exhaust gas decreased with time from 2% to 1% CO_2.

THE FATE OF PLANT PATHOGENS DURING COMPOSTING OF CROP RESIDUES

G.J. BOLLEN
Department of Phytopathology, Agricultural University Wageningen
The Netherlands

Summary

A review is made of survival of 25 plant pathogens including fungi, bacteria, nematodes, and viruses. Inactivation is caused by (a) heat generated during the thermophilic phase, (b) microbial antagonism, or (c) toxicity of conversion products formed during decomposition of organic material. This implies that inactivation can occur after exposure to temperature-time dosages below those needed for thermal death.
The majority of pathogens did not survive efficient composting. Sclerotial fungi were highly susceptible. No definite conclusions could be drawn for *Pyrenochaeta lycopersici*, whereas little or no data are available on inactivation of *Olpidium* spp. and the various *formae speciales* of *Fusarium oxysporum* being the most heat-resistant fungal pathogens. Viruses with high inactivation temperatures, e.g. tobacco rattle virus (TRV), can survive composting. It is unlikely that TRV will cause disease in crops on soils amended with properly prepared compost because of the feeding habits of its vectors. Resting spores of *Plasmodiophora brassicae* were destroyed during the heat phase but survived the maturing phase without loss of viability. Probably, heat is the major factor in inactivation of obligate fungal pathogens.

1. INTRODUCTION

Preparation and use of compost has always been an intrinsic part of organic gardening. For the last few years, its utilization on a large scale in conventional horticulture has also been investigated. Composting of crop residues may provide an appropriate solution for the disposal in areas with intensive horticulture, especially where glasshouse crops are grown (14).

Manuals on organic gardening (4) and compost making in general seldom provide information on the risk of spreading plant diseases by the use of compost made from plant residues. A few authors advise growers to avoid incorporation of infested plant material, e.g. of cabbage with clubroot (24). However, infested plants do not always show symptoms, especially when root-infecting pathogens, like *Olpidium* and some *Fusarium* species, are involved. Moreover, when composting is to be practised on a large scale, separation of diseased material from bulk quantities of crop residues is not economically feasible. Knowledge of the risk of spreading diseases by the use of compost is important for workers in organic gardening and for growers of horticultural crops in general.

In contrast to the vast literature on survival of human pathogens during composting of municipal wastes and sewage sludge (10, 12, 30), there are few data on survival of plant pathogens during preparation of compost.The present interest of plant pathologists in compost is mainly on the disease-suppressing properties of the product. Amendment of soil and container media with special composts, e.g. composted hardwood bark, shows great promise for control of root-infecting fungi in ornamentals (8, 15). This subject has been reviewed recently (15). The scope of the present review is the elimination of pathogens during composting of residues of infected crops.

2. SURVIVAL OF PATHOGENS IN COMPOST HEAPS

Some of the few reports on this topic are abstracts of papers given at meetings, and details needed for a further analysis of the data are often lacking, e.g. the organic materials and other ingredients used, the temperatures reached during peak heating and the duration of the period of elevated temperatures. The pathogens include fungi, bacteria, viruses and nematodes. Fungi received more attention than the other pathogens, probably because of their relative importance in horticulture.

In most of the trials, infected plant residues were placed in nylon fabric nets (7, 16, 29) or in perforated metal containers (tea infusers) (19), buried in compost heaps. Upon completion of composting, the samples were retrieved and assayed for survival of pathogens. Viability was tested by using test plants or, in the case of fungal pathogens, also by plating out particles of the composted material on selective media.

The data have been summarized in Table I. Obviously, the chances of survival of fungal pathogens are very low. Even pathogens like *Sclerotium cepivorum*, *Stromatina gladioli* and *Verticillium dahliae*, known for their formation of resistant structures (sclerotia) which can persist for more than 10 years in soil, were eradicated. For *Pyrenochaeta lycopersici*, the corky root pathogen of tomato, results were inconsistent. In a trial where the temperature exceeded 44 °C, depending on the location within the compost bin, the test plants grown on compost samples from the various locations became infected. Because complete destruction of the pathogen was obtained in earlier trials, the survival of the pathogen is presently studied in more detail.

Nematodes seem to be very susceptible to composting. Even cyst--forming species and root-knot nematodes did not survive the process. The data on the potato cyst nematode were derived from investigations on survival of the cysts in sewage sludge of the potato processing industry. The temperatures in the sludge did not exceed 32-34 °C, nevertheless the cysts were killed (26).

It must be realized that detection of the viability of cysts is a problem apart. Cysts and eggs of the potato cyst nematode sometimes fail to hatch even in the presence of highly susceptible test plants although they may appear to be viable after many years of storage in soil. However, from the available data together with arguments given later in this paper, it can be concluded that survival of these nematodes is very unlikely when material is properly composted.

There are too few data on the survival of viruses to draw definite conclusions, but it seems that these pathogens are more resistant to composting than fungi and nematodes. Activity of tobacco mosaic virus (TMV) was considerably decreased, but not inactivated completely by extracts of various composts (2). The virus was not detected in compost made from diseased crops (13). In the 'Multibacto' procedure for rapid composting of municipal waste, samples of plant material infested with tobacco rattle virus (TRV) were kept for 6 days at ca. 68 °C. This procedure resulted in an inactivation of the virus in only one out of

Table I. Literature data on survival of plant pathogens

Pathogen	Host plant[1]	Disease	Temp.[2] (°C)	Sur-vival[3]	Refer-ence
Fungi					
Armillaria mellea	not mentioned	honey fungus, root rot	50	-	32
Botrytis allii	onion	neck rot	47-73	-	29
B. cinerea	geranium	grey mould	40-60	-	16
Didymella lycopersici	tomato	stem rot	39	-	21
Fusarium oxysporum	chinese aster	wilt	47-73	-	7
Phomopsis sclerotioides	gherkin	black root rot	47-73	-	29
Phytophthora cinnamomi	rhododendron	root rot	40-60	-	16
P. cryptogea	chinese aster	foot rot	47-74	-	29
P. infestans	potato	late blight	44-65	-	7
Plasmodiophora brassicae	chinese cabbage	clubroot	47-73	-	7
Pyrenochaeta lycopersici	tomato	corky root	47-73	-	29
	tomato	corky root	44-65	+	7
Pythium irregulare	rhododendron	root rot	40-60	-	16
Rhizoctonia solani	potato	black scurf	67	-	18
	potato	black scurf	47-73	-	29
	sugar beet	damping off	40-60	-	16
	not mentioned	-	50	-	32
Rhizoctonia sp.	tobacco	damping-off	49-63	-	13
Sclerotinia sclerotiorum	lettuce	foot rot	47-73	-	29
Sclerotium cepivorum	onion	white rot	47-73	-	29
S. rolfsii	not mentioned	-	50	-	32
Stromatinia gladioli	gladiolus	dry rot	47-73	-	29
Thielaviopsis basicola	tobacco	black root rot	49-63	-	13
Verticillium dahliae	hop	wilt	57-70	-	25
	hop	wilt	34	+[4]	25
	not mentioned	-	50	-	32
Bacteria					
Erwinia chrysanthemi	chrysanthemum	bacterial blight	40-60	-	16

Table I. continued.

Pathogen	Host plant[1]	Disease	Temp.[2] (°C)	Survival[3]	Reference
Nematodes					
Globodera rostochiensis	potato	potato cyst nematode	32-34	-	26
Meloidogyne incognita var *acrita*	red pepper, cucumber	root-knot nematode	57	-	19
Viruses					
TMV	tobacco	mosaic	49-63	-	13
	tobacco	mosaic	-	(+)[5]	2
TRV	tobacco	rattle virus	68	+	19

[1] Host plants of which residues were used for composting.
[2] Maximum temperatures reached at various sites inside the compost heap or sludge from where samples were retrieved.
[3] - and + means complete inactivation and survival of inoculum.
[4] Survival at only one out of the three sites included in the trial.
[5] Not completely inactivated by extracts of composts.

five trials (29). The virus is an important pathogen of various crops, e.g. potato, tobacco, tulip and gladiolus. In soil, it is transmitted by free-living nematodes and, as will be explained later on, it is unlikely that these vectors will survive composting.

3. MECHANISMS OF INACTIVATION

Three factors are involved in the eradication of pathogens through composting: (a) heat evolved during the thermophilic phase, (b) microbial antagonism, and (c) toxicity of conversion products formed during decomposition of organic material.

3.1 Inactivation by heat

Thermal death is probably the most important cause of destruction of pathogens in compost heaps. Supporting evidence was obtained for human pathogens (10) and plant pathogens as well (7, 19, 21, 25). Most data on heat resistance of fungi apply to short exposure periods, mostly 30 min. and varying from 10 min. to 4 h. The majority of soil-borne pathogens is eliminated by a treatment at 55 °C for only 30 min. (1, 5, 6). Few pathogens are slightly more resistant. One of them is *Fusarium oxysporum*. Its various *formae speciales* are a serious threat to a large number of crops, causing wilting that is difficult to control once it has been established. Unfortunately, there is little information on their fate during composting. Complete eradication requires treatments at 57.5 or

60 °C for 30 min. (6). In compost heaps, such high temperatures are not always reached, but exposure to elevated temperatures lasts from 2 days to several weeks depending on the method used.

The temperature-time dosages needed for complete inactivation is dependent on the inoculum density. The exposure time required for thermal death through treatment at a given temperature is characterized by the decimal reduction value (D-value), which is the time needed for reduction to 10% of the number of viable propagules in the untreated sample. D-values were assessed for *F. oxysporum* f. sp. *melongenae*, the pathogen causing wilt in eggplants. At 50 °C, a temperature commonly attained in composting processes, the D-value was 2.0 h. As the heat resistance of this fungus is typical for that of wilt-causing fusaria, it can be inferred that there is little chance that these pathogens occur in properly prepared compost. This assumption does not change the need for more attention to the fusaria in studies on pathogen survival in compost. The same applies to *Olpidium* species, not for their importance as pathogens, but because of their role in the transmission of viruses. The resting spores of *O. brassicae* are even more resistant to heat than the chlamydospores of *F. oxysporum* (6).

The literature on control of soil-borne pathogens by solar heating of mulched soils is a new source of data on thermal death (17). The data apply to long periods (2-4 weeks) of exposure at temperatures of 40 to 52 °C; therefore these data provide valuable information for composting processes. An example is the study on four common soil-borne pathogens by Pullman et al (23).

Bacteria, except for spore-forming species, are sensitive to heat. It seems unlikely that plant pathogenic bacteria will survive composting. With the exception of *Bacillus cereus* as an exopathogen of tobacco (31), spore-forming bacteria are not known as plant pathogens.

Nematodes are more sensitive to heat than most fungal pathogens. None of the eleven species mentioned by Wallace survived exposure to 50 °C for 1 h. (28). The root-knot nematodes were among the less sensitive ones, but both of the species tested were killed by treatment at 44 °C for 4 h. Thus problems with nematodes in well-prepared compost are not to be expected.

The temperatures required for the inactivation of some viruses like TMV and TRV are higher than those reached during the composting process and this probably accounts for the survival of TRV as mentioned above. However, the nematodes needed for transmission of TRV are killed during composting. It seems unlikely that problems arise when the compost is used for amendment of field and garden soils that harbour nematodes because of the feeding habits of these vectors. The species transmitting TRV are parasites that feed on living plants and not on dead substrates like compost. Since the virus is transmitted through seed, this assumption only applies to properly prepared compost that does not contain viable seeds of infested crops or weeds.

3.2 Inactivation by microbial antagonism

A few cases are known where pathogens are eliminated in spite of the fact that temperature-time dosage remained below the dosages needed for inactivation (25, 26, 32). In such examples, the result is attributed to antagonism of the microflora or to toxic products from the decomposing organic matter as it is difficult to study these factors separately. Yuen and Raabe (32) found that sclerotia of *Sclerotium rolfsii* in sealed vials remained viable, whereas those that were placed in mesh bags were inactivated in compost at sublethal temperatures. The inactivated sclerotia were colonized by bacteria that formed fungitoxic substances.

The sensitivity to antagonism observed for sclerotial fungi and other pathogens does not apply to obligate parasites with hardy resting spores. An example is the clubroot fungus *Plasmodiophora brassicae*. The following results show that the microbial activity during the maturing phase has no detrimental effect on the pathogen. Samples of cabbage with clubroot were incorporated in and then removed from a compost heap before and after the two major phases. During the heat phase, temperatures above 40 °C were recorded for periods varying from 4 to 10 days, depending on the site. Maximum temperatures ranged from 43 °C in outer layers to 58 °C in the centre of the heap. Viability tests with chinese cabbage as a test plant showed that the fungus was destroyed during the heat phase but survived the maturing phase without loss of viability (Table II).

Table II. Survival of composting by *Plasmodophora brassicae* during the two major phases of the process (7).

Period of exposure (weeks)	Phase of composting	Number of test plants	
		Total[1]	With clubroot
0	-	53[2]	46
3	heat phase	89	1
27	maturing phase	90	90

[1] Each analysis included samples from six locations in the compost heap.
[2] Among the control plants seven out of 60 were killed by other pathogens.

The discovery of disease-suppressive properties of special composts in the last decade has led to comprehensive studies on the mechanisms involved. Remarkable results were obtained with composted hardwood bark and it is used on a commercial scale as an amendment in container media for control of root-infecting fungi of ornamentals (8, 15). There is substantial evidence that its suppression of pathogens is induced by an antagonistic microflora (20). When the microflora confers suppressive properties to mature composts, it can be assumed that it is also active in the last phase of the composting process. It would be interesting to investigate whether obligate parasites, like *Plasmodiophora*, are affected by composted hardwood bark.

3.3 Inactivation by toxic conversion products

Leachates and extracts from compost can inhibit the development of fungal pathogens (27) or decrease the viability of TMV (2). These effects need not be due to conversion products only, as the leachates also contain microbial products.

Volatiles formed during decomposition of crop residues suppressed spore germination of *Verticillium dahliae* and inhibited sporulation of other fungi, both parasites and saprophytes (3). Methanol was one of the components of these volatiles.

Ammonia is formed from protein amino acids during the first phase of the compost process. In relatively high concentrations it has a detrimental effect on some fungi, e.g. *Phytophthora cinnamomi*, a common pathogen on roots of various crops (11). Ammonia may play a role in the inactivation of fungal pathogens during decomposition of organic products rich in nitrogen, but supporting data are lacking. The ammonium ion is utilized for protein synthesis by microorganisms and in later stages of the compost process it is also converted to nitrate (22).

The inactivation of potato cyst nematodes in waste sludge from potato processing factories may be caused by toxic fermentation products (26). The temperatures were 34 °C at most and this is too low for thermal inactivation.

4. FINAL REMARKS

The present knowledge on pathogen control by composting reveals that many of the plant pathogens are destroyed during the process. This statement requires some comments.

Efficient composting is associated with a heat phase during which the temperature-time dosages exceed those that are needed for inactivation of pathogens, except a few viruses. Contrary to expectation, the survival of the most heat-resistant fungal pathogens, viz. *Olpidium* species and the *formae speciales* of *Fusarium oxysporum*, has hardly been studied. Data on the fate of these pathogens during composting, also in piles where the heat generated has been insufficient to reach lethal temperatures, are needed as a basis for providing advice.

At the periphery of compost heaps lethal temperatures are never reached; although inoculum can be reduced considerably, it is not inactivated completely (25). Therefore, either the compost from the outer layers should be mixed with fresh residues in new heaps or the material should be turned and mixed during the early stages of composting in order to expose all infected residues to the heat inside the heap.

Obviously, the conclusions do not apply to refuse dumps of crop residues in gardens and small holdings. In such 'mixed heaps' the heat generated and the concentration of toxic products are usually insufficient for inactivation of pathogens. The use of the final product can lead to increased spread rather than to control of plant diseases.

The inactivation during the process of composting is the result of various factors operating at the same time. It depends on the pathogen involved which of these factors is the primary one. Heat might have been the sole factor causing eradication of the clubroot pathogen in the example mentioned above, although an effect of toxic decomposition products cannot be excluded. On the other hand, inactivation of the potato cyst nematode in sludge was attributed to toxic products. In some cases, evidence was obtained for a combined effect. Exposure to sublethal temperatures may weaken the pathogen so that it is more susceptible to microbial antagonism or toxic products. This kind of effect plays an important role in biological control of plant pathogens (9).

5. REFERENCES

1. Baker, K.F. (1957). The U.C. System for producing healthy container--grown plants. Calif. Exp. Stn Manual 23: 126-129.
2. Bartels, W. (1956). Untersuchungen über die Inaktivierung des Tabaksmosaikvirus durch Extrakte und Sekrete von höheren Pflanzen. Ein Beitrage zur Frage der Kompostierung tabaksmosaikhaltigen Pflanzenmaterials. Phytopath. Z. 25: 113-152.
3. Berestetskii, O.A., Kravchenko, L.V. and Makarova, N.M. (1982). The effect of volatile products of decomposed plant residues on the development of fungal spores. Mikologiya i Fytopatologiya 16: 126-129. Abstract in Rev. Pl. Path. 61 (1982) no 6289.
4. Bockemühl, J. (1978). Vom Leben des Komposthaufens. Elemente der Naturwissenschaft 29 Heft 2: 1-67.
5. Bollen, G.J. (1969). The selective effect of heat treatment on the microflora of a greenhouse soil. Neth. J. Pl. Pathol. 75: 157-163.
6. Bollen, G.J. (1984). Lethal temperatures of soil fungi. In: C.A. Parker et al (Eds.). Ecology and management of soil-borne plant pathogens. Am. Phytopathol. Soc., St. Paul, Minnesota. In press.
7. Bollen, G.J., Wijnen, A.P. and Volker, D. (1984). Inactivation of fungal plant pathogens during composting of crop residues. Submitted to Neth. J. Pl. Path.
8. Chef, D.G., Hoitink, H.A.J. and Madden, L.V. (1983). Effect or organic components in container media on suppression of *Fusarium* wilt of chrysanthemum and flax. Phytopathology 73: 279-281.
9. Cook, R.J. and Baker, K.F. (1983). The nature and practice of biological control of plant pathogens. Am. Phytopathol. Soc., St. Paul, Minnesota, 550 pp.
10. Finstein, M.S., Wei-Ru Lin, K. and Fischler, G.E. (1978). Sludge composting and utilization: review of the literature on temperature inactivation on pathogens. Report New Yersey Agric. Exp. Stn, Project 03543, 29 pp.
11. Gilpatrick, J.D. (1969). Role of ammonia in the control of avocado root rot with alfalfa meal soil amendment. Phytopathology 59: 973-978.
12. Golueke, C.G. (1976). Composting: a review of rationale, principles and public health. Compost Sci. 17: 11-15.
13. Grushevoi, S.E. and Levykh, P.M. (1940). The possibility of obtaining seed-bed soil free from infection in compost heaps. The A.I. Mikoyan pan-Soviet sci. Res. Inst. Tob. and Indian Tob. Ind. (VITIM) 141: 42-48. Abstract in Rev. appl. Mycol. 20 (1941): 87.
14. Hoek, K.W. van der, and Oosthoek, J. (1984). Composting: odour emission and odour control by biofiltration. This volume.
15. Hoitink, H.A.J. (1980). Composted bark, a light weight growth medium with fungicidal properties. Pl. Disease 64: 142-147.
16. Hoitink, H.A.J., Herr, L.J. and Smitthenner, A.F. (1976). Survival of some plant pathogens during composting of hardwood tree bark. Phytopathology 66: 1369-1372.
17. Katan, J. (1981). Solar heating of soil for control of soil-borne pests. Ann. Rev. Phytopathol. 19: 211-236.
18. Martin, P. (1966). Plant pathology problems in refuse composting. Compost Sci. 6: 23.
19. Menke, G. and Grossmann, F. (1971). Einfluss der Schnellkompostierung von Müll auf Erreger von Pflanzenkrankheiten. Z. PflKrankh. PflSchutz 71: 75-84.
20. Nelson, E.B. and Hoitink, H.A.J. (1983). The role of microorganisms in the suppression of *Rhizoctonia* in container media amended with

composted hardwood bark. Phytopathology 73: 274-278.

21. Phillips, D.H. (1959). The destruction of *Didymella lycopersici* Kleb in tomato haulm composts. Ann. appl. Biol. 47: 240-253.
22. Poincelot, R.P. (1975). The biochemistry and methodology of composting Conn. Agri. Exp. Stn, New Haven, Bull. 754, 18 pp.
23. Pulmann, G.S., DeVay, J.E. and Garber, R.H. (1981). Soil solarization and thermal death: a logarithmic relationship between time and temperature for four soil-borne plant pathogens. Phytopathology 71: 959-964.
24. Schilthuis, W.Th. (1977). Nederlandse praktijk van de biologisch--dynamische land- en tuinbouw. Vrij Geestesleven, Zeist, 110 pp.
25. Sewell, G.W.F., Wilson, J.F. and Martin, D.G. (1962). Machine-picking in relation to progressive *Verticillium* wilt of the hop. II The effect of composting on the infectivity of machine-picked hop waste. Ann. Rep. East-Malling Res. Stn 1961, pp. 102-106.
26. Sprau. F. (1967). Das Verhalten von Zysten des Kartoffel-nematoden in Kläranlagen. Mitt. biol. BundAnst. Ld- u. Forstw. Heft 121: 39-43.
27. Spring, D.E., Ellis, M.A., Spotts, R.A., Hoitink, H.A.J. and Smitthenner, A.F. (1980). Suppression of the apple collar rot pathogen in composted hardwood bark. Phytopathology 70: 1209-1212.
28. Wallace, H.R. (1963). The biology of plant-parasitic nematodes. Edw. Arnold Ltd., London, 280 pp.
29. Wijnen, A.P., Volker, D. and Bollen, G.J. (1983). De lotgevallen van pathogene schimmels in een composthoop. Gewasbescherming 14: 5.
30. Wiley, B.B. and Westerberg, S.C. (1969). Survival of human pathogens in composted sewage. Appl. Microbiol. 18: 994-1001.
31. Woltz, S.S. (1978). Non-parasitic plant pathogens. Ann. Rev. Phytopathol. 16: 403-430.
32. Yuen, G.Y. and Raabe, R.D. (1979). Eradication of fungal plant pathogens by aerobic composting. Phytopathology 69: 922.

DISCUSSION

LYNCH: You have addressed the question of organic gardening rather than organic farming. What advice would you give to the organic farmer who is likely to have only one type of plant residue in the compost heaps? Should certain diseased residues be avoided?

BOLLEN: When composting is done properly, most pathogens will not survive. It will not be economically feasible to examine the crop residues for diseases and, moreover, many pathogens can be present in roots of crop that do not show clear symptoms.

STENTIFORD: I would like to ask about the vectors for viruses. Do nematodes in composts carry viruses?

BOLLEN: Yes. Viruses will only infect in the presence of nematodes.

PLANT PATHOGEN SURVIVAL DURING THE COMPOSTING OF AGRICULTURAL ORGANIC WASTES

J. LOPEZ-REAL and M. FOSTER
Department of Biological Sciences
Wye College, University of London

Summary

Composting of organic wastes requires strict monitoring of potential pathogenic microorganisms in order to minimise spread of disease. Criteria that have been established for the control and monitoring of human pathogens, in the processing of sewage sludge, have not been tested for plant pathogenic organisms likely to be present in wastes of agricultural origin. A preliminary study was made of the viability and survival of *Pseudomonas phàseolicola*, *Botrytis cinerea*, *Plasmodiophora brassicae* and *Tobacco Necrosis Virus* (TNV) in laboratory thermal inactivation experiments and in 'in situ' natural non-aerated compost piles. Both *P. phaseolicola* and *B. cinerea* were inactivated in cool composts while *P. brassicae* and TNV survived. At median temperature levels (average 54°C) both organisms were inactivated within 72-96 hours of exposure. At high temperature levels TNV and *B. brassicae* were not isolated after 24 hours of exposure. All test pathogens showed greater temperature tolerance to laboratory induced conditions than when exposed to the temperature, moisture and microbial diversity of the compost piles.

Introduction

The composting of municipal organic wastes has attracted considerable attention in recent years, most notably in the USA with the development of the Beltsville (BARC) and Rutgers aerated static piles for the treatment of sewage sludge. The advantages of composting such wastes as a waste treatment process have been summarised and include principally volume reduction, water removal and the production of a stabilised aesthetically acceptable finished product for landfill or other uses. Such materials contain varying amounts of pathogenic organisms which may possibly survive the composting process. In recent years a great deal of research has been carried out to determine the fate of human pathogens during the composting process (Wiley & Westerberg 1969; Cooper & Golueke 1979; Golueke 1983). The data has been reviewed recently by Finstein, Wie-Ru & Fischler (1982). As a result, standards based on the attainment and maintenance of high temperatures have been proposed for judging pathogen kill in turned windrows and static pile systems. In windrows these standards range from "significant reduction (50°C/5 days + 4 hrs at 55°C) to "further reduction" (55°C/5 days) and turned five times (U.S. EPA 1981). Other authorities have recommended that a seven-day period at 55-60°C be required (Ware 1980). The suggested standard applied to static pile systems is that pathogen elimination will have occurred with the attainment and maintenance of a temperature higher than 55°C over a three-day period (U.S. EPA 1981).

An alternative criterion is the use of heat resistant f2 bacteriophage as an indicator organism and the temperature-time relationship based on that required to inactivate 15 logs of the bacteriophage (Burge 1983). The long-term reliability of all these standards has, however, yet to be demonstrated and supported (Golueke 1983).

The composting of agricultural wastes may present similar problems if animal manures and materials are utilised. Much of the data outlined above is of direct and immediate relevance. Such wastes are, however, more likely to be recycled directly into agricultural production sites and soils for crop and horticultural uses. An extra problem therefore arises in this context with respect to the transfer of plant residues infected with plant pathogenic microorganisms. Plant pathogen survival through such systems and its subsequent utilization in arable sites would provide an inoculum source for the subsequent crop. Data on the survival of plant pathogens during composting or related waste processing systems is limited though Hales (1974) noted the elimination of *Heterodera rostochiensis* (potato cyst eelworm) during anaerobic digestion of sludges. Clearly some evaluation and determination of minimum operating standards are necessary before large-scale agricultural waste composting can be undertaken.

Two key factors that bear an important relationship to pathogen destruction are that pathogenic organisms vary with respect to temperature tolerance and that the thermal death point may vary with the stage in the life cycle of the pathogen. Normal compost heaps will generate temperatures of up to 50-65°C well above the thermal death point of mesophilic pathogens. In studies on composted bark wastes, *Phytophthora cinnamomi*, *Pythium irregulare*, *Rhizoctonia solani*, *Botrytis cinerea* and *Erwinia carotovora var chrysanthemi* could not be recovered from infected plant material buried in a compost pile (40-60°C internal temperature) for 10-12 weeks (Hoitink 1976). The differences exhibited between thermal tolerance possessed by animal and plant fungal pathogens have been tabulated from laboratory 'in vitro' experiments (Biology Databook 1972). In addition to those listed, the viruses present further problems. Certain viruses can withstand high temperatures, notably tobacco mosaic virus. Extracts from infected tobacco leaves and stems are found to be infective after burial for six weeks in a compost pile at 50-75°C. Knowledge of thermal inactivation from laboratory studies may give a useful guide to the limits of tolerance but are likely to overestimate the survival capacity of a pathogenic propagule when confronted with a composting 'milieu' - temperature, moisture, pH change, competition and antibiosis all interacting on the pathogen.

A preliminary investigation was therefore undertaken to establish the ability of representative plant pathogens from bacterial, fungal and viral groups to survive a range of composting temperatures. The paper that follows is an initial report on this investigation.

Materials and Methods

The methods of inoculum preparation and inoculation of test plants for assessment and viability are considered below for each pathogen. *Plasmodiophora brassicae* infected soil was obtained from a naturally contaminated source (ADAS, Wye). The material was macerated in distilled water and the presence of resting spores of *Plasmodiophora* checked microscopically. Cabbage seedling test plants approximately 2-3 weeks old were dipped into inoculum suspension before transplanting to pots containing sterilised peat. Plants were watered daily and observed weekly for development of club root symptoms. Subsequent inoculum operations were prepared from roots of these infected plants. The infected cabbage root material was used for the investigation of heat inactivation on infected material.

Tobacco necrosis virus (TNV, strain 3) was obtained from culture stock in the Biology Department, Wye College. The concentration of inoculum utilised was one part infected leaf material to four parts water (weight basis). Cellite was added to the inoculum for abrasive purposes and the mixture smeared over the upper leaf surface of the test plants. French beans (Phaseolus vulgaris - 'Prince') at two-leaf stage was used as test plants for the virus throughout the investigation. Characteristic small brown necrotic lesions of TNV were observed between 2-3 days after inoculation.

Pseudomonas phaseolicola (Race 2) was obtained from stock cultures and maintained on nutrient agar. Following 72 hours of incubation at 20°C inoculum suspensions were prepared by adding 5 ml of sterile distilled water to the surface of the plates and mixing with a sterile glass rod. The bacterial suspension was decanted into a glass vial and injected, utilising a fine sterile hypodermic needle, into the air spaces of the upper leaf surface of French bean plants. Four 'water soaked' patches each more than 1 cm square were obtained per leaf. The characteristic brown rotting lesion of P. phaseolicola was obtained 3-4 days after inoculation.

Botrytis cinerea (Persoon ex Persoon) was maintained on potato dextrose agar at 20°C. A conidial inoculum suspension was obtained by pouring 5 ml of sterile distilled water onto the surface of a 5-day old culture. The suspension was decanted and excised surface leaves of the French bean inoculated by pricking leaf surface with a sterile needle and adding a drop of the suspension to the damaged site. The petiole of the excised leaves were wrapped in moist cotton wool and the whole leaf placed in a humidity chamber consisting of a plastic box whose surfaces were lined with well-fitting filter paper.

Temperature inactivation: Laboratory 'in vitro' tests.

Temperature inactivation of the four test pathogens was carried out under both laboratory 'in vitro' and compost 'in vivo' conditions. Inoculum suspensions of club root, TNV and P. phaseolicola were filtered through muslin gauze to remove plant debris. For each pathogen 5 ml replicates of the filtered suspensions were placed in thin walled test tubes and incubated in a water bath at pre-set temperatures covering a range from 35-75°C at 10°C intervals. Cooled treated suspensions were inoculated into test plants following the procedures as described earlier. All inoculated test plants were maintained in the glasshouse. Plant development was monitored daily and results recorded as presence or absence of disease symptoms on inoculated plants.

Compost inactivation tests:

Exposure of infected plant material to 'in vivo' conditions was carried out through the construction of simple 'Indore'-type compost piles. Technical limitations precluded the use of a controlled 'Rutgers' aerated pile and exposure of infected material was therefore monitored and related to the natural fluctuating temperature obtained from static non-aerated piles. Compost piles (1.5 cubic metres) consisted of grass cuttings, hop waste and either dairy cattle manure or ammonium nitrate as nitrogen activator sources. Layers of vegetation (15 cms approx) were alternated with the cattle manure (5 cms) or ammonium nitrate sprinkled over surface layers Wooden pallèts were used as a base in the construction of the compost piles to aid ventilation. Temperatures within the compost piles were monitored daily using thermistor probes buried in the middle zone of the pile adjacent to the exposed infected plant material.

The four pathogens were incubated in infected test plant material at the centre of the compost piles inside 5 cm square nylon mesh bags. Such bags containing plant material infected with the same pathogen were linked together with 0.5 m nylon cord so that single bags could be pulled out from the centre of the heaps at time intervals of 1-2 days dependent on temperature reading. Removal in this manner did not disturb the bags remaining in the centre of the compost heaps. Each nylon bag contained 1 g of infected plant material for each pathogen. The infected plant material had previously been chopped into small pieces and air dried at room temperature overnight. On removal, the infected material was used in the manner described previously and inoculated into indicator host plants. Control bags were left at the compost site exposed to ambient air temperatures in order to monitor the natural decline of the pathogen. Uninfected plant material controls were also incubated in the compost piles.

To limit symptom variation, all inoculant suspension on the nylon bag contents were stored at -24°C until sampling was complete. Test plants of the same age were then inoculated. In order to accommodate variability associated with lesion count methods for TNV, the half leaf method was used to compare controls with treatments. Opposite half leaves of the same plant were inoculated with control and treatment inoculum respectively. The control inoculum was prepared from an infected leaf material with a standardised lesion count of 305 per half leaf. Inoculum material was prepared to give a 1:4 dilution immediately on excising the infected leaf material from test plants and stored at -24°C. Average lesion counts were based on six replicate half leaves for both control and treatments.

Results

Temperature inactivation under 'in vitro' conditions

The results obtained for laboratory water bath incubation of sap extracts and conidial suspensions of test pathogens are shown in Table I. Under these conditions, all four pathogens showed considerable resistance to temperature, especially TNV (65°C) and *Plasmodiophora brassicae* (65°C).

Temperature inactivation following compost incubation

In the limited time available for this study a range of compost heaps built and the elimination of the test pathogens related to naturally occurring temperatures in the heaps. Quite fortuitously, three heaps constructed provided a range of temperature curves giving a low, median and high internal temperature peak. The temperatures recorded during the three runs carried out can be seen in Figure I.

Compost pile A:

A gradual and low temperature regime was established in this compost pile. The maximum temperature reached at the centre of the pile was 35°C. The temperature did not rise above the mesophilic range for the whole period of sampling. Despite this, conditions were severe enough to inactivate *Pseudomonas phaseolicola* and *Botrytis cinerea* after a four-day period of incubation. Both the *P. brassicae* and TNV remained viable at 16 days of incubation when the run was discontinued.

Compost pile B:

A rapid but moderate temperature regime was established in this run. A maximum temperature of 56.9°C was attained in the centre of the pile after three days of incubation. An average temperature of 54°C was recorded for the duration of the entire sampling period after an initial reading of 15.5°C on construction of the pile. The results with TNV (Table II) show a considerable decrease in lesion counts per half leaf for the treated samples. The infected leaf material showed a decrease of 234 lesions per half leaf within the first 24-hr period; thereafter the lesion count decreased to a negligible level and could not be detected after 96 hours incubation at an average temperature of 54°C. *Plasmodiophora brassicae* was not detected after 24 hours of incubation at this temperature.

Compost pile C:

A rapid and high temperature regime was established in this pile. A maximum temperature of 75°C was reached after three days of incubation. A minimum temperature of 65°C was recorded for the whole period of sampling and an average temperature of 70.5°C for the run. Results of samples of both *P. brassicae* and TNV taken after 24 hours showed inactivation had occurred. The maximum temperature reached during this period was 68°C. As continuous temperature recording was not used, it is not known for what period of time the pile had been at or close to this level.

Discussion and conclusions

The overall results showed that temperature inactivation of the test plant pathogens was achieved at considerably lower temperatures in the compost piles than that indicated by laboratory conditions. There are clearly many inactivating factors other than temperature operating within a compost pile that militate against the survival of plant pathogenic propagules. The compost pile possesses enormously diverse and extremely high population levels of micro-organisms and there are undoubtedly factors such as antibiosis, competition, parasitism in addition to fluctuating moisture, pH and unknown volatile compounds that will affect propagule survival. Temperature as a major process control parameter is the simplest to relate to pathogen survival. The laboratory results showed that temperature, in isolation, can be tolerated to much higher levels by the test pathogens.

The bacterial pathogen *Pseudomonas phaseolicola* and the fungal pathogen *Botrytis cinerea* were both highly intolerant of elevated temperatures and compost pile conditions. Both test pathogens were easily eliminated in a cool pile run. The remaining test runs (B and C) were directed to establishing the resilience of TNV and *P. brassicae*. In terms of average temperature attainment, both runs B and C could be superficially compared to the Rutgers and BARC alternatives that have been extensively utilised in the USA. In both instances the pathogens were eliminated with broadly speaking a 72-96 hour requirement at 55°C level and 24-48 hours at the 70°C level.

Thermal inactivation of plant viruses has attracted considerable interest in the past for considerations of plant propagation and plant vigour. The majority of these studies have been strictly concerned with thermal inactivation in sap extracts and not in leaf tissue. A detailed study of the inactivation of TNV has been reported in the literature (Babos & Kassania 1963). They reported that TNV is inactivated only at high temperatures of 80-95°C in subextract. Thermal inactivation depended on the virus strain and duration of exposure to temperature. Thermal inactivation points of 90-95°C, 85-90°C and 80-85°C were reported for strains D and G; A, F and S; and B respectively, at 10 minutes exposure. The same authors showed that thermal inactivation of T and V was influenced by pH with higher values ($>$ pH7) causing precipitation of the virus from solution. It is possible that similar shifts in the pH in composts could cause similar decreases in viral effectivity.

Plasmodiophora brassicae is spread mainly as resting spores produced in the infected root where all tissues are liable to be attacked. Inactivation of these spores in diseased tissue is therefore critical for control of pathogen dissemination. The resting spores of P. brassicae are reported to possess a longevity of from five to seven years in the soil. Temperature inactivation of such spores has apparently not been examined.

This preliminary investigation indicates that the standards set for elimination of human pathogens in the composting of sewage sludge may also be applied to agricultural wastes and plant pathogens. Further and more extensive surveys of the important plant pathogens and their ability to survive composting is however a necessary precaution. A comparative evaluation of static and aerated systems would also be useful in this context.

Fig 1. Temperature curves of non-aerated static piles.

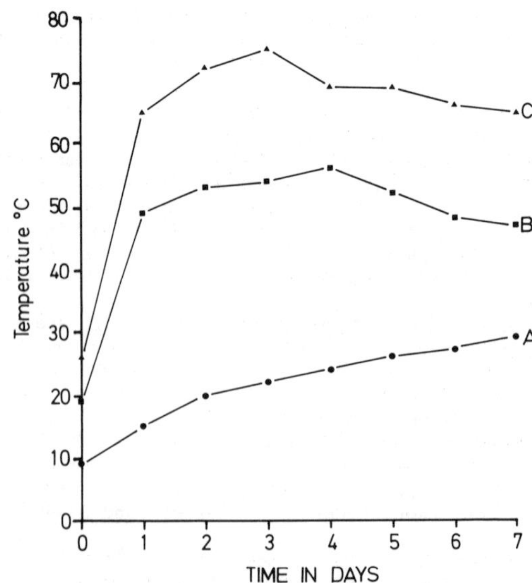

TABLE I

Survival of test pathogen in macerated host tissue exposed to a range of temperatures

Pathogen	Temperature °C (10 mins exposure)				
	35	45	55	65	75
Plasmodiophora brassicae	+	+	+	+	-
Tobacco necrosis virus	+	+	+	+	-
Pseudomonas phascolicola	+	+	+	-	-
Botrytis cineria	+	+	+	-	-

(+) = pathogen survival indicated by viability in vivo test

(-) = inactivation indicated by failure to infect test plant

TABLE II

Tobacco Necrosis Virus inactivation during composting of infected leaf tissue in compost heap (B)

Treatments	Time (hours) after incubation in compost heap (Average lesion counts per half leaf at 3 days.)											
	24 hrs		48 hrs		72 hrs		96 hrs		120 hrs		144 hrs	
	$\overline{x}$	SE	$\overline{x}$	SE	$\overline{x}$	± SE	$\overline{x}$	± SE	$\overline{x}$	SE	$\overline{x}$	SE
Control	313	± 9.54	207.33	± 7.46	288.67	± 8.65	247.0	± 5.62	192.67	± 5.72	254.67	± 4.27
Exposed Leaf tissue	65.67	± 5.37	C.33	± 0.62	0.67	± 0.88	0	-	0	-	C	-

BABOS, P. & KASSANIS, B. 1963 'Thermal inactivation of Tobacco Necrosis virus'. Virology 20 490-497.

Biology Data Book 2nd Ed (1973) Vol II p848-864. Eds P. Altman and D. Dittmer, Federation of American Societies for Experimental Biology, Bethesda, Maryland.

COOPER, R.C. & GOLUEKE C.G. (1979) 'Survival of enteric bacteria and viruses in compost and its leachate'. Compost Science/Land Utilization 20 (2) 29-35.

FINSTEIN, M.S., WEI-RU LIN, K. & FISCHLER, G.E. (1982) Sludge composting and utilization: Review of the literature on temperature inactivation of pathogens. New Jersey Agricultural Experiment Station, Cook College, Rutgers.

GOLUEKE, C.G. (1983) Epidemiological aspects of sludge handling and management. Biocycle 24 (4): 50-59.

HALES, D. (1974) Salmonellae in dried sewage sludge. Environmental Health 82 (11): 213-215.

HOITINK, H.A.J., HERR, L.J. & SCHMITTHENNER (1976) Survival of some plant pathogens during composting of hardwood tree bark. Phytopathology 66: 1369-1372.

U.S. EPA (1981) Technical Bulletin: Composting processes to stabilize and disinfact municipal sewage sludges. Office of Water Program Operations. Washington D.C. EPA 430/9-81-011.

WARE, S.A. (1980) 'A survey of pathogen survival during municipal solid waste and manure treatment processes'. EPA 600/8-80-034. Municipal Environmental Research Laboratory, Office of Research & Development, U.S. EPA, Cincinnati.

WILEY, J.S. & WESTERBERG, S.C. (1969) Survival of human pathogens in composted sewage. Applied Environmental Microbiology 18: 994.

DISCUSSION

BOLLEN: Dr Lopez-Real has made a valuable contribution to the knowledge on elimination of pathogens in compost by showing the TNV is inactivated after exposure from 3 to 4 days. The mechanisms involved may be heat-inactivation or attack of the virus protein. The latter will also occur in field soil or mature compost. Are there any data on longevity of TNV in these habitats?

LOPEZ-REAL: I have no data.

LYNCH: Composting eliminates plant pathogens by heat. What is the effect on beneficial micro-organisms?

LOPEZ-REAL: "Beneficial" organisms such as N_2-fixers become re-established by recolonisation. Attempts to inoculate with *Trichoderma harzianum* during the composting phase have failed.

BOLLEN: *Trichoderma* survives heat quite well, for example 30 minutes at 60°C. *Rhizobium* is thought to be eliminated, as are *Nitrobacter* and *Nitrosomonas*, but they probably re-invade the cooled compost.

POSTERS

Utilization of heat recovered from low temperature waters in agricultural production and fish culture

Heating of greenhouses by trickling water

Conversion of liquid animal slurry to a stackable solid material

Preparation of compost for the production of mushrooms in the Netherlands

The place and potential of waste as organic amendments and substrates

Composting of food factory, fruit and vegetable waste, tannery sludge and cork waste

A mobile laboratory for composting facilities

UTILIZATION OF HEAT RECOVERED FROM LOW TEMPERATURE WATERS IN AGRICULTURAL PRODUCTION AND FISH CULTURE

J DELMAS, J MARQUET AND Ph LEMAITRE
CEA, CADARACHE, 13115 SAINT PAUL LES DURANCE, FRANCE

Studies on the use of low temperature water for agricultural and fish production have been conducted at the CEA since 1974. The CEA (or Centre d'Etudes Nucleaires) of Cadarache has developed, in cooperation with the appropriate industries, techniques of full scale plant and fish production used by professional market-gardeners and fish farmers.

At Pierralatte, water at 20°C is being used to increase growth of eels in ponds extending over $3000m^2$; an area soon to be increased to more than one hectare.

It was demonstrated that heat derived from water between 15 and 40°C could be used for agricultural and horticultural production. Processes use combined techniques that have proved successful for many years with new floor and enclosure heating techniques. In particular, with market-gardeners, forcing of crops such as tomatoes, melons, pimentos, cucumbers, aubergines, marrows, ... was made possible by employment of a low-cost plastic double wall greenhouse with water trickling between the walls. Production was obtained either with heat recovered from industrial waste water at temperatures ranging between 25 and 40°C, or with heat recovered from the ground-water (15°C), solar energy, or urban and agriucultural composts.

HEATING OF GREENHOUSES BY TRICKLING WATER

J DELMAS, J MARQUET AND Ph LEMAITRE
CEA, CADARACHE, 13115 SAINT PAUL LES DURANCE, FRANCE

The greenhouse has water trickling within a double wall heated by nautrally available heat and not by industrial wastes. Such heat is recovered by heating the ground-water with solar energy and biomass fermentation. The greenhouse frame is conventional; the outer wall is of polyethylene, 180 micron thick, the inner wall is of thin PVC. Water is distributed throughout the double wall by black helical tubles for drip irrigation to heat up the air contained within the walls. The water flowrate ranges between 8 $l/h/m^2$ and 16 $l/h/m^2$, and 40 $kcal/h/m^2$ are dissipated at maximum performance. Thus radiative heating of the inside of the glasshouse is obtained through the wall. The water trickling along the inner wall is recovered in a gutter and is stored in reservoirs. There it receives heat during the sunny periods. Such reservoirs also store the heat developing from biomass fermentation. During cold periods, hot water from the reservoirs will be supplied to trickle between the walls and will heat up the glasshouse.

Over a 500 h period of use in Spring, 7 l/m^2 of fuel are saved because of the heat recovered from solar energy and biomass. Close to the vernal equinox, 500 $kcal/m^2$ of trapped and stored solar heat will ensure an integral antifreeze protection with a temperature higher by 10° to 14°C than the atmospheric.

Two types of biomass were tested, viz prunings of grape vines and urban compost after hot-state treatment. A 20-ton compost heap produces 10 to 15 thermal units during 5 hours every 2 days and raises the circulating water temperature by 1°C to 1.5°C. The heap temperature must not drop below 45°C. Such heat drawn from the biomass makes it possible to save from 1 to 1.5 litre of fuel during 500 h glasshouse heating time.

CONVERSION OF LIQUID ANIMAL SLURRY TO A STACKABLE SOLID MATERIAL

M.K. GARRETT and H. LOGAN
The Queen's University of Belfast and
Department of Agriculture for Northern Ireland

A pilot plant study has established the feasibility of removing approximately 30% of the water from pig slurry through optimised drainage using barley straw admixed (1% ww^{-1}) with the slurry and drainage pipes traversing a specially designed settlement unit. This unit consists of fixed parallel walls through which the drainage pipes pass and removable timber slats at either end. The unit was designed to hold the effluent from 100 pigs for a 9 month period.

Liquid which drains from the unit is highly polluting (BOD = 15000 mg l^{-1}, P = 400 mg l^{-1}) but may be recirculated to accomplish zero effluent production. Whilst this is readily achieved when the potential evapotranspiration exceeds 3.75 cm $month^{-1}$, conditions which occur typically in Northern Ireland between the months of April and September, liquid reduction in the winter period is relatively slow and a storage capacity equivalent to 30% of the total volume of slurry handled is required to retain liquid effluent for recirculation when conditions permit. Alternatively drained liquid may be treated with lime and an anionic flocculent to produce a liquid potentially suitable for irrigation (pH = 10, P $<$ 1 mg l^{-1}, BOD = 2000 mg l^{-1}) or it may be irrigated in an untreated state when conditions permit. Stackable solids are readily removed from the unit using normal farm equipment.

After approximately two weeks settlement within this unit the temperature of the settled solid material rises between 10 and 15° above ambient temperature and this is maintained for a period of approximately 8 weeks after addition ceases.

The system presents a possible alternative to conventional liquid storage systems for pig slurry but may be difficult to scale up.

PREPARATION OF COMPOST FOR THE PRODUCTION OF MUSHROOMS IN THE NETHERLANDS

J.P.G. GERRITS
Mushroom Experimental Station, Horst (L.)
The Netherlands

Summary of poster

In 1983 80,000 tonnes of mushrooms (mainly Agaricus bisporus) were produced in the Netherlands by a total of 820 growers. All the necessary compost is made by two firms. The total production is about 10,000 tonnes per week of which 80% is produced by the Co-operation and 20% by a private enterprise. The most important basic material is straw-bedded horse manure. If this is not available in sufficient quantity, straw compost is used. In winter this amounts to 20%, in summer to 50%. In the past a great many materials were added to a compost such as malt sprouts, cotton seed meal, urea, ammonium sulphate, calcium carbonate, superphosphate and gypsum. Straw compost was additionally supplemented with a range of minor elements. Research has fundamentally changed this formulation. Nowadays chicken manure is used exclusively as the organic N source. The only inorganic supplement left is gypsum which turns out to be of utmost importance. Two formulae are in use now, one for horse manure compost and one for synthetic (= straw) compost. A tonne of horse manure is supplemented with 100 kg of chicken manure and 25 kg of gypsum. The amount of water to be added varies from 200-800 l/t. This results in 900-1300 kg of compost. Prewatering takes about a week followed by the proper composting in windrows, which takes another week with turning every 2-3 days. On average an optimum ammonia content in the compost pile is achieved with 100 kg of chicken manure. If horse manure contains much straw, more chicken manure has to be added and vice-versa. Mushroom production is best when the NH4 content of the compost is 0.4% at the end. In that case the N content (determined in dried samples) is fairly constant from the beginning to the end.
For synthetic compost, wheat or rye straw is preferred. Per tonne, 600-900 kg of chicken manure, 75 kg of gypsum and about 5000 l of waters are added, resulting in 3000 kg of compost. The straw is first mixed with 2/3 of the chicken manure, watered and allowed to heat for 7-10 days. The purpose of this pretreatment is to make a product similar to horse manure as it leaves the stables.
Gypsum decreases the pH of the compost. This is particularly important if the NH4 content is high. The role of ammonia and consequently ammonification and N immobilization in the compost needs further study.
After the outdoor composting in windrows the compost is pasteurized and conditioned in layers (in a mushroom house) or in bulk (in tunnels). The time of this process depends on the NH4 content of the compost.

REFERENCES

1. GERRITS, J.P.G. (1974). Development of a synthetic compost for mushroom growing based on wheat straw and chicken manure. Neth.J.agric.Sci. 22: 175-194.
2. GERRITS, J.P.G. (1978). The significance of gypsum applied to mushroom compost, in particular in relation to the ammonia content. Neth.J.agric.Sci. 25: 288-302.

THE PLACE AND POTENTIAL OF WASTE AS ORGANIC AMENDMENTS AND SUBSTRATES

B. ROBLOT (presented by J.M. MERILLOT)
ANRED, 22 Square la Fayette
49004 Angers Cedex, France.

The study was carried out by the national agency for the recovery of waste as organic amendments and substrates. The inquiry started with a questionnaire, then French firms manufacturing and selling these products were visited.

Organic amendments (Standard AFNOR U 44.051) are meant to maintain the organic matter content of a soil. France does not import any organic amendments but exports them to countries whose soils are low in organic matter. Many firms serve the market, the four biggest ones realizing 30% of the sales, the rest being small enterprises owned by local manufactures

About 200,000 t of organic amendments are sold a year, from animal manures, household compost, prunings, mushroom bed compost...

Substrates (standard U 44.551) such as vegetable mould, leaf mould, peat, generally differ from organic amendments in that their organic matter is more inert.

In 1982, France used 155,000 t peat (39% more than in 1981).

80% (about 400,000 tons) of the moulds used in France are imported which shows that the French market for substrates is dependent on foreign firms.

Conclusion:

This study illustrates the importance of foreign firms in our market. Because of their commercial organization and technical level they hold a dominant place on the French market. French manufacturers, aware of this dependence on peat, are more and more interested in substitute products.

The use of composted barks as a substitute for peat offers interesting possibilities of utilization for a few by-products.

More attention must be given to the organization of the industry to promote the investigations and development of information among horticulturists and amateurs.

COMPOSTING OF FOOD FACTORY, FRUIT AND VEGETABLE WASTE, TANNERY SLUDGE AND CORK WASTE

G. Vallini, M. L. Bianchin and A. Pera

Centro di Studio per la Microbiologia del Suolo, C.N.R., Via del Borghetto n. 80, I-56100 Pisa (Italy)

Agricultural wastes consist mainly of crop residues and livestock effluents and by-products of food manufacturing and forestry.

Disposal of these wastes, owing to their organic and putrescible nature, poses problems for environmental pollution. Recycling agricultural waste reduces potential pollution and also offers the opportunity of recovering and re-using part of these wastes. Composting is one of the most feasible and economic methods for treating agricultural waste and the compost produced can be utilized as an organic soil fertilizer.

Research on the process of composting has shown that most biodegradable organic wastes can be composted. Previous conditioning may be required to enable the development of the microbial reactions which carry out the process of composting.

However, although composting is suggested here as a safe method of eliminating polluting organic wastes, it is not the only way of managing organic waste from crop or animal production. In many cases a more defined recovery of energy is to be preferred when the characteristics of the wastes and other conditions permit. In such cases, combustion, pyrolysis, gasification or biochemical conversion to produce ethanol, biogas and biomass may be used. Other processes may allow recovery of valuable materials such as proteins, starch and fats.

Composting may sometimes be necessary as a way of treating organic waste before releasing it safely into the environment. All organic waste, whether originating from livestock production or agro-industrial processes, requires a stabilizing process before disposal even when the end product serves no useful purpose, such as discharges from canning factories, olive oil and wine production as well as technological processes like tanning. These wastes which are often in a highly diluted liquid form can only be processed cheaply by something like composting.

Some experiments were described on composting: 1) a mixture of wheat straw and tomato processing waste, 2) sawdust from cork processing mixed with urban sewage sludge, 3) a mixture of olive-husks and sewage sludge, 4) a mixture of vegetable tannery sludge and the organic fraction of solid urban waste. Some data obtained with agronomical use of these composts are also given.

Some experiments were done with the static pile composting system using forced pressure ventilation in conjunction with temperature-feedback control. The main physical, chemical and biological factors which influence microbial activity were discussed.

A MOBILE LABORATORY FOR COMPOSTING FACILITIES

J-P. VELLAUD and Ph. THAUVIN,
SATUC (technical advisory service for composting facilities),
ANRED, 22 Square la Fayette, 49004 Angers Cedex, France

In France 10% of the household waste is treated in about 100 composting facilities, producing 800,000 t of compost per year.

In order to assess the performance of these plants and to improve the quality of the end-products, ANRED set up in 1981 a technical advisory service (SATUC) provided with a mobile laboratory. The equipment on the vehicle allows the determination of the main physical and chemical properties of the compost.

Since 1981, half of the composting plants have been visited. The technical advisory service (SATUC) prepared a report on each plant visited, mentioning at least the following points:

- maintenance and operating conditions of the plant,
- physical and chemical qualities of the end-products,
- operating costs,
- marketing and distribution of the compost.

A further study in the laboratory of ANRED at Angers is now developing methods for measuring the level of impurities such as plastics and glass.

The experience gained after three years of operation enables the SATUC to make recommendations to the plant operators and builders on the following points:

- the plant in its environment,
- the sorting processes and their effectiveness,
- the benefit of the shredding systems,
- the fermentation monitoring,
- the quality of the end-products required by the users (maturity, particle size...),
- the treatment of residual material.

CONCLUSIONS AND RECOMMENDATIONS

A number of themes have recurred throughout the Seminar which allow some general conclusions to be drawn, and recommendations to be made. The main subject areas may be defined as engineering aspects, the composting process and utilization of the heat produced, the product and its utilization, and economic considerations.

Engineering and the Composting Process

Compost is made from bulky materials, which, in total in agriculture, represent large amounts of material to be handled. Therefore, at all stages of composting consideration needs to be given to the design of plant and equipment, so that the operations may be carried out efficiently and effectively at least cost and with minimum use of fossil fuels. During composting dry matter is lost and heat is generated; water also is usually lost. Control of composting may be exercised mainly in two ways, by turning the heap or by ventilating. Turning does not lend itself to exact control, but allows comminution of the material and the distribution of the desired micro-organisms throughout the heterogeneous matrix.

A suggested definition of composting was by using the equation

$$Q = m\,(h_{out} - h_{in})$$

where Q = rate of heat generation; m = rate of ventilation;

h = heat content of the air

out = outlet air

in = inlet air

the heat content depends mainly on the temperature and the relative humidity of the air.

The objective should be to maximise the rate of heat generation, unless there are other overriding factors.

Recommendations For agriculture and the food processing industries there are identified needs for further work on:

i) Operational Research on the composting process.

ii) Development of plant and machinery to improve the efficiency and economy of composting for various starting materials.

iii) The development and application of existing microbiological knowledge of the process.

Temperature and Pathogen Control

The major factor influencing the death of pathogens is the temperature - time function during composting. Thermophilic temperatures are in general more efficient in controlling pathogens than mesophilic ones. Other factors involved are microbial antagonism and toxic products formed during decomposition of organic wastes such as crop residues.

Incomplete heat treatment may enhance the effects of these factors. Complete control is required for some specialized pathogens like obligate parasites of plants which can rapidly spread in susceptible crops. This aim is achieved by efficient composting for the majority of the pathogens. More information is needed for a few relatively heat-resistant pathogenic fungi and plant viruses.

Recommendation More work is needed on the effects of composting on pathogenic organisms and the importance of residual populations to provide a valid assessment of the acceptable level for various types in finished compost.

Heat Production and Recovery

Heat production is accepted as a valid description, particularly of the earlier stages, of composting. This heat can be used, for example for soil heating and for space heating but it should be used efficiently. This may require new equipment and new approaches to the concept of heat recovery systems.

Recommendation More development work is required on integrated systems which allow the efficient recovery and use of heat produced during composting.

Control of Composting

The control of composting requires methods of assessment to determine the stage the process has reached. Two types of method are required. Some methods should be simple for on farm use or at an industrial composting plant. Other methods, usually more sophisticated, are required for research and development projects. When compost is applied to soil its effects need to be known, for example the available N it contains and its effects on soil structure. Accepted standard methods are needed for these and other measurements.

Recommendations Further work on methods for assessing the state of compost and its likely effects on being added to soil is needed. The projects should be co-ordinated so that agreed standard methods are developed.

Use of Compost

Most compost is used for agricultural purposes or amenity work such as tree planting. These uses pose few problems. Horticultural crops and particularly container grown plants are more sensitive to the presence of toxic compounds (sometimes present in the starting material; some are produced ephemerally or for longer periods during composting) or growth promoting compounds. The latter effect was particularly noticed in plants grown in worm-worked waste. For some plants, the form was altered, and in some flowering plants the time to flowering was shortened. Composting can affect beneficial as well as pathogenic organisms and may influence the uptake of iron by plants which is partly controlled by rhizosphere microbiological activity and also the presence and activity of mycorrhiza, which are essential for the optimum growth of some species.

Recommendations The value of compost at various stages of maturity should be evaluated for different uses. Particular attention should be paid to effects of both phytotoxicity and regulatory aspects of plant metabolism.

Economic Considerations

Economic appraisals are required both of broad research issues, such as composting as an alternative to straw burning, and of specific systems which are developed as a result of research work. The former ensure that research is directed to areas which offer the possibility of substantial economic returns, and the latter that the systems which emerge are financially attractive to those who are expected to adopt them. In both cases, economic analysis can highlight areas where data are inadequate for proper appraisal, and show the sensitivity of the outcome to variations in assumed costs and benefits. This should lead to better decisions on where to allocate resources. Appraisal should be revised as new information becomes available.

Economic appraisal on a national scale raises the problem of taking into account the cost of decisions made on political or environmental grounds. For example, limitations on straw burning or changes in odour control regulations could greatly alter the viability of composting compared with other forms of agricultural waste treatment. Not everything can be expressed in financial terms, but the most cost-effective method can still be determined within the limitations of such non-economic conditions.

Economic appraisals of specific systems may only be relevant for particular types of waste material in particular locations, and caution must be shown when drawing conclusions about the viability of similar systems where conditions differ. So that, even if composting is found to be economically viable there may still be barriers to its recognition and acceptability as a commercial process. Only in Italy does it seem to be accepted at present as a feasible waste management system.

Recommendation There is an urgent need for more and better use to be made of economic appraisal and analysis to guide programmes on composting agricultural and food wastes. This should be co-ordinated at the national level.

LIST OF PARTICIPANTS

BAINES, S.
Dept. of Microbiology
The West of Scotland Agricultural College
- AUCHINCRUIVE, AYR
UNITED KINGDOM

BAUDUIN, F.
Faculté des Sciences Agronomiques de l'Etat

5800 - GEMBLOUX
BELGIUM

BIDDLESTONE, A.J.
Department of Chemical Engineering
University of Birmingham
Edgbaston
B15 2TT - BIRMINGHAM
UNITED KINGDOM

BIDLINGMAIER, W.
Institut für Siedlungswasserbau,
Wassergüte- und Abfallwirtschaft
Abt. Abfalltechnik
Bandtäle 1
7000 - STUTTGART 80
GERMANY

BOLLEN, G.J.
Laboratorium voor Fytopathologie
P.B. 8025
6700 EE - WAGENINGEN
THE NETHERLANDS

BRUCE, A.M.
Water Research Centre
Elder Way
SG1 1TH - STEVENAGE HERTS
UNITED KINGDOM

CHAPMAN, S.J.
Agricultural and Food Research Council
Letcombe Laboratory
Wantage
OX12 9JT - OXON
UNITED KINGDOM

DE BERTOLDI, M.
Istituto di Microbiologia Agraria
Università di Pisa
Via del Borghetto 80
56100 - PISA
ITALY

DEMUYNCK, M.
Unité de Génie Biologique
Université Catholique de Louvain
Place Croix du Sud, 1
Bte 9
1348 - LOUVAIN-LA-NEUVE
BELGIUM

EDWARDS, C.A.
Rothamsted Experimental Station

- HARPENDEN, HERTS
UNITED KINGDOM

EVANS, M.R.
Microbiology Dept.
The West of Scotland
Agricultural College
- AUCHINCRUIVE AYR
UNITED KINGDOM

FERRANTI, M.P.
Commission of the European Communities
200, rue de la Loi
1049 - BRUSSELS
BELGIUM

FERRERO, G.-L.
Commission of the European Communities
200, rue de la Loi
1049 - BRUSSELS
BELGIUM

FIELDSON, R.S.
National Institute for Agricultural
Engineering
Wrest Park
MK45 4HS - SILSOE, BEDS
UNITED KINGDOM

FINSTEIN, M.S.
Department of Environmental
Science Rutgers
The State University
P.O. Box 231
08902 - NEW BRUNSWICK - N.J.
USA

FLEGG, P.B.
Glasshouse Crops Research Institute
Worthing Road
BN16 3PU - LITTLEHAMPTON, WEST SUSSEX
UNITED KINGDOM

GARRETT, M.K.
Agricultural and Food
Chemistry Research Division
Department of Agriculture
Newforge Lane
BT9 5PX - BELFAST
UNITED KINGDOM

GASSER, J.K.R.
Agricultural and Food
Research Council
160, Great Portland Street
W1N 6DT - LONDON
UNITED KINGDOM

GERRITS, J.P.G.
Mushroom Experimental Station

- HORST
THE NETHERLANDS

GODIN, P.
Direction de la Prévention des Pollutions
Service des Déchets
14-16 Bd du Général-Leclerc
92524 - NEUILLY-SUR-SEINE
FRANCE

HALL, J.E.
Water Research Centre
Medmenham Laboratory
Henley Road
Medmenham
P.O. Box 16
SL7 2HD - MARLOW, BUCKS
UNITED KINGDOM

HARPER, S.H.T.
Agricultural and Food Research Council
Letcombe Laboratory
Wantage
OX12 9JT - OXON
UNITED KINGDOM

HOVSENIUS, G.

Tunavägen 32
194 51 - UPPLANDS VÄSBY
SWEDEN

LANDER, R.A.
Agricultural and Food Research Council
Letcombe Laboratory
Wantage
OX12 9JT - OXON
UNITED KINGDOM

LANG, J.
Ministry of Agriculture, Fisheries
and Food
Great Westminster House
Horse Ferry Road
SW1P 2AE - LONDON
UNITED KINGDOM

LE ROUX, N.W.
Warren Spring Lab.
Gunnels Wood Road
SG4 7AE - STEVENAGE, HERTS
UNITED KINGDOM

L'HERMITE, P.
Commission of the European Communities
200, rue de la Loi
1049 - BRUSSELS
BELGIUM

LOPEZ-REAL, J.M.
Biological Sciences Dept.
Wye College
University of London
Wye
TN25 5AH - ASHFORD, KENT
UNITED KINGDOM

LYNCH, J.M.
Glasshouse Crops Research Institute
Worthing Road
BN16 3PU - LITTLEHAMPTON, WEST SUSSEX
UNITED KINGDOM

MERILLOT, J.-M.
A.N.R.E.D.
2, Square la Fayette
B.P. 406
49004 - ANGERS CEDEX
FRANCE

MONACO, A.
University of Naples
Istituto di Microbiologia Agraria
80055 - PORTICI
ITALY

MOREL, J.L.
E.N.S.A.I.A.
38, rue Ste. Catherine
54000 - NANCY
FRANCE

NAVEAU, H.
Unité de Genie Biologique
Université Catholique de Louvain
Place Croix du Sud, 1 - Bte. 9
1348 - LOUVAIN-LA-NEUVE
BELGIUM

NICKOLS, S.G.
Agricultural and Food Research Council
Letcombe Laboratory
Wantage
OX12 9JT - OXON
UNITED KINGDOM

OOSTHOEK, J.
N.V. VAM
Marijkeweg 11
6709 PE - WAGENINGEN
THE NETHERLANDS

PERA, A.
Centro di Studio per la
Microbiologia del Suolo
C.N.R.
Via Borghetto 80
56100 - PISA
ITALY

PHILLIPS, V.R.
National Institute for Agricultural
Engineering
Wrest Park
MK45 4HS - SILSOE, BEDS
UNITED KINGDOM

ROCCHI, A.
Ministero Industria
Direzione Generale Miniere
Via Molise 2
 - ROMA
ITALY

ROUSSEL, F.
DERS/SERE
CEN - Cadarache
13115 - ST PAUL LEZ DURANCE
FRANCE

SIPPOLA, J.
Agricultural Research Centre
M.T.T.K.
31600 - JOKIOINEN
FINLAND

SMITH, V.G.
Agricultural and Food Research Council
Letcombe Laboratory
Wantage
OX12 9JT - OXON
UNITED KINGDOM

STENTIFORD, E.I.
Dept. of Civil Engineering
The University
LS2 9JT - LEEDS
UNITED KINGDOM

SUPERSPERG, H.
Universität für Bodenkultur Wien
Institut für Wasserwirtschaft
Gregor-Menzel-Strasse 33
1180 - WIEN
AUSTRIA

SVOBODA, I.F.
Dept. of Microbiology
The West of Scotland Agricultural
College
- AUCHINCRUIVE, AYR
UNITED KINGDOM

THOSTRUP, P.
Crone & Koch
Faelledvej 1
8800 - VIBORG
DENMARK

TUNNEY, H.
The Agricultural Institute
Johnstown Castle Research Centre
- WEXFORD
IRELAND

VALLINI, G.
Centro di Studio per la
Microbiologia del Suolo
C.N.R.
Via Borghetto 80
56100 - PISA
ITALY

VAN DER HOEK, K.W.
Rijks Agrarische Afvalwater Dienst
Kemperbergerweg 67
6816 RM - ARNHEM
THE NETHERLANDS

VAN FAASSEN,
Institute for Soil Fertility
P.O. Box 30003
9750 RA - HAREN (Gr.)
THE NETHERLANDS

VERDONCK, O.
Faculty of Agriculture
University of Ghent
Coupure Links 653
9000 - GENT
BELGIUM

VIGERUST, E.
Agricultural University
Boks 28
1432 - AS-NLH
NORWAY

WOOD, D.A.
Glasshouse Crops Research Institute
Worthing Road
BN16 3PU - LITTLEHAMPTON, WEST SUSSEX
UNITED KINGDOM

ZOGLIA, M.
Danieli Ecologia S.p.A.
San Giovanni al Natisone
33100 - UDINE
ITALY

ZUCCONI, F.
Ist. Coltivazioni Arboree
Università di Napoli
Via Università 100
80055 - PORTICI-NAPOLI
ITALY